THE SECOND BRAIN

THE SECOND BRAIN

A Groundbreaking
New Understanding of
Nervous Disorders of the
Stomach and Intestine

MICHAEL D. GERSHON, M.D.

An Imprint of HarperCollinsPublishers

A hardcover edition of this book was published in 1998 by HarperCollins Publishers.

HarperCollins books may be purchased for educational, business, or sales promotional use. For information please write: Special Markets Department, HarperCollins Publishers Inc., 10 East 53rd Street, New York, NY 10022.

First HarperPerennial edition published 1999.

Reprinted in Quill 2003.

Designed by Charles Kreloff

The Library of Congress has catalogued the hardcover edition as follows:

Gershon, Michael D.
The second brain : a groundbreaking new understanding of nervous disorders of the stomach and intestine / Michael D. Gershon.
p. ill. cm.
ISBN 0-06-018252-0
Includes index.
1. Gastrointestinal system—Innervation. 2. Autonomic ganglia.
3. Gastrointestinal system—Diseases—Psychosomatic aspects. 4. Irritable colon—Psychosomatic aspects. I. Title.
QP145.G33 1998
612.3'2—dc21 98-28577

ISBN 0-06-093072-1 (pbk.)

07 08 09 SPS / RRD 20 19 18 17 16 15 14 13 12 11

To the United States Government and its politicians,
whose faith in science and unflagging support of research
have made possible the discoveries described in this book.

CONTENTS

ACKNOWLEDGMENTS

WHEN I WAS YOUNG I used to imagine that authors of books were intrepid souls who worked long hours, alone, as their work flowed smoothly from pen to page. Although I admired the putative writers that I conjured up, I never hoped to emulate one. I knew my limitations. The ability to work alone on anything significant was not my strong suit. If authorship meant the single-handed production of a finished volume, I realized that I would not be destined to produce books. I need to interact. Feedback and help from compatriots, therefore, are for me, necessary for success. Fortunately, my immature musing about the creative process turned out to be incorrect. Books do not have to be written in absolute solitude. It is possible to obtain help and, if one is as lucky as I have been, it is possible to find gracious and talented people who are willing to provide assistance. In short, I wrote this book, but not without aid when I critically needed it. I would thus like to take this opportunity to express my gratitude and dedicate this book to those individuals whose efforts on my behalf made it possible to get it written.

First, I would like to thank my wife, Anne Gershon, for her willingness to provide frank and sometimes painful criticism. She never tired of listening to me babble on about my ideas, and she was always willing to read and referee whatever I wrote. Above all, I cannot even begin to list the number of concepts that she added to the text, because they are too numerous to remember. These valuable and sometimes critical additions were always offered delicately as "suggestions" for me to consider. Finally, whether or not authors are intrepid, their wives have to be. Loneliness, as I have said, does not have to be the fate of a writer, but it does seem to be the fate of the writer's wife. Long hours are involved in the creative process and these hours are time taken away from the joyous pursuit of life together, which is the essence of a successful marriage. Not only, therefore, do I wish to thank Anne for her help, I also want to apologize to her in this very public place for hibernating as I have while writing; *mea culpa*.

Second, I will, for the rest of my days, be grateful to my editor, Nellie Sabin. Nellie acted as literary arbiter, grammarian, and psychotherapist. She kept me going when I thought all would be lost, and she helped me know when I began to write in English instead of in "Science." The rapidity and professionalism Nellie brought to her work never failed to astonish me. Her cheerfulness and willingness to listen to me rant were equally amazing. Nellie clearly made it possible for me to write this book; I could not have done it without her.

Third, there are many additional individuals whose assistance allowed me to write a book. My agents, Herb and Nancy Katz, for example, convinced me to do it in the first place. They, and they alone, were responsible for prodding and enticing me sufficiently to overcome an initial period of massive resistance to the idea. Nancy and Herb had faith in me even before I had faith in myself. They were also invaluable as critics and taskmasters, seeing to it that the work was completed in good time; moreover, I am thoroughly indebted to Nancy and Herb for effectively insulating me from the business of authorship and allowing me to concentrate instead on its creative aspects. My executive editor, Larry Ashmead, and editors Jason Kaufman and Allison McCabe, all of HarperCollins deserve to be acknowledged for their bravery in being willing to engage a writer whose previous published output consisted entirely of technical articles (and grant applications) written for the most specialized of audiences. Their vision, that a wider and non-specialized audience (the public) needs to learn about the second brain, and their willingness to make it possible for this audience to do so is every bit as commendable as their bravery. I would like to thank my colleague, Michael Camillieri of Mayo Foundation, for providing me with an impromptu lecture (during a meeting of the council of the American Gastroenterological Association) on functional bowel disease and for patiently explaining the Rome criteria for diagnosing that illness. Finally, I would like to express my appreciation to my secretary, Helena Leiter, who ceaselessly monitored my peripatetic wandering all over the globe. No matter how remote the destination, or how exotic the scientific meeting, Helena was able to get me there and back, and to enable people who had to reach me to do so.

My last words in this section are another apology. These are directed to my co-workers, both in the Department of Anatomy and Cell Biology at Columbia, and in my laboratory. I am sorry for my long absence from their lives. I would like to tell them what Evita said to Argentina, "I never left you." I have now completed a book that I hope will bring new hope and knowledge to many who need both. The writing of it, however, is over and I am back. I promise once again to devote myself to research, education, and administration (in that order). After all, no matter what I have said about the second brain, there is still much more to tell, and I have to try to learn what that is.

PREFACE

HUMANS ARE A self-centered bunch. Evolution—or if one is a fundamentalist, creation—is viewed as a story with a happy ending. The process is seen as having culminated in the production of the ultimate species, the only one that reflects the image of God. Because we see ourselves as special, whatever detracts from the centrality of the human condition is inevitably viewed with suspicion, if not outright hostility. Copernicus and Galileo did not receive the plaudits of their contemporaries when they suggested that the sun, the stars, and the planets might not revolve around the Earth. That was because humans live on Earth. It seemed totally unreasonable to believe that God would place the ultimate species on a backwater planet in a third-rate galaxy. To deny the centrality of the human position is tantamount to denying God, a venture never to be embarked upon without risk.

Science often interferes with the human self-image. Its observations are made without regard to their potential impact on human feelings. Nature is nature. Scientists do not make or invent principles, they discover them. The profession is thus a dull one. Science is not creative, as, for example, is art. An artist produces an opus, a scientist merely observes the facts and communicates them. Bliss to a scientist is to be right, while bliss for an artist is to be beautiful, whimsical, and imaginative. The plodding nature of their profession often leads scientists to grief. They follow their trails of discovery wherever those trails lead, which sometimes is to trouble.

Consider the lowly gut and its nervous system. The bowel just is not the kind of organ that makes the pulse race. No poet would ever write an ode to the intestine. To be frank, the popular consensus is that the colon is a repulsive piece of anatomy. Its shape is nauseating, its content disgusting, and it smells bad. The bowel is a primitive, slimy, snakelike thing. Its body lies coiled within the belly and it slithers when it moves. In brief, the gut is despicable and reptilian, not at all like the brain, from which wise thoughts emerge. Clearly, the gut is an organ only a scientist would love. I am such a scientist.

Actually, I am a neurobiologist. Most of my colleagues study the brain. The few who do not investigate the spinal cord or "models," the nervous systems of more primitive beasts, which they hope will help them to understand how the brain works. Those of us whose trails of discovery have led to the gut are beyond rare. We are just a little on the common side of unique. I have become accustomed, at meetings of the Society for Neuroscience, to being a house novelty. Until a recent revolution led to the establishment of the new field of neurogastroenterology, the nervous system of the gut was not to be taken seriously. I suffered.

Still, there I am, a neurobiologist who has devoted the whole of his career to the part of the nervous system that runs the bowel. My route to the gut was a tortuous one. It began years ago, in 1958, when I was a student at Cornell. I was taking a course on the neurobiology of behavior and became interested in what was then a newly discovered chemical of the body, serotonin. Serotonin had generated a great deal of excitement at that time because its ability to contract a rat's uterus had just been found to be blocked by LSD. Now before you laugh and dismiss that observation as the kind of thing that would only agitate a professor of something or other, remember that this was the age of Timothy Leary. The mind-altering effects of LSD and the hallucinations it causes were big news. Serotonin was known to be present in the brain. If LSD could block the action of serotonin on a rat's uterus, people reasoned, it seemed logical to assume that LSD would also block the action of serotonin in the brain. If serotonin's action was important in brain function, which was likely, then the antagonism of serotonin by LSD could be the basis of the hallucinatory effects of LSD. The mental state brought about by LSD, moreover, was considered to be similar to schizophrenia. Perhaps, therefore, schizophrenia could be understood as a disease of serotonin deficiency.

While I wanted to learn more about serotonin, the brain scared me. It is a very complicated organ and was, I thought, too daunting. I longed for a simple nervous system, one that I might be able to understand. When I learned that over 95 percent of the body's serotonin is made in the bowel, therefore, I decided that the organ had promise. In fact, I now know that my original concept of a "simple" nervous system was wrong. A simple nervous system is an oxymoron, like jumbo shrimp; nevertheless, the enteric nervous system, the nervous system of the gut, is simpler than the brain, and its study has served to keep me off the streets. Despite the trouble it has occasionally brought me, the enteric nervous system has provided a wonderful life, packed with surprise, excitement, and a degree of interest that has even attracted the media. Although the gut may be reptilian, people are fascinated by reptiles. The lines at the reptile house at the zoo are long, and no museum ever went broke pushing its dinosaur exhibit. Neurobiologists, like pre-Copernican theologians, may once have failed to look beyond the universe that they could see, but the discoveries of science, even the most outrageous, are eventually recognized if they are correct.

The Thoughtful Bowel

We now know that there is a brain in the bowel, however inappropriate that concept might seem to be. The ugly gut is more intellectual than the heart and may have a greater capacity for "feeling." It is the only organ that contains an intrinsic nervous system that is able to mediate reflexes in the complete absence of input from the brain or spinal cord. Evolution has played a trick. When our predecessors emerged from the primeval ooze and acquired a backbone, they also developed a brain in the head and a gut with a mind of its own. The organism could thus attend to more attractive things, like finding food, escaping destruction, and having sex with other organisms. All this could occur while the bowel handled digestion and absorption beyond the pale of cognition. It was not necessary to devote cerebral energy to visceral matters because the viscera took care of themselves.

That primitive nervous system is still with us. In fact, as animals became more complicated, so too did the enteric nervous system. That is nature's trick. The brain in the bowel has evolved in pace with the brain in the head. Our enteric nervous system is not even small. There are more than a hundred million nerve cells in the human small intestine, a number roughly equal to the number of nerve cells in the spinal cord. Add on the nerve cells of the esophagus, stomach, and large intestine and you find that we have more nerve cells in our bowel than in our spine. We have more nerve cells in our gut than in the entire remainder of our peripheral nervous system. The enteric nervous system is also a vast chemical warehouse within which is represented every one of the classes of neurotransmitter found in the brain. Neurotransmitters are the words nerve cells use for communicating with one another and with the cells under their control. The multiplicity of neurotransmitters in the bowel suggests that the language spoken by the cells of the enteric nervous system is rich and brainlike in its complexity. Neuroscientists, whose horizon ends at the holes in the skull, are continually amazed to find that the structure and component cells of the enteric nervous system are more akin to those of the brain than to those of any other peripheral organ.

The enteric nervous system is a curiosity, a remnant of our evolutionary past that has been retained. That certainly does not sound like something that would excite anyone, but it should. Evolution is a powerful editor. Body parts that are frivolous or not absolutely necessary have little chance of making it through the rigors of natural selection. An enteric nervous system, however, has been present in each of our predecessors through the millions of years of evolutionary history that separate us from the first animal with a backbone. The enteric nervous system thus has to be more than a relic. In fact, the enteric nervous system is a vibrant, modern data-processing center that enables us to accomplish some very important and unpleasant tasks with no mental effort. When the gut rises to the level of conscious perception, in the form of, for example, heartburn, cramps, diarrhea, or constipation, no one is enthused. We want our bowel to do its thing, efficiently and outside our consciousness. Few things are more distressing than an inefficient gut with feeling.

Surveys have shown that over 40 percent of patients who visit internists do so for gastrointestinal problems. Half of those have "functional" complaints. Their gut is malfunctioning, but no one knows why. No anatomical or chemical defects are obvious. Physicians become angry. Patients who present themselves to doctors with problems that are insoluble are perceived as threatening and are often dismissed as mentally unbalanced, with the epithet "crocks" whispered behind their backs. They are considered to be examples of poor protoplasm whose neurotic thought processes are communicating themselves to their bowel. Their gut is thus acting up in such a way as to defy the best that modern medicine has to offer, which in this case is ignorance compounded by lack of compassion. While it is indeed true that the brain can affect the behavior of the bowel, the gut can also manage to get along without hearing from the brain. Only one to two thousand nerve fibers connect the brain to the hundred million nerve cells in the small intestine. Those hundred million nerve cells are quite capable of carrying on nicely, even when every one of their connections with the brain is severed; nevertheless, physicians have only recently begun to believe that it just might be possible for diseases of the bowel to arise within the gut.

Since the enteric nervous system can function on its own, it must be considered possible that the brain in the bowel may also have its own psychoneuroses. That new concept, simplistic as it may be, is likely to turn out to be as revolutionary and hopeful as Copernicus's discoveries. Cures come when diseases are understood. Malfunction of the enteric nervous system may be resistant to therapies aimed at the head, but therapies aimed at the gut just might work.

The Magnetism of the Undiscovered

The fate of the bowel's own nervous system has until recently been a cruel one. Ignored, despised, and troublesome, its inner workings (both normal and abnormal) have escaped discovery. Its microcircuits have yet to be mapped, the chemical symphony played by its neurotransmitters still has not been heard, and even the scope of the behaviors it controls remains unknown. The status of our knowledge of the enteric nervous system has been, until recently, positively medieval. Medieval ignorance, however, yielded once before to the Renaissance and the Renaissance led eventually to the Enlightenment. A renaissance of the gut is under way. Therein lies the marvel of this system. It is an uncharted frontier. Could any curious person resist, let alone one who calls himself a scientist, or better yet, an investigator? The shackles of scientific resistance to the obvious are disappearing. To paraphrase President Reagan, it is morning in the abdomen.

It is reasonable to ask why anyone should care that we all have a second brain where we least expected to have one. The answer, of course, is that we should care about our second brain for the same reasons we care about our first. Descartes may have said, "I think, therefore I am," but he only said that because his gut let him. The brain in the bowel has got to work right or no one will have the luxury to think at all. No one thinks straight when his mind is focused on the toilet.

Here is a new field, a new horizon, and a new science. It is enticing. For me, the presence in the gut of an ancient second brain is not merely the stuff of science, it is also an intriguing and surprising story of discovery that I want the world to know. Wonderfully, there is help for many on the way, and it is exciting to be one of those able to sound the trumpets heralding its arrival. Serotonin hooked me and set me on a course that was soon to upset many of my scientific elders. I became involved in a scientific war that raged over the enteric nervous system and its serotonin content. That conflict was ultimately resolved with an extraordinary deus ex machina in, of all places, Cincinnati. The resolution of that very personal little war, however, did not end the tale. In fact, the story is still unfolding, and becoming more interesting as it does so.

This book tells the beginning of the story of the second brain. I wish it were possible to include the denouement. That, however, is coming in the future. The beginning of this beginning, Part I, provides the background of this particular storyteller and introduces some of the other scientists whose work rescued this topic from scientific obscurity. Also included in Part I is necessary information about how the nervous system is constructed and how it works.

Part II presents a mouth-to-anus travelogue of the inner sanctum of the gut. This section essentially follows the food chain from ingestion to egestion and covers, as it goes, the critical processes of digestion and absorption. This section also deals with threats to the body and describes the cooperation between the second brain and the immune system in defending us against an evil army of microbes that seeks eternally to turn the bowel into a route of invasion.

Part III covers the results of modern research into the development and disorders of the second brain. Some of these disorders, in fact, need all the coverage they can get because they are frequently overlooked by physicians in a rush to attribute gastrointestinal symptoms to psychoneurosis of the brain in the head. Taken as a whole, the book provides a history of scientific discovery and some insight into how these discoveries are made. It tells of a process brought about not by magic, nor by prayer, but by the hard, rational work of a great many ordinary people.

Finally, the book ends on the note of hope that the new understanding of the second brain holds for millions of people, particularly for the 20 percent of Americans who suffer from functional bowel disease.

Part I

THE EARLY BREAK-THROUGHS

1

THE DISCOVERY OF THE SECOND BRAIN

THOSE OF US WHO deal in science, even the most enlightened of us, have a strong and objectionable tendency to hubris. Hubris for scientists comes from an inadequate knowledge and appreciation of the past. Discoveries are thus made and claimed that are really rediscoveries—not new advances at all, but history lessons.

Neurogastroenterology: A Rediscovery Brings Hope for the Future

Not long ago, the *New York Times* ran an article about the second brain in its science section. David Wingate, a gastroenterologist with an academic practice in London, was cited as a source for a comment that identified me as the father of the new field of neurogastroenterology. I admit to being the father only of three children. Clearly, David did not insult me by attributing to me the fatherhood of a field. In fact, I would love to be able to just send him a note saying something nice, like "David, you've noticed." Unfortunately for my ego, however, I know better.

I have made discoveries in my scientific career, but the basic principles on which my work is based are about to celebrate their one hundredth anniversary. That bit of information is very good for putting down my own particular brand of hubris. I am not really disappointed, because

I have to concede priority to people who came before me. Rediscovery is every bit as good as discovery, if what is rediscovered is important and was forgotten. It is better still when the rediscovered information has the capacity to improve the lives of those around us.

Neurogastroenterology began when the first investigators determined that there really is a second brain in the bowel. The seminal discovery that established its existence was the demonstration that the gut contains nerve cells that can "go it alone"; that is, they can operate the organ without instructions from the brain or spinal cord. Those of us who qualify as neurogastroenterological fathers in David Wingate's estimation are really children. None of us discovered the existence of the second brain. That discovery had passed, however, like the Roman Empire, into oblivion. What I have done, with a great deal of help from colleagues around the world, is to find it again and return it to scientific consciousness. To me, that accomplishment, which will soon provide relief to millions of people suffering from the misbehavior of an ill-tempered bowel, is sufficient. *Dayenu.*

Ecclesiastes Was Right: There Really Is Nothing New Under the Sun

Neurogastroenterology really started with Bayliss and Starling, two investigators whose work in nineteenth-century England established them as immortals in the Pantheon of Physiology. I love to envision what life must have been like in the laboratories of England at the turn of the century. It was a time when notorious fogs descended on London and mixed with the smoke of thousands of coal stoves to clog lungs and blot out the view. This was the time of Jack the Ripper, David Copperfield, and Ebenezer Scrooge. I had experienced an update of an English laboratory in 1965–1966, when I was a postdoctoral fellow at Oxford. One needs to read Dickens to appreciate the conditions under which English scientists work.

Winters in England are not usually very cold. Certainly, New York is colder than Oxford or London. The trouble with England is that there is very little difference in most of the country between the indoor and the outdoor temperatures. I worked at my laboratory bench with my scarf on, and I wore gloves with the fingertips cut off so that I could feel what I touched. The benches tended to be high, and we sat on backless wooden stools. Since these laboratories were the result of half a century of progress, the working conditions faced by Bayliss and Starling must have been almost penal. Whatever their laboratories were like, the accomplishments of Bayliss and Starling were quite startling. In fact, until Margaret Thatcher became prime minister and suffocated British science, the physiological laboratories of England were a match for those of any other nation.

The Law of the Intestine

Bayliss and Starling worked with dogs. They isolated a loop of intestine in anesthetized animals and studied the effects of stimulating the bowel from within its internal cavity, thereby mimicking the effects that normal intestinal contents might exert on the wall of the gut. In their critical experiments, Bayliss and Starling increased the pressure within the loop of bowel. The gut responded with a stereotyped behavior that, in its reproducibility, caught their attention. Whenever its internal pressure was raised sufficiently, the bowel would exhibit muscular movements that had the effect of propelling the contents of the intestine in a startlingly one-way direction. The propulsive movements consisted of a coordinated descending wave of oral contraction and anal relaxation that forced the intestinal contents relentlessly and inevitably in an anal direction.

Bayliss and Starling called this response of the gut to increased internal pressure "the law of the intestine." Bayliss and Starling were very much into "laws." Their physiological legacy includes a "law of the heart" and a "law of the circulation" as well as the "law of the intestine." They probably used the word "law," which now seems quaint, to imply that they had discovered an everlasting principle that governs the behavior of a biological system. Perhaps it was the contemporaneous notoriety of the case of Jarndyce versus Jarndyce in Dickens's great book *Bleak House* that focused their terminology in such a legal direction. In any case, the "law of the intestine," despite its catchiness as a phrase, failed to persist. Not that Bayliss and Starling were wrong. In fact, their work has stood up well over time and is easily reproduced. The "law of the intestine" that Bayliss and Starling formulated still describes the behavior of the bowel, but the name of the activity has changed. The "law of the intestine" is now known as the *peristaltic reflex,* a much more prosaic term but one that is more descriptive of what the gut is actually doing, and certainly less evocative of an unknown higher power. After all, if there are laws, surely there must also be law enforcement, which as a scientific concept leaves something to be desired.

Bayliss and Starling correctly associated the coordinated nature of the "law of the intestine" with nerves. A surprising finding, however, was made when they cut all of the nerves entering or leaving the loops of dog bowel that exhibited the "law of the intestine." They knew that if they were to cut all the nerves to limbs or other organs, reflexes would be lost. Reflex behavior anywhere but the gut always involves the participation of the brain or spinal cord. Other organs do not make decisions for themselves; instead they inevitably follow the instructions they receive from the central nervous system. Cutting the nerves that connect these organs to the brain or spinal cord deprives them of their directions and the organs become paralyzed, like an airline ticket agent whose computer has crashed.

Bayliss and Starling reasoned that severing all of the nerves entering or leaving a loop of bowel would cut all nerve-mediated communication between the gut and the central nervous system. When they did this, however, the "law of the intestine" still prevailed. Increases in internal pressure continued to be followed by the same descending wave of oral contraction and anal relaxation that they saw before the nerves were cut. Since a reflex behavior could thus be elicited in segments of bowel after all input from the brain or spinal cord had been eliminated, Bayliss and Starling attributed the "law of the intestine" to what they called the "local nervous mechanism" of the gut. In other words, if outside nerves are not required, then inside nerves must be the ones that do the job.

There Is a Nervous System Inside

The conclusion that intrinsic nerves are responsible for the "law of the intestine" was a reasonable one for Bayliss and Starling to reach because they were aware, even before they began their studies, that a complicated nervous system is embedded in the wall of the bowel. The existence of the enteric nervous system had been discovered in Germany while the Civil War was raging in America. Working with a primitive optical microscope, a German scientist by the name of Auerbach had found that the bowel contains a complex network, or *plexus,* of nerve cells and fibers. This plexus, wedged between the two layers of muscle that encircle the gut, is still called *Auerbach's plexus,* as if he owned it. Since some scientists hate to include a person's name in the nomenclature of body parts, Auerbach's plexus is also known as the myenteric (my = muscle; enteric = gut) plexus.

After Auerbach's discovery, another smaller plexus was found in a layer of the bowel called the *submucosa.* The submucosa gets its name from its location, which is just beneath the lining of the gut's internal cavity, where the business of digestion takes place. The inner lining is called the *mucosa*; therefore, the layer under the mucosa, logically enough, is the submucosa. The submucosa is a region of dense connective tissue that is so tough and resistant to stretch that it enables gut, to literally be used to make surgical sutures and strings for tennis rackets. The second network of intestinal nerve cells is called *Meissner's plexus,* by those who like to award structures to their discoverers, and the *submucosal plexus* by others who hate people's names in the nomenclature. Since Bayliss and Starling knew that the gut contains a large nervous system, they felt free to postulate that this system, the "local nervous mechanism" of the bowel, could provide the gut with the means to manifest reflex activity, even in the absence of external nervous input.

It Works All by Itself

Eighteen years after Bayliss and Starling first published their observations, Ulrich Trendelenburg, on the German side of the trenches that divided Europe from the Swiss border to the English Channel, mounted an isolated loop of guinea pig's intestine on a hollow J-shaped tube. This experiment, published in 1917, turned out to be a critical one. Trendelenburg suspended the bowel, on its tubular support, in a test tube containing a warm nutritive solution, which he supplied with oxygen. The bowel survived well in this artificial environment. The apparatus, within which living organs survive for several hours, is called an *organ bath.*

When Trendelenburg blew through the J-shaped tube into the bowel, the gut blew back. This experiment sounds terribly simple, and it is. The consequences of the phenomenon Trendelenburg observed and recognized, however, are profound. In order to blow back, the segment of guinea pig intestine isolated in an organ bath had to display the same reflex behavior that Bayliss and Starling had observed many years previously in an intact dog. To do this, the segment of gut needed to be able to detect the increase in pressure that blowing into the tube had caused to occur within its internal cavity. Then, to blow back, the isolated bowel had to respond with a coordinated descending wave of oral contraction and anal relaxation, mimicking Bayliss and Starling's "law of the intestine." In Trendelenburg's laboratory, moreover, this behavior did not occur in an intact animal. The brain, spinal cord, and sensory ganglia had all been discarded with the rest of the guinea pig. There was nothing in the organ bath but gut.

The modern term "peristaltic reflex" was introduced by Trendelenburg to describe the pressure-induced propulsive activity of the gut. This caused the equivalent phrase, "the law of the intestine," to fade from use. The observation that the peristaltic reflex could be elicited in a segment of gut isolated in an organ bath confirmed that Bayliss and Starling had been correct in attributing the pressure-induced descending wave of oral contraction and anal relaxation to the "local nervous mechanism" of the bowel. For the reflex to take place in a system that contains no other organ but the intestine, all of the necessary elements have to be intrinsic components of the wall of the gut. That they all should be there is striking, because a similar neural apparatus does *not* exist in any other organ. Cut the connections between the bladder, the heart, or the skeletal muscles and the central nervous system, and all reflex activity ceases. Trendelenburg's simple experiment, therefore, was nothing short of revolutionary. Trendelenburg had demonstrated that the intrinsic nervous system of the gut actually has properties that are like those of the brain and its subservient appendage, the spinal cord. To a neurobiologist, this is like saying that the bowel is close to God.

2

THE AUTONOMIC NERVOUS SYSTEM AND THE STORY OF CHEMICAL NEUROTRANSMISSION

THE STORY OF THE enteric nervous system picks up again at the next landmark, which was reached in 1921. The setting returned to England, where, in Cambridge, J. N. Langley published his great book, *The Autonomic Nervous System.* Most practicing doctors, even some who have heard of the enteric nervous system, think they know about Langley's classification of the divisions of the autonomic nervous system. To tell a physician about the definition of this system should be like telling Michael Jordan about James B. Naismith's definition of basketball. Very few modern doctors, however, have actually read Langley's book. Langley published *The Autonomic Nervous System* almost eighty years ago. What most doctors know about Langley's book is thus based on what they have read *about* it in textbooks, which is both inadequate and wrong.

Whether people have read his work or not, Langley remains the single individual most responsible for our current understanding of the autonomic nervous system. Langley's definition of the autonomic nervous system was that it is an entirely motor system of nerves that control the behavior of the visceral muscles, blood vessels, heart, and glands. A *motor system* is one that carries information away from the brain and/or spinal cord. There was nothing in Langley's definition about the exercise of free will or volitional control over the effectors (targets) of autonomic nerves. Langley also envisioned his system as a one-way street. The word of the brain is passed out along the nerves of this system, and what comes back to the brain is received via a different system. Many textbooks tell you that the activities under the control of the autonomic nervous system are involuntary and that there are autonomic nerves that sense what is happening in the periphery and so inform the brain. These views are those of the authors of these books, which would be all right if they were not wrong.

It is true that the activities directed by the autonomic nervous system are not usually under conscious voluntary control. The involuntary nature of most autonomic behavior is the reason the system was called "autonomic" in the first place. Another system of motor nerves controls the behavior of skeletal muscles, which are usually operated voluntarily. This system is called the *skeletal motor system.* The volitional difference between the two systems, however, is not absolute. On the one hand, some people can be conditioned to will changes to occur in autonomic activities, such as the rate of their heartbeat or their blood pressure. On the other hand, none of us can make certain skeletal muscles, such as those in the middle ear, contract whenever we want them to. The middle ear muscles are activated only as an involuntary reflex response to loud noise.

Peripheral Synapses: To Have or Not to Have

There is a major anatomical difference between the nerves that go to skeletal muscles and autonomic nerves. All of the nerves to skeletal muscles run directly from the central nervous system to their skeletal muscle targets. In contrast, autonomic nerves never run directly from the central nervous system to their effectors (muscles, blood vessels, or glands).

The autonomic nervous pathway is always interrupted by at least one

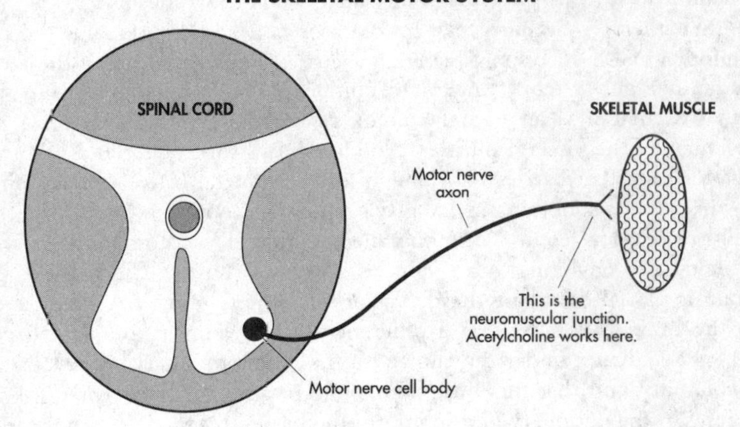
THE SKELETAL MOTOR SYSTEM
SPINAL CORD
SKELETAL MUSCLE
Motor nerve axon
This is the neuromuscular junction. Acetylcholine works here.
Motor nerve cell body

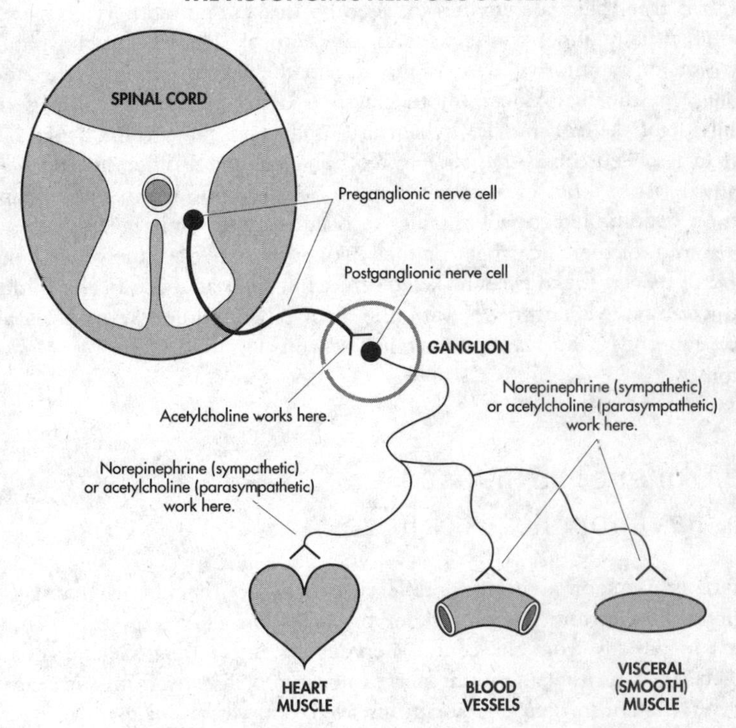
THE AUTONOMIC NERVOUS SYSTEM
SPINAL CORD
Preganglionic nerve cell
Postganglionic nerve cell
GANGLION
Acetylcholine works here.
Norepinephrine (sympathetic) or acetylcholine (parasympathetic) work here.
Norepinephrine (sympcthetic) or acetylcholine (parasympathetic) work here.
HEART MUSCLE
BLOOD VESSELS
VISCERAL (SMOOTH) MUSCLE

junction between nerve cells, called a *synapse.* An autonomic signal from the brain to an effector thus must be carried by two or more nerve cells, while a signal going from the brain to a skeletal muscle requires only one.

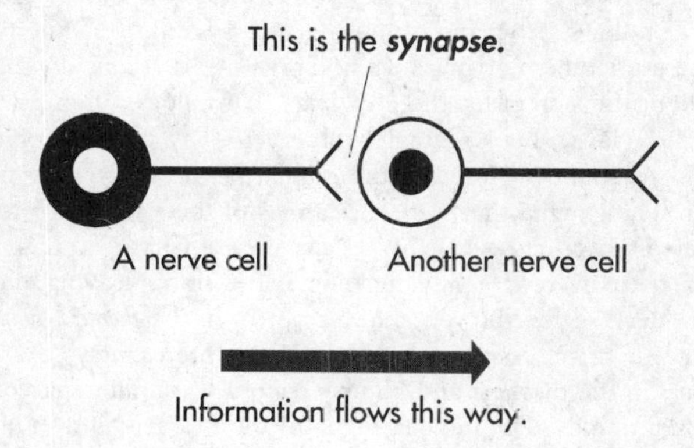

The consequences of the anatomical difference between nerves are actually quite profound. A signal leaving the central nervous system en route to a skeletal muscle either gets there, intact and unchanged, or it is not received at all. Signaling is a simple "all or none" phenomenon. In a sense, it is black and white. There is no gray. In contrast, an equivalent signal leaving the central nervous system en route to a blood vessel, the heart, or a gland may be amplified, weakened, or otherwise modulated by processes that occur at the autonomic synapses. The activation of autonomic effectors thus is infinitely more subtle than that of skeletal muscle. There is room in the process of autonomic innervation for shades of gray. This subtlety in the autonomic nervous system reaches a crescendo in the bowel. Subtlety is important when it comes to nerves. Not only is there a nice ring to it, but it also provides for instant adaptation to changing circumstances.

Langley's classification of the divisions of the autonomic nervous system proceeded from his realization that the autonomic innervation of an effector involves a chain of nerve cells. The first nerve cell in the chain, which initially carries the commands of central processing centers, is located within the brain or spinal cord. This cell passes the instructions via a synapse to the second nerve cell, which is located in a *ganglion* (a regional aggregate of nerve cell bodies). The first nerve cell carries signals *to* this ganglia and so is called, logically enough, *preganglionic.* The second nerve cell, which lives in the ganglia, is called *postganglionic,* because its processes lead *away* from the ganglia to the effectors waiting in the outlying districts of the body.

Sympathetic and Parasympathetic Divisions

Langley realized that he could use the locations of the preganglionic nerves, or "outflows" from the central nervous system, to define two divisions of the autonomic nervous system. These two divisions dominated all thought about the system for the fifty years that followed Langley's publication of his classic book. Langley observed that preganglionic nerve fibers could be found in some of the cranial nerves emanating from the brain itself. The ganglia that were the targets of these nerve fibers tended to be located in, or close to, the organs they innervated. Langley also noted that there were no preganglionic nerve fibers leaving the spinal cord in the neck region; however, lower down, in the *thoracic, lumbar,* and *sacral* levels, these nerves were again present. Interestingly, the preganglionic fibers in the thoracic and lumbar regions were different from those in the cranial nerves. The target ganglia of the thoracic and lumbar outflows of preganglionic nerves were not in, or near, the organs they innervated. Instead, these ganglia were all located in prominent groups near the vertebral column, a considerable distance from their effectors. At the sacral level, the preganglionic fibers again resembled those of the brain, in that their target ganglia were once more located far from the central nervous system, near the effectors.

Unfortunately, most people still divide the autonomic nervous system into two parts. Langley used the similarity of the cranial and sacral outflows to define one of these parts as the *parasympathetic division,* and that of the thoracic and lumbar outflows to define the other as the *sympathetic division.* In Langley's view, the critical difference between the sympathetic and the parasympathetic divisions was an anatomical one, based on the locations of the respective outflows of preganglionic nerves, and nothing else. Other differences exist, but these are not absolutes, and thus do not help in distinguishing one system from the other.

Langley included two groups of ganglia in his sympathetic division. Both are supplied by preganglionic nerve fibers that exit from the central nervous system at thoracic or lumbar levels (which is what makes them sympathetic), but the two groups differ somewhat in their location. One set of sympathetic ganglia is concentrated in two long chains extending on either side of the spinal column, from the neck to the tail (even if the tail is reduced, as it is in humans, to rudimentary bones that cannot be seen from the outside). Since these ganglia lie next to the vertebrae, they are called *paravertebral.* Because they connect to one another, they are also called the *sympathetic chain ganglia.* Another set of sympathetic ganglia is located in front of the vertebrae and is thus called *prevertebral.* These ganglia, which supply the gut with sympathetic nerves, are composed of clusters of nerve cells that encircle the abdominal branches of the aorta, the great artery that carries blood away from the heart.

In contrast to the well-delineated sympathetic ganglia, parasympathetic ganglia are harder to find because they are all situated within, or just outside, the organ they innervate. As a result, the preganglionic nerves to the parasympathetic ganglia are long, and the postganglionic nerves are short. This is just the opposite of the sympathetic division, where preganglionic nerves are short and postganglionic nerves are long. This anatomical difference is functionally significant. In both the sympathetic and the parasympathetic systems, the preganglionic nerves are well insulated with a fatty sheath (called *myelin*) and conduct rapidly, while the postganglionic nerves are not ensheathed and conduct slowly.

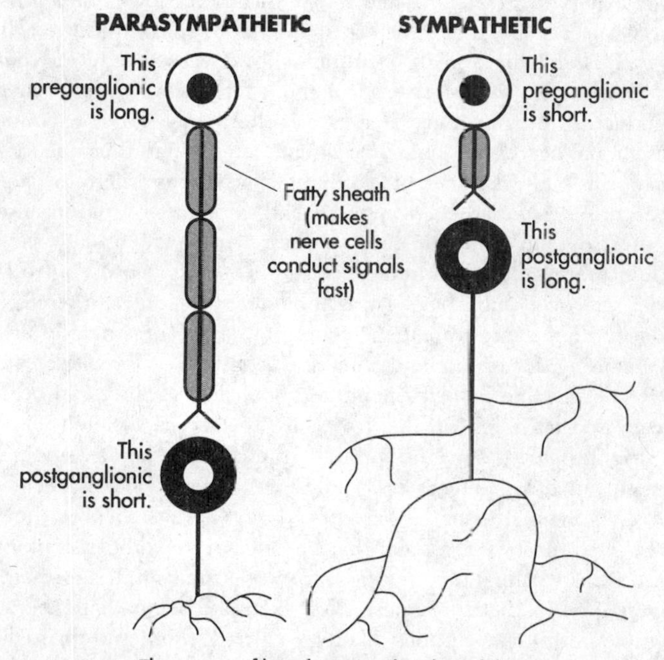

Because of their anatomy, parasympathetic responses tend to be faster and more precise than sympathetic responses, which are usually slower and more diffuse!

The system can be thought of as an operational, if not an anatomical, relative of Amtrak's trains on the Northeast Corridor. The fast-conducting preganglionic nerve fibers have their counterpart in Metroliners, while the slow-conducting postganglionic nerve fibers are analogous to commuter locals. If you happened to be taking some of these trains to carry a message from New York City to Paoli (a Philadelphia suburb that is the effector cell in this metaphor), you would leave Pennsylvania Station on a Metroliner, no matter what. Somewhere along the way, however, you would have to switch (synapse) to a commuter local, because Metroliners do not go to Paoli. Where you switch trains makes a big difference in the amount of time it will take you to reach Paoli. If you get off the Metroliner in Newark and take locals from there to Paoli, the trip will be much longer than if you ride the Metroliner to Philadelphia and just take a local for the short hop to Paoli.

In the parasympathetic system, the ensheathed, fast-conducting preganglionic nerves are long, while the unsheathed, slow-conducting postganglionic nerves are short. In this system, so to speak, the Metroliner is taken to Philadelphia and the commuter local is used only for the short hop to Paoli. The reverse is true of the sympathetic system, where the fast-conducting preganglionic nerves are short and the slow-conducting postganglionic nerves are long. The sympathetic system thus abandons the Metroliner in Newark and takes commuter locals from there to Paoli. For this reason, parasympathetic responses tend to be more rapid in onset and more precise than sympathetic responses, which are usually slower in onset and more diffuse (commuter locals not only go slowly but they also meander all over suburbia). Parasympathetic responses, therefore, are more likely than their sympathetic counterparts to be restricted to a single organ—causing, for example, the pupils to constrict or the bladder to contract. Responses that involve the entire body, such as the rapid beating of the heart and elevation of blood pressure that are associated with stress (flight, fright, or fight), are the specialty of the sympathetic system.

These functional differences, while real, are tendencies, not absolute distinctions. Sympathetic responses need not always involve the whole body, or even most of it, a point that introductory textbooks of physiology often overlook. Sympathetic responses can be just as limited in their scope as any parasympathetic response. For example, sympathetic nerves are responsible both for our pupils dilating in a dark room and for male ejaculation during orgasm. Happily, these responses can occur individually, independently of one another. It is thus not necessary for a male to ejaculate in order for his pupils to adjust to the dark. Since functional differences between the sympathetic and parasympathetic nervous systems do not always hold true, the two systems cannot be differentiated on the basis of function alone. Langley's anatomical definitions thus, even today, remain the only foolproof means of distinguishing sympathetic and parasympathetic nerve cells.

The Enteric Division

Given the importance that Langley attached to the locations of the outflows of preganglionic nerves in establishing the divisions of the autonomic nervous system, it was apparent to him that the enteric nervous system could be considered neither sympathetic nor parasympathetic. Clearly, Langley was aware of the work done earlier by Bayliss and Starling and by Trendelenburg, which showed that, alone among the organs, the gut can manifest reflex activity independently of input from the central nervous system. Langley also appreciated the fact that, in comparison to the number of nerve cells in the bowel, the number of motor nerve fibers connecting the brain or spinal cord to the gut is incredibly small. In humans, for example, there are only about two thousand preganglionic nerve fibers in the *vagus* nerves (the large cranial nerves that connect the brain to the bowel) at the point where those nerves enter the abdomen. These preganglionic fibers are all that exist. In contrast, there are over *one hundred million* nerve cells in the human small intestine.

It is doubtful that Langley knew the exact number of nerve cells in the bowel of any species of animal. The first relatively accurate estimate of the number of enteric nerve cells (in the guinea pig intestine) did not appear until ten years after the publication of Langley's opus on the autonomic nervous system. Actually, the issue is still a matter of contention. These cells are hard to count, and many people are still fighting over how to obtain an accurate enumeration. Langley, however, did not need to know the precise number. He knew that there was an overwhelming disparity between the number of nerve cells in the bowel and the number of nerve fibers that were available to provide them with a preganglionic innervation. This disparity indicated to Langley that the majority of nerve cells in the gut probably do not receive any input at all from the central nervous system. He did not seem to consider it likely that the small number of preganglionic nerve fibers from the brain could divide into enough branches to make contact with all the nerve cells in the gut.

Subsequent investigations, carried out many years after Langley's death, have confirmed that his supposition about the innervation of enteric nerve cells was basically correct. Most enteric nerve cells probably are not directly connected to the central nervous system. Not that Langley's view is free of criticism from modern revisionists. Terry Powley, a Canadian anatomist, shows beautiful pictures of vagal nerve endings in the rodent bowel at meeting after meeting and exults at how many there seem to be. In fact, there are many thousands. However, thousands of nerve endings still pale in comparison to the millions of nerve cells in the bowel and the many hundreds of millions of intrinsic nerve fibers that these cells put out to talk to one another. What makes Powley's pictures so beautiful, in fact, is the relative rarity of vagus nerve fibers in the gut. Since they are labeled in his pictures, the vagus nerves stand out in individual splendor, looking like varicose snakes winding tortuously through the nervous strands that interconnect enteric ganglia. The nerve fibers of the vagus are seen because the intrinsic nerve fibers are not seen. If, in the same pictures, the intrinsic nerve fibers of the enteric nervous system were to be stained, the fibers of the vagus would vanish, covered over by an obscuring vast broad swipe of color.

Certainly, the voice of the brain is heard in the bowel, but not on a direct line to every member of the enteric congregation. If the majority of enteric nerve cells receive no direct connections from the brain or spinal cord, it is clearly impossible to call them, by Langley's rules, sympathetic or parasympathetic. For this reason, Langley included a third division, the enteric, in his definition of the autonomic nervous system. The enteric nervous system differs from the sympathetic and the parasympathetic in its anatomical and functional independence from the brain and spinal cord. It is this separate-but-equal classification of the enteric nervous system that still surprises audiences of doctors and even neuroscientists.

Langley's work has turned out to be of lasting importance because it provides the basis for understanding the independence of the enteric nervous system and why it is actually a second brain. Everyone and everything needs a place in the grand scheme of affairs, and Langley found one for the enteric nervous system in the scheme of the grand innervator.

Despite the overwhelming complexity of the vertebrate nervous system, its basic plan is simple. There are two major divisions, a *central nervous system,* known as the CNS, and a *peripheral nervous system,* called the PNS. The central nervous system consists of the brain (referring now to the one in the head) and spinal cord, while the peripheral nervous system includes everything else. These two systems are thoroughly interconnected and they work together, so their separation might appear to be somewhat arbitrary; nevertheless, the central nervous system clearly outranks the peripheral nervous system, as there is no question about which system gives the orders and which system follows them. Commands flow *from* the brain and spinal cord via the nerves of the peripheral nervous system *to* muscles and glands (the effectors of the body). Information detected by the body's sensory receptors travels back to the brain and spinal cord, again via the nerves of the peripheral nervous system. There is thus a hierarchy of function and a necessary linkage. The brain is at the top, the effectors and sensory receptors are at the bottom. Effectors do what the central nervous system tells them to do, and the information gathered by the sensory receptors is sent upstairs to be appreciated by a CNS audience.

Rebellion in the Bowel: The Brain Below

The enteric nervous system, uniquely, can escape from this functional hierarchy. Technically, the enteric nervous system is a component of the peripheral nervous system, but it is so only by definition. Everything is part of the peripheral nervous system as long as it is nervous and is not brain or spinal cord. In contrast to the remainder of the peripheral nervous system, however, the enteric nervous system does not necessarily follow commands it receives from the brain or spinal cord; nor does it inevitably send the information it receives back to them. The enteric nervous system can, when it chooses, process data its sensory receptors pick up all by themselves, and it can act on the basis of those data to activate a set of effectors that it alone controls. The enteric nervous system is thus not a slave of the brain but a contrarian, independent spirit in the nervous organization of the body. It is a rebel, the only element of the peripheral nervous system that can elect *not* to do the bidding of the brain or spinal cord.

Independence

The enteric nervous system is thus an independent site of neural integration and processing. This is what makes it the second brain. The enteric nervous system may never compose syllogisms, write poetry, or engage in Socratic dialogue, but it is a brain nevertheless. It runs its organ, the gut; and if push comes to shove (as it does in the millions of people who have had their vagus nerves surgically interrupted), it can do that all by itself. When the enteric nervous system runs the bowel well, there is bliss in the body. When the enteric nervous system fails and the gut acts badly, syllogisms, poetry, and Socratic dialogue all seem to fade into nothingness.

All of this, of course, is more than Langley knew, but I doubt it is more than he suspected. The precocious discoveries of Bayliss and Starling and the work of Trendelenburg can be explained by nothing else. Even so, many years after Langley published his great work, another generation of scientists, of which I am one, would be out discovering it all over again and spreading the information to an audience that would greet the rediscoveries, first reluctantly, and finally with a collective "WOW!" as if these verities had just been revealed for the first time. Fortunately, at least some of the WOW comes not just from surprise but from the potential the rediscovery of the second brain holds for clinical medicine. To paraphrase the commercial for Tums, understanding spells relief.

Eclipse

Langley achieved a fair amount of eminence during his lifetime. He was not only the editor of the prestigious *Journal of Physiology,* he owned it. Unfortunately, as an editor Langley had all the lovable characteristics of Saddam Hussein. Investigators are not amused by imperious editors who, without reason, alter their work. Langley was not known for his well-reasoned criticism. Acceptance of editorial change, however, is facilitated when criticism is well reasoned and subject to rebuttal and appeal. Unanticipated alterations, moreover, in one's manuscript can cause some antipathy. Antipathy may simmer internally, but whether it simmers internally or erupts externally, it is unlikely that the editor who makes changes by fiat will be remembered with fondness.

Langley's nasty reputation and the rumors that swirl around his memory help to make sense of what happened to his concept of the autonomic nervous system as a troika. When I was a medical student, I was taught authoritatively (albeit erroneously) that there are only two divisions of the autonomic nervous system, the sympathetic and the parasympathetic. I suspect that most doctors are still taught that this is so. Since no scientific assertion can be made in a scholarly work without a citation, Langley's book, *The Autonomic Nervous System,* was cited as the reference for this binary division. It therefore came as no small surprise to me when I actually read what Langley wrote. There were *three* divisions, not two.

Poor old Langley. After his death, he was the victim of irony. In life he had been an imperious editor, but later the imperious editor had been edited. A change had been made in his legacy with which he would certainly have disagreed. The revision occurred, moreover, after Langley could no longer defend himself or his ideas. The deletion of the enteric nervous system as a division of the autonomic nervous system might have occurred independently of Langley's personality; nevertheless, it is conceivable that Langley's ideas might have had a stronger hold on his scientific constituency if he had been nicer.

Whatever the reason, the concept that the autonomic nervous system is an entity with three parts virtually disappeared in the years that followed Langley. The accomplishments of Bayliss and Starling, Trendelenburg, and Langley were eclipsed by the dazzling discovery of neurotransmitters that could explain the effects of stimulation of parasympathetic and sympathetic nerves. The enteric nervous system faded from view as first the parasympathetic and then the sympathetic neurotransmitters were identified. These radiant discoveries seemed to blind our more recent predecessors.

The enteric nervous system, when it was thought of at all, was considered to be parasympathetic. Even its name fell into disuse. The nerve cells in the bowel were written off as the postganglionic links in the parasympathetic nervous pathway to gastrointestinal smooth muscle and glands. There was a rationale, albeit a flawed one, for this point of view. The vagus nerve, which innervates the entire gut from the esophagus all the way down to the middle of the colon, is, after all, a cranial nerve. Where the vagus nerve's territory ends, sacral nerves take over. The parasympathetic definition is met, in the sense that the nerves that connect the gut to the central nervous system are cranial and sacral. Like other parasympathetic ganglia, moreover, the enteric ganglia are embedded within the innervated organ. It all fits perfectly well, if one is ignorant of, or wishes to overlook, the fact that most enteric nerve cells are not directly innervated by either the vagus or the sacral nerves. The smooth muscle and glands of the gut are not supplied by a chain of two nerve cells but by complex intrinsic enteric neural circuits that may involve many nerve cells. In fact, since reflexes occur when the gut is actually cut off from the central nervous system, the vagus nerves and the sacral nerves may be totally irrelevant to many of the behaviors of the bowel.

It is not easy to explain why understanding of the enteric nervous system went into a decline. Willful ignorance of history and established biological truth is not a common practice among scientists. Nothing was discovered that proved, or even suggested, that Bayliss and Starling, Trendelenburg, and Langley had been wrong. It is hard to believe that the Dark Ages of the enteric nervous system were precipitated only by our predecessors' love of simplicity and order, but that seems to be the most charitable explanation. Any other would involve bizarre theories of conspiracy, which even if true, are both unknown and unappealing to me. I thus attribute the lapse in knowledge of the enteric nervous system to the emergence of the two chemical neurotransmitters that came to dominate all scientific thought about the autonomic nervous system. Once it became known that there were two chemical neurotransmitters, the temptation to have only two autonomic systems, one for each, became too great to resist. Unfortunately, the yielding of our predecessors to this temptation led to the Dark Age of enteric neurobiology. Fortunately, the work of Langley, like the writings of the Venerable Bede, survived to keep alive the flame of enteric learning and served as a beacon to guide the Renaissance of the bowel.

The functioning of any part of the vertebrate nervous system involves the communication of nerve cells with one another. For the most part, messages are delivered in the form of chemicals; therefore, it was necessary to understand what neurotransmitters are and how they act as they do before scientists could hope to understand how any part of the nervous system functions. For the enteric nervous system, the history of chemical transmission is particularly poignant because the discovery of chemical neurotransmission served initially to mask knowledge about the second brain. Recently, however, as the true abundance of neurotransmitters has finally become apparent, it has become possible to use the actions of these substances to unlock the secrets of the mind of the bowel and to use this information to bring solace to the troubled gut.

Perspicacious Poisons

Curiously, the story of chemical neurotransmitters can be thought of as a series of episodes, in which the unlikely hero of each is a hideous poison. Toxins and poisons, often the instruments of death, have in fact taught us much about life.

The story of our understanding of chemical neurotransmission begins in France in the nineteenth century with the investigations carried out by Claude Bernard on *curare,* a deadly plant toxin used by native peoples of South America. These peoples are hunter-gatherers for whom the ability to bring down fleeing game is a serious business. The hunters use curare to coat the tips of darts they shoot from blowguns at animals in the forest. The curare quickly paralyzes any animal hit by the dart, so that the hunters can collect the beasts at their leisure and prepare them for one or more meals. Fortunately for all concerned, except of course the animals, curare is not absorbed orally, so the poisoned meat can be happily consumed without danger.

Bernard worked not with the autonomic nervous system but with the nerves that supply skeletal muscle. He found that curare blocked the ability of these muscles to respond to stimulation of their motor nerves; nevertheless, even though the curare-treated muscles sat flaccidly in their chambers in the face of what should have been irresistible nerve stimulation, they were not paralyzed. Bernard observed, probably to his delight, that despite the presence of curare, both the nerves and the muscle still worked perfectly well if they were stimulated individually. This observation demonstrated that there is a specialized region, the *neuromuscular junction,* between nerve and muscle that is the site of the action of curare. The equivalent region in the autonomic nervous system where smooth muscle, cardiac muscle, or glands are innervated (the *neuroeffector junction*), is unaffected by curare, indicating that the mechanism of autonomic neurotransmission must be different from that which occurs at the skeletal neuromuscular junction.

THE AUTONOMIC NERVE-EFFECTOR JUNCTION

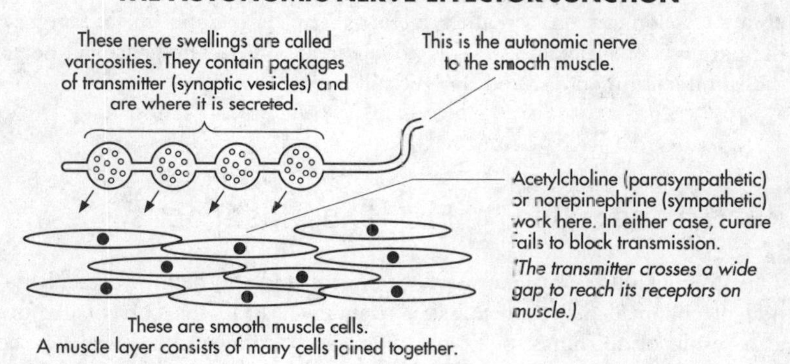

THE SKELETAL NEUROMUSCULAR JUNCTION

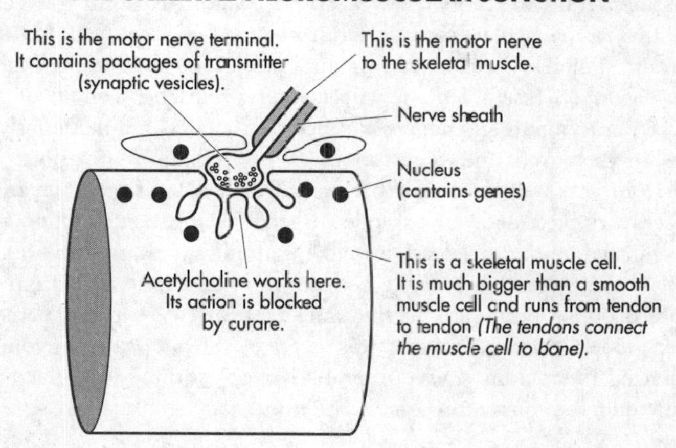

Modern surgeons still use curare for a purpose that is not very different from that of the hunters of South America. Naturally, surgeons do not shoot curare into patients from blowguns, but they do drip curare into patients' veins for its original intent: to paralyze muscles. This enables much less anesthetic to be used during surgery. In contrast to the animals poisoned by curare, artificial respiration is provided for the human patients who receive it. These patients need a respirator to breathe for them, but they survive because the cardiac muscle of the heart is different from the skeletal muscles, which are responsible for breathing. When nerves stop talking to skeletal muscle, the muscle becomes flaccid and lies still in the hammock of its bony attachments. Cardiac muscle, however, has its own intrinsic rhythm and beats regularly, even if it receives no information at all from the nerves that supply it. Transmission from nerves to cardiac and smooth muscle is not, in any case, affected by curare. As long as one breathes for a patient, as is done during surgery, the paralysis of skeletal muscles is not a threat to life. The heart still beats, the brain still functions, and the gut still moves.

Muscarine: Don't Eat the Mushrooms

The story of the autonomic nervous system and the definitive confirmation of chemical neurotransmission resumes early in the current century as a result, of all things, of English picnics. Once again, a poison figured heavily in the events. Picnics are popular in England, and wild mushrooms are considered by many to be the perfect, if not always safe, accompaniment. Sir Henry Dale, a don at Oxford early in this century, noted that a particular clinical syndrome, which occurred in unwary mushroom fanciers who gathered and ate a fungus by the name of *Amanita muscaria,* resembled the effects that would be seen if all of the patient's parasympathetic nerves had been stimulated simultaneously. The victims' eyes teared and their pupils constricted into tiny dots. Sweat oozed from pores all over their bodies, saliva drooled from their mouths, and feces dribbled or even exploded uncontrollably out of their anuses. Guts churned and moved so strongly that intestinal motion was often painful to the patient and visible beneath the skin of the abdomen. Blood pressure dropped ominously as the patients' hearts slowed and sometimes stopped, albeit briefly. Survival was not a sure thing, and anyone who experienced the condition was miserable. The syndrome was called "muscarism" after the mushroom that caused it.

The responses of effectors (smooth muscles, glands, the heart, and blood vessels) to muscarine, the poison extracted from *Amanita muscaria*, were all found to be identical to those elicited by the parasympathetic neurotransmitter, the identity of which was not known before Dale did his work. In contrast to curare, muscarine is absorbed when one eats it. Muscarine does not occur naturally in the body, so it was obvious that it was not itself the parasympathetic neurotransmitter; however, Dale found that all of the effectors that respond to muscarine respond identically to *acetylcholine,* which does occur naturally. Stimulated parasympathetic nerves, moreover, were demonstrated first by Otto Loewy, and then by Dale, to release acetylcholine. These observations suggested to Dale and others that acetylcholine might actually be the neurotransmitter of parasympathetic nerves.

Scopolamine and Atropine: Hot, Dry, and Wide-Eyed

Another set of plant poisons helped to confirm the suggestion of Loewy and Dale that acetylcholine is the parasympathetic neurotransmitter. Responses of effectors to muscarine, acetylcholine, and parasympathetic nerves are all specifically blocked by toxins that are present in an extract of belladonna and other plants. These compounds are relatively simple molecules and can be easily purified or synthesized commercially. The pure *antagonists* of responses to acetylcholine, like the belladonna extracts, also specifically block parasympathetic responses. One such antagonist is called *hyoscine* in Britain and *scopolamine* in America. Another, now used extensively by ophthalmologists to dilate pupils for eye exams, is called *atropine.*

The toxicity of scopolamine and atropine follows from their ability to block the actions of acetylcholine. Victims are said to become blind as a bat (the lenses of their eyes cannot accommodate), hot as Hades (they cannot sweat or spit), red as a beet (the blood vessels in their skin dilate), and mad as a hatter (the drugs enter the brain and do the same thing there that they do in the periphery). Their mouths become dry, their pupils open wide, and their hearts race. Both scopolamine and atropine thus abolish the responses of effectors to parasympathetic nerve stimulation, but they leave sympathetic responses intact. It seemed clear to Dale that acetylcholine must be the final parasympathetic neurotransmitter, and that the sympathetic neurotransmitter had to be something else. Dale's contemporaries agreed, and his conclusion has not since been challenged.

Dale won the Nobel Prize for his discoveries. The place in history, moreover, of the invidious mushroom *Amanita muscaria* and its repulsive toxin, muscarine, were also marked by the scientific establishment. That place is now memorialized in the official nomenclature. Those responses of effectors to acetylcholine that mimic parasympathetic nerve stimulation are called "muscarinic"; the *receptors* (molecules expressed on cell surfaces that act as switches that can be turned on by signaling molecules, such as neurotransmitters or hormones) upon which acetylcholine acts are known as "muscarinic receptors"; and the compounds, scopolamine and atropine, which prevent the activation of these receptors by acetylcholine, or compounds like it, are called "muscarinic antagonists." This nomenclature did not evolve because scientists wished sentimentally to remember the mushroom. The nomenclature evolved because acetylcholine has a much greater degree of versatility in its actions than its mimicry of the effects of muscarine might suggest. The actions of acetylcholine are not *all* muscarinic, and names had to be invented to distinguish those effects of acetylcholine that are muscarinic from those that are not.

Nicotine: Another Reason Not to Smoke

The junction between the postganglionic nerve cell and a smooth muscle, cardiac muscle, blood vessel, or gland is only the final link in a chain. You will remember that the neural pathway in the autonomic nervous system involves at least two nerve cells. The first cell must communicate with the second. Transmission from the first to the second nerve cell in a ganglion is thus just as important as the final (neuroeffector) junction in getting the message of the central nervous system to the effectors. Surprisingly, muscarine, which mimics the effect of acetylcholine at the end of the chain of nerve cells, was found to have no effect on transmission at the junction in the middle. When applied to either parasympathetic or sympathetic ganglia (the locations of the junction in the middle), muscarine neither mimicked nor blocked the action of the ganglionic neurotransmitter. In contrast, both of these ganglia were found to respond to nicotine, the very same nicotine that corrupts cigarettes.

Nicotine potently stimulates both sympathetic and parasympathetic postganglionic nerve cells. In fact, Langley used nicotine to map out the autonomic nervous system. He did this simply by painting nicotine onto ganglia and watching the parasympathetic or sympathetic responses. Whether the nicotine paint elicited a parasympathetic or a sympathetic effect was determined by which ganglion he painted. The response, in turn, established the division to which the ganglion painted with nicotine belonged.

In contrast to its striking actions on ganglion cells, nicotine was found to be powerless at the end of the autonomic chain of nerve cells. Nicotine thus does virtually nothing when it is applied to either parasympathetic or sympathetic neuroeffector junctions. For example, when nicotine is painted onto a ganglion that innervates a gland, the nerve cells in the ganglion become excited by the nicotine. They, in turn, excite the gland that they innervate, and the gland secretes. When nicotine is painted directly onto the same gland, however, nothing happens. With muscarine, just the opposite is true. If muscarine were to be painted onto the same ganglion that responded to nicotine, nothing would happen. On the other hand, if muscarine were to be applied to the gland that ignored the direct application of nicotine, the gland would secrete.

At first glance, these phenomena can seem to be dizzyingly inconsistent and even self-contradictory. After all, acetylcholine is the neurotransmitter that excites nerve cells in ganglia, and it is also the neurotransmitter that excites secreting cells in the gland. Why, then, should the ganglionic actions of acetylcholine be mimicked by nicotine and not by muscarine, while at the neuroeffector junction, the actions of the same molecule, acetylcholine, are mimicked by muscarine and not by nicotine?

Receptors: Ears for Chemical Words

The answer to this apparent paradox lies in the receptors, the molecules expressed on cell surfaces that enable cells to respond to signaling substances in their environment. Nerves talk with a chemical language. To hear a chemical language, one needs the molecular equivalent of ears. The "ears" that respond to chemical "words" are the receptors present in the membranes of responding cells. Signaling molecules, which may be quite discriminating in their molecular taste, bind to these receptors. After they have bound a signaling molecule, the receptors go on to initiate the Rube Goldberg–like "transduction machinery" that converts the binding of the signaling molecule into a physiological event. Drugs and toxins, like nicotine and muscarine, are able to take over the molecular switches that activate cells by mimicking natural signaling molecules and binding to receptors.

Many receptor molecules, which are totally different from one another, share the ability to bind acetylcholine. As a result of this shared property, and despite profound molecular differences, these receptors will cause any cell that expresses them to respond to acetylcholine. The nerve cells that are excited by acetylcholine in autonomic ganglia thus express receptor molecules that bind acetylcholine, as do the effectors in organs that respond to parasympathetic nerves; however, these cells each express a completely different type of acetylcholine receptor molecule. These receptor molecules can be distinguished from one another pharmacologically by the actions of nicotine, muscarine, or other drugs. Nicotine binds only to some of the receptors that recognize acetylcholine, while muscarine binds to others, and none of the receptors that bind nicotine bind muscarine. Acetylcholine can thus be thought of as a gourmand that binds happily to any type of acetylcholine receptor. In contrast, nicotine and muscarine are gourmets, molecules that are far more discriminating in their taste for receptors than is acetylcholine.

The nicotine-binding type of acetylcholine receptor is called *nicotinic.* The muscarine-binding type of acetylcholine receptor is, as we've seen, called *muscarinic.* Since nerve cells in ganglia express nicotinic receptors, they respond both to nicotine and to acetylcholine. Similarly, since effectors express muscarinic receptors, they respond both to muscarine and to acetylcholine. Muscarine ignores the nicotinic receptors in ganglia, and nicotine ignores the muscarinic receptors on smooth muscle, cardiac muscle, or glands. Modern investigators, using techniques of molecular biology, have recently uncovered the existence of a previously unsuspected cornucopia of acetylcholine receptors; nevertheless, the old toxins, nicotine and muscarine, have left their mark, in that they successfully distinguished the major categories of acetylcholine receptor and associated their names with them.

The fact that nerve cells in autonomic ganglia express nicotinic receptors and thus respond to acetylcholine does not, by itself, establish that acetylcholine is the ganglionic neurotransmitter. In the world of cells, news can be carried by more than a single messenger. Even worse, as we have just seen, molecular mimics, which impersonate the true messenger, are abundant, and cells may not be able to tell the difference between them. T. R. Elliot, long before Dale worried about mushrooms at Oxford picnics, had suggested that if a natural substance mimicked the effects of a neurotransmitter, then that substance must *be* the transmitter. Dale never accepted this argument, which in fact he exploded with his investigation of muscarine. Additional evidence, however, confirming that acetylcholine did double duty, transmitting at autonomic ganglia as well as at the neuroeffector junction, followed soon after the early work of Loewy and Dale.

Once again, curare, the very same South American dart poison that Claude Bernard had used to demonstrate the special nature of the skeletal neuromuscular junction, played an important role. The acetylcholine receptors at the neuromuscular junction, like their counterparts in parasympathetic and sympathetic ganglia, can be activated by nicotine (although nicotine binds somewhat more readily to the acetylcholine receptors in ganglia than to those at the neuromuscular junction). The acetylcholine receptors of the neuromuscular junction, therefore, are nicotinic. Since curare blocks the effects of acetylcholine at the neuromuscular junction, it follows that curare is a nicotinic *antagonist*. Antagonists, like neurotransmitters and their chemical mimics, bind to receptors. Unlike the molecules that activate receptors, which are called *agonists*, antagonists simply sit on the receptors without turning them on. By sitting on the receptors, the antagonists block the actions of neurotransmitters or other agonists simply by preventing the activators from binding to their chosen receptors.

As a nicotinic antagonist, curare would be expected to inhibit the action of acetylcholine and nicotine on ganglionic acetylcholine receptors, in the same way as curare prevents their action at the neuromuscular junction. To an antagonist, in a perversion of the words of Gertrude Stein, a nicotinic receptor is a nicotinic receptor is a nicotinic receptor. In fact, curare was found to do everything to ganglia that was expected of it. When curare is applied to autonomic ganglia, it stops acetylcholine and nicotine in their tracks and halts communication between pre- and postganglionic nerve cells. Years after Dale, the release of acetylcholine by stimulated parasympathetic and sympathetic nerves was confirmed by chemical means. This confirmation, when it eventually came, was the necessary and ultimate piece of evidence that acetylcholine is really the neurotransmitter in all autonomic ganglia. Few eyebrows were raised by the demonstration, however, because by that time it seemed like nothing more than the final crossing of the last "t."

We now accept, almost as dogma, that in the parasympathetic division of the autonomic nervous system, acetylcholine is the neurotransmitter at both of the synapses in the pathway. It functions in the ganglia to link the first nerve cell of the parasympathetic pathway to the second, and it also functions at the junction where the second nerve cell sends signals to the effectors under its control.

The effectiveness of curare as an antagonist at both the ganglionic and the skeletal neuromuscular junction demonstrates the similarity between the acetylcholine receptors found at these two synapses. Both are nicotinic; however, modern research, which has deciphered the genes and the messages that encode the actual receptors, has now revealed that the muscular and ganglionic nicotinic receptors are complexes of several individual molecules that are similar, but not identical, to one another. It is thus possible to design drugs that take advantage of these differences. In contrast to curare, these new compounds act as antagonists whose efficacy is tailored to be greater at ganglia than at the neuromuscular junction, or the reverse. This selective difference in potency at the two types of nicotinic receptor enabled a class of drugs known as *ganglion blocking agents* to be produced. The ganglion blockers have now been replaced by better drugs, but they were once very useful for the control of malignant high blood pressure. In contrast to curare, ganglion blockers can stop neurotransmission through autonomic ganglia without inducing a neuromuscular paralysis.

Since the activity of sympathetic nerves raises the blood pressure, ganglion blocking drugs can reduce nerve-induced high blood pressure (hypertension) by turning off the nerves. To be sure, the effects of ganglion blocking agents are not limited to just those sympathetic nerve cells that control blood pressure, nor do these drugs allow blood pressure to be regulated once they have been given. A treated person may thus stand up and faint because his/her blood pressure falls as blood pools in the legs and the blood pressure becomes too low to adequately support the brain. Still, while they were in use, the ganglion blocking agents probably did save the lives of many patients with malignant hypertension. Malignant hypertension and the stroke that he suffered because of it were what killed Franklin Delano Roosevelt. When I visited his grave at Hyde Park last summer, I contemplated how the history of the twentieth century might have been changed if ganglion blocking drugs had been available to his physicians.

Botulinum Toxin: The Wonder of Home Cooking

One more poison figured in establishing that acetylcholine serves as the transmitter at three major peripheral synapses: the parasympathetic neuroeffector junction, the nerve-nerve junction in parasympathetic and sympathetic ganglia, and the skeletal neuromuscular junction. This poison, like that of *Amanita muscaria,* also frequents picnics. Unlike muscarine, however, this poison is not an intrinsic constituent of food that is consumed by mistake. This poison is added by the chef. The chefs do not, like Lucrezia Borgia, deliberately poison their victims. Instead, they do their deadly work without a lethal intent. In fact, the killers who distribute this poison are often some of the sweetest and nicest of folks.

The poison is botulinum toxin. It is made by a bacterium, *Clostridium botulinum,* and causes an often fatal syndrome known as botulism. These germs cannot live in oxygen. They are thus said to be *obligate anaerobes,* which means simply that they are under an obligation to live in places that have no air. Since we live in air, and our blood distributes oxygen to our internal tissues, *Clostridium botulinum* normally poses no problem to us. There is not going to be a *Clostridium botulinum* plague occurring anytime soon. On the other hand, the organism makes spores, which enable it to persist even when growing conditions are not right. The spores are remarkably resistant and long-lasting. When the spores reach an environment favorable for growth, they germinate, like the seeds of a plant, and give rise to a portentous crop of living *Clostridium botulinum.* As the bacteria grow, they secrete their toxin and produce gas.

Spores can contaminate food. When food is canned or put up as preserves, therefore, the spores had better be killed by the canning or preserving process. There is no oxygen in a sealed can. As Giuseppe Verdi dramatized in the last act of *Aida,* when he essentially canned Radames and Aida, oxygen becomes depleted inside of a sealed vessel and is not restored. Organisms trapped inside that depend on oxygen, like Radames and Aida, are doomed, but organisms like *Clostridium botulinum,* which cannot stand oxygen, flourish.

Home-canned preserves that were not sufficiently sterilized by their unwitting makers are frequent offenders. They are sealed and lack internal oxygen. If spores are included in the cans, conditions inside are perfect for *Clostridium botulinum.* When opened, the *Clostridium botulinum*–infested preserves disgorge a burst of gas, which may not be noticed by the cook before he or she presents a delicious-tasting mixture of botulinum toxin and preserves to his/her unintended victims. In addition to preserves, improperly canned tuna is another threat. A can that bulges because it contains gas is a can that is not to be consumed. In evaluating cans, bulges must be distinguished from dents. A dent may be only the mark of a Teamster at work. A bulge, however, is likely to be the mark of *Clostridium botulinum* at work.

Botulinum toxin kills because, like curare, it abolishes transmission at skeletal neuromuscular junctions. The poisoned individual becomes gradually but inexorably weaker as nerves lose their ability to induce muscular contraction. Muscles that are used most frequently fail first, and the more a victim struggles, the faster he or she becomes paralyzed. Eyelids droop, the head sags, and wrinkles disappear from the face, along with the activity of the facial muscles that caused them. As the muscular partition that separates the cavity of the nose from that of the mouth ceases to function, speech becomes nasal. The weakness of the tongue causes words to slur. The legs lose their ability to support the body, and the arms can no longer reach out for help. Worse yet, breathing stops, which is why botulism can kill. Salvation lies in a respirator, but since botulinum toxin is not as reversible as curare, artificial respiration may have to be provided for months. Respirators, furthermore, are not found at most picnic sites.

Botulinum toxin is the most toxic of toxins, the most poisonous of poisons. Weight for weight, nothing is nearly as deadly. The toxicity of cyanide is trivial in comparison to botulinum toxin. The fact that Saddam Hussein was strongly suspected to have vats of the stuff in Iraq before the Gulf War was thus quite unsettling to the allied armed forces.

Botulinum toxin does more damage than curare because its mechanism of action is different. Curare interferes with the action of acetylcholine at nicotinic receptors. Botulinum toxin does not. Add acetylcholine locally to a junction that has been poisoned by botulinum toxin and it works as well as if nothing was wrong. Stimulate the nerve and the nerve conducts well. Stimulate muscles, glands, or blood vessels . . . everything works. The nerves, the effectors, everything looks beautiful, even when examined at ultrahigh resolution with an electron microscope. The problem is that botulinum toxin gets into nerve endings and digests critical proteins that enable the little internal packets of neurotransmitter to fuse with the cell membrane and release acetylcholine. A botulinum-poisoned nerve thus is unable to secrete its neurotransmitter. The nervous signal arrives at the terminal, which stands mute and unable to respond. All of the junctions at which acetylcholine is the neurotransmitter thus cease to function.

Unlike curare, which blocks only those junctions that utilize nicotinic receptors, botulinum toxin inactivates every nerve it can reach that utilizes acetylcholine as its neurotransmitter. Not only do the skeletal muscles fail in patients with botulism, but the pupils no longer contract, sweat ceases to form, and saliva stops flowing. Botulinum toxin thus provides the ultimate proof that acetylcholine is a neurotransmitter. Junctions that should release acetylcholine no longer do so after they have become poisoned with botulinum toxin. Like the laundry where the motto is no ticket, no wash, the rule of the nervous system is, no transmitter, no function.

Surprisingly, botulinum toxin can actually be useful. Not all muscle contractions are desirable. Muscles sometimes contract in disturbing, uncomfortable, or even unsightly ways. The most trivial use of botulinum toxin is its use as the ultimate cosmetic. Facial wrinkles are caused by the contractions of the face; therefore, wrinkles can be banished simply by paralyzing the facial musculature. Cutting the nerves to the muscles is a bad idea because the nerve trunks that are accessible and easily found by a surgeon would, when severed, cause the paralysis of the muscles responsible for smiles and the eyelids shutting. A little botulinum toxin, given the familiar and cheery name "Botox," can be applied to just the crease-forming muscles. More seriously, botulinum toxin works wonders for patients with serious medical conditions involving excessive or spontaneous muscle contractions. It is wonderful, for example, in the prevention of *belepharospasm,* the spontaneous and uncontrollable contraction of the eyelids. With respect to the gut, botulinum toxin is used to prevent the spasm of the muscle that forms the gate between the esophagus and the stomach. If this muscle stays shut, there is obviously a major problem because food will become jammed in the esophagus, unable to go down the gullet. In all of these cases, botulinum toxin is a wonder drug; however, it is only a short-lived wonder. Eventually a new apparatus for transmitter release is produced, the toxin wears off, and it has to be administered again . . . and again . . . in an endless cycle.

Sounds Great: It's ALL NATURAL!

It is interesting to think of muscarine, botulinum toxin, and nicotine when one sees advertisements touting products as ALL NATURAL, as if that property somehow confers a special blessing upon them. The assumption behind the ads is that the chemicals added to foods to preserve freshness, retard spoilage, and inhibit the growth of bacteria must be bad for you. The advertisers who push ALL NATURAL products want you to distrust those unseen gray men of government who make the rules about what goes into our food. An ALL NATURAL product, the ads imply, must be free of the artificial stuff that these conspiratorial men allow to be placed in what we eat. The mere fact that something is ALL NATURAL, however, should not be perceived as a recommendation. Some of the deadliest toxins known to humankind are ALL NATURAL. Muscarine, botulinum toxin, and nicotine are ALL NATURAL. They are not made by people but by mushrooms, bacteria, and tobacco plants.

Personally, I am very happy to buy my mushrooms from the Korean grocer on the corner. When my mood turns to nature, I can think of this store as a magical forest where all the mushrooms are edible and come in clean packages. The ease of gathering the mushrooms and my confidence in surviving their consumption more than compensates me for having to ingest whatever chemicals the mushroom suppliers may have added during handling. I also appreciate the industrial processing and quality control that goes into canned tunafish. Since *Clostridium botulinum* is a normal resident of the intestines of some of the fish, it is comforting to know that the improper canning of tuna is frowned upon, not only for aesthetic and moral reasons but also for legal ones. When it comes to cans of tunafish, it is good to have the government on your side.

My distrust of wild mushrooms and bulging cans of tunafish is an abstract response. I have learned what could be in them and am dispassionately cautious as a result. Tobacco, however, is personal. This product is something that I rigorously avoid in all its forms. Fourteen years ago, tobacco killed my brother, and just this year it killed my father-in-law. My niece was only a year old when her father died. My mother-in-law, who needs chronic care, was left alone by her husband's death and now lives with us. Tobacco is bad, not only because it may cause cancer, but also because its nicotine is a great stimulator of the nicotinic receptors in ganglia. That means that when you light up a cigarette, you also light up your autonomic nervous system. Fortunately, the concentration of nicotine in tobacco smoke is not high enough to cause serious or immediate harm to most people. Long-term problems, however, are another issue. Hypersensitive people are yet one more. I vividly remember such a person. A hideous medical condition involving hypersensitivity to nicotine is called *Berger's disease.* I met an individual with advanced Berger's disease when I was a medical student. He was smoking in bed with a mechanical device to hold his cigarette. He needed the device because he had no fingers. The nicotine from the cigarettes to which he was addicted had stimulated his sympathetic ganglia to close down his blood vessels so severely that his fingers and toes had become gangrenous and all had to be amputated. Small comfort to him that tobacco and nicotine are ALL NATURAL.

It is a terrible shame that some of the worst, as well as some of the best, things in life are free. The natural world provides us with good, and in equal measure provides us with bad. Natural, therefore, does not mean good for you, and artificial does not mean bad for you. The decline in stomach cancer that began in the 1940s, for example, may well have been caused by the addition of antioxidants such as *tertiary butylatedhydroxy-toluene* to cereal to prevent spoilage. There is nothing natural about tertiary butylatedhydroxy-toluene. The name cries out as if it were the personification of chemical evil, yet it prevents the formation of carcinogens in our stomach that occurs through the oxidation of foods that we eat. What we need, therefore, is to know the nature, not the source, of what we consume. We can learn, and have learned, a lot about the enteric and other parts of the nervous system from the study of poisons, but of course it is best not to derive this information from the investigation of their unwitting human victims. Experimental subjects are much better.

The Sympathetic Transmitter: Acetylcholine's Elusive Partner

The sympathetic nervous system turned out to be more resistant than the parasympathetic to scientific analysis. Acetylcholine was eventually established as the transmitter in sympathetic ganglia, but the transmitter that activated the effectors at the end of the chain of sympathetic nerve cells was clearly not acetylcholine and needed to be identified. The difficulty encountered in finding the final sympathetic neurotransmitter, however, was not troubling to our scientific forebears because they did not notice it. Shortly after the turn of the century, in 1904, T. R. Elliott, in England, had shown that a hormone extracted from the adrenal gland called *adrenaline* exerted effects that closely resembled those brought about by stimulating sympathetic nerves. The feeling thus became established that "adrenaline" was the final sympathetic neurotransmitter. This feeling sufficed, in lieu of definitive evidence, to give rise to the dominant two-neurotransmitter/two-division view of the autonomic nervous system. "Adrenaline" was equally accepted as the sympathetic neurotransmitter on the American side of the Atlantic, but we Americans insisted on using the rather less picturesque name, *epinephrine,* to distinguish the neurotransmitter from Adrenalin, the trade name of a drug.

Terminology is idiosyncratic. At the time the name was chosen, Americans, who often display a puritanical streak, evidently preferred a politically correct scientific lexicon to linguistic euphony. Personally, I think that the choice of epinephrine over "adrenaline" was unfortunate. A challenge that gets the "adrenaline" flowing seems formidable, but when the same challenge only gets the epinephrine flowing, it seems to be hardly a challenge at all. Still, as Shakespeare put it in another context, a rose by any other name would smell as sweet. Whatever they called it, few people on either side of the Atlantic fretted about identifying the sympathetic neurotransmitter. Like pornography, they thought they knew what it was when they saw it.

In T. R. Elliott's day, and for a long time afterwards, the tricks the nervous system can play on the unwary were not adequately appreciated. Consequently, no one demanded that a candidate neurotransmitter satisfy the rigorous set of criteria that we now use for neurotransmitter identification. In fact, these criteria, which occupied my full attention for a decade many years later when I applied them to serotonin, had not yet been formulated. That was a pity, because the identity of the transmitter at sympathetic neuroeffector junctions is *not* epinephrine, as everyone now knows, but its biosynthetic precursor, *norepinephrine.*

The role of norepinephrine as the final sympathetic neurotransmitter was finally discovered by a Swedish scientist, U. S. Von Euler, after World War II. The fact that the actual neurotransmitter is norepinephrine (and not epinephrine) is obviously an important detail. That detail, however, was not disquieting to Von Euler's predecessors, who had been convinced, albeit wrongly, that epinephrine mimicked the effects of sympathetic nerve stimulation. Unfortunately, epinephrine provides only a close facsimile; it does not actually mimic the real thing. The receptors that respond to epinephrine and norepinephrine—which are called in the trade *adrenoceptors*—respond differently to the two chemicals. As a result, the effects of epinephrine are not the same as those of norepinephrine or, therefore, of sympathetic nerve stimulation. The differences in the responses of organs to epinephrine and norepinephrine are easily overlooked. Before people knew what to look for, and more importantly, before they had the tools to look with, the effects of epinephrine and norepinephrine were not easily distinguished.

The hindsight provided by modern knowledge makes it easy for to us to criticize earlier workers, who lacked our information. It is hubris, however, for anyone who is engaged in scientific exploration to do so. Every year, I tell my students in my first lecture that at least half of what I am about to teach them will eventually be shown to be wrong. The trouble is that I do not know which half. The future is a rough taskmaster. Nevertheless, a herd instinct often grips the imaginations of scientists. Like lemmings, we are prone to charge over cliffs when a large enough pack of us moves in that direction. The idea that epinephrine was the sympathetic neurotransmitter was accepted before Von Euler, more because the concept had become dogma than because a good case had been made for it.

We can, however, easily understand why the mistake was made. The chemical structures of epinephrine and norepinephrine are very similar, and the amounts of neurotransmitter released by nerve stimulation are very small. Prior to the development of modern analytical techniques, made possible by the postwar emergence of the National Institutes of Health as a source of biomedical funding, epinephrine and norepinephrine could not be detected by precise chemical measurements. They were studied instead by *bioassays,* which measure not the hormones themselves but the responses the hormones evoke in a living system. The bioassays were sufficiently sensitive to detect and quantify the total content of epinephrine *plus* norepinephrine in body fluids and tissue extracts, but the bioassays then in use did not reveal that epinephrine and norepinephrine exerted different effects.

Only a few of the most highly perceptive and critical of observers, such as Walter Cannon in America, noticed differences in the responses of organs to epinephrine and norepinephrine. Unfortunately, although

Cannon actually determined that the effects of epinephrine and norepinephrine were not identical, he was unwilling to jettison the idea that epinephrine was a sympathetic transmitter. Cannon thus proposed that there was not one, but two, transmitters, which he called "sympathin I" and "sympathin II." This proposal is now history; however, there is no second sympathetic transmitter.

As we have already seen, poisons were critically important in determining the effects of acetylcholine in the body. With the exception of nicotine, which helped to define and locate sympathetic ganglia, there were no sympathetic equivalents of the poisons that revealed so much about the parasympathetic functions of acetylcholine. The investigators struggling with the sympathetic neurotransmitter were on their own. No deus ex machina dropped out a mushroom to save them. Since neither selective antagonists nor chemical methods of assay were available to Von Euler's predecessors, it is wrong to fault them for not correctly identifying the sympathetic neurotransmitter as norepinephrine. Even so, their acceptance of the wrong substance without sufficient evidence is not a practice that should be emulated. Fortunately, science is self-correcting. Whatever the herd is doing, truth eventually becomes known and stops the stampede. That happened with norepinephrine, and it was later to happen to me.

The Butterfly

The history of research on the enteric nervous system had been characterized by discovery and eclipse. By the time I began to study it, no field could have been more perfect for a young scientist. Almost anything I found had to be revolutionary. So many of the prevailing concepts about the enteric nervous system were misconceptions that every new observation I made would both advance the field and dispel an old belief. Many areas of biology were then unexplored and ripe for discovery. This one, however, was cloaked in denial and thus needed more than exploration. The enteric nervous system required both a reality check and public relations. Investigation was required, not only to find the truth but to dispel the widespread acceptance of untruth.

To me, research on the enteric nervous system seemed like the life cycle of a butterfly. The eggs that were to give rise to enteric neurobiology were planted by Auerbach and Meissner, who, in the middle of the nineteenth century, showed that a complex system of ganglia is present in the gut. These eggs incubated and then, as the twentieth century began, an independent nervous system, the equivalent of a second brain, emerged in the work of Bayliss and Starling, who demonstrated that the bowel can manifest reflexes even in the absence of nervous input from the brain or spinal cord. Like a lively caterpillar, the field voraciously consumed information and enlarged with the contributions of Trendelenburg and Langley until it peaked in Langley's 1921 book, *The Autonomic Nervous System*. This book was, for enteric neurobiology, a cocoon, the covers of which had hidden the field from sight for fifty years, even though the pages of the book contained a clear definition of the system.

By 1981, the cocoon was ripe for opening. A metamorphosis of the concealed and pupating field of enteric neurobiology—or, to use the term coined by David Wingate and given to me as a field to father, neurogastroenterology—was about to occur. The enteric nervous system, forgotten since the time of its definition in Langley's book, was on the verge not only of rediscovery but of reacceptance by the scientific establishment. The wonder of neurotransmission had been digested, powerful new drugs to manipulate neurotransmitters were under development, and new methods of investigation were at hand. A butterfly was thus about to take wing, in the form of a new field of research, and that event was scheduled to occur in Cincinnati.

3

THE TURNING POINT

MY OWN INTRODUCTION to the second brain came slowly and indirectly. It began with a love affair, not with the bowel or my wife, Anne, but with serotonin. In 1958, as an undergraduate at Cornell, I learned during a course called "The Physiology of Behavior" that serotonin was likely to be a neurotransmitter and that problems with serotonin might lie at the core of schizophrenia and other mental diseases. I resolved to do some work on serotonin myself as soon as I could.

The opportunity presented itself in medical school, now no longer far above Cayuga's waters but only a few feet above the East River. In those days, medical students were encouraged to take time off to do research, which is what I did during summers and a wonderfully productive year. When I first began, I never suspected that a neurotransmitter encountered during a course on physiology could lead me to the gut, but that is what happened.

First, Some History

My first experiments were designed to locate the sites in the body where serotonin is made. To do this I injected mice with a radioactively tagged form of serotonin's immediate chemical precursor. This molecule has an unhappily tongue-twisting chemical name, *5-hydroxytryptophan*, but, as is also true of southerners who dislike their given names, the compound is familiarly known by its initials, 5-HTP. As expected, the mice that I had injected with radioactive 5-HTP rapidly converted the injected molecules into radioactively labeled serotonin that I could easily detect and quantify. Locating the tagged serotonin, however, turned out to be somewhat more difficult, because it did not stick around very long. I thus needed to develop a method to hold the radioactive serotonin in place so that I could find it.

Serotonin is naturally broken down in the body at a relatively fast rate. For serotonin to function, this is all to the good; zip in, zip out, and never accumulate. For me, in my efforts to try to discover, and actually visualize, the sites where the radioactive serotonin was located, the fast natural rate of breakdown was counterproductive. It was also expensive, because no one gives radioactive chemicals away. I needed to preserve the radioactive serotonin intact, so that I could locate the sites where the radioactively tagged 5-HTP I had injected was converted to labeled serotonin. I reasoned that this could be done if I also injected the mice with a drug that prevented the breakdown of serotonin.

Fortunately for me, a drug that exerted exactly this action had recently been introduced for the treatment of clinical depression. Serotonin is broken down in the body mainly by an enzyme with the obligatory long name *monoamine oxidase.* Drugs that inhibit this enzyme, which are classified, naturally enough, as *monoamine oxidase inhibitors,* tend to preserve serotonin and cause it to accumulate within the cells that make it. The first of the monoamine oxidase inhibitors to be used clinically, a drug called *iproniazid,* was initially given a trial as an agent for the treatment of tuberculosis. Although iproniazid was not as good against tuberculosis as its relative, *isoniazid,* the clinical investigators who were studying iproniazid unexpectedly observed that many of the sick and depressed patients who received the drug were no longer depressed, even if iproniazid failed to cure their tuberculosis. On the basis of these observations, iproniazid was dropped as an antitubercular agent but was tested to see if it would work as an antidepressant. Iproniazid passed that test, and although the risk of giving iproniazid (it occasionally, and for no obvious reason, destroyed the livers of unlucky patients) eventually caused it to vanish from the market (to be replaced by the antidepressant monoamine oxidase inhibitors Marplan, Nardil, and Parnate), iproniazid left behind two great legacies. One was the idea that drugs could really be used effectively to combat mental illness, and the other was that serotonin was a substance that played a critical role in the creation of happiness. A theory that the malfunctioning of serotonin as a neurotransmitter in the brain caused depression was launched. This theory is still in vogue.

The ability of the monoamine oxidase inhibitors to alleviate depression was not what interested me in them. What really excited me about the monoamine oxidase inhibitors was that they might make it possible for me to actually carry out the experiments that I wanted to do. The monoamine oxidase inhibitors, I realized, had the ability to prevent the destruction of the radioactive serotonin I had produced, at great expense, in mice. I thus injected iproniazid simultaneously with radioactive 5-HTP. The monoamine oxidase inhibitor performed as advertised, and the radioactively labeled serotonin remained intact long enough for me to look for it.

While iproniazid took care of one of the biological obstacles that stood in the way of my experiments, there were others. Radioactivity is easily detectable, but simply finding radioactivity in a sample does not, by itself, provide a clue as to what substances in that sample are radioactive. The analysis of radioactive material is a little like the analysis of the use of a bank's ATM machine. It is easy to look at the bottom line and determine how much money was withdrawn in a given period of time, but to discover the identity of the people who obtained the money, one would have to ascertain the PIN code of each user and then decode them to produce the names. I had injected radioactive 5-HTP into mice, and I found radioactive serotonin, as I thought I would, but I still needed to know what other radioactive compounds the mice had made and how to distinguish radioactive serotonin from them. Equally necessary, I needed to distinguish radioactive serotonin from any residual radioactive 5-HTP that had not yet been converted to serotonin. Essentially, a molecular PIN code was required. I also needed to find a method to hold the radioactive serotonin in place while I went about looking for it. Inhibition of monoamine oxidase would be helpful in this regard, in that it would prevent the formation of radioactive breakdown products of serotonin. The innovation, however, that made my experiment feasible came unexpectedly and easily from my studies of how to preserve or "fix" radioactive serotonin in place.

I was testing the adequacy of a variety of preservatives or "fixatives" of the kind that are generally used for the microscopic examination of tissues. Formaldehyde seemed almost to work. When I added the formaldehyde, there was a brief moment during which the outflow of radioactive serotonin increased, but then it stopped totally. Except for the initial moment of disaster, this outcome would have been perfect. The formaldehyde must have chemically coupled the radioactive serotonin to protein successfully, because nothing short of incineration could extract it. The initial brief increase in the outflow of radioactive serotonin, however, was a problem that had to be prevented. If radioactive serotonin was allowed to dance in whatever direction the tides of molecular motion took it, there would be no point in finding out where it ultimately came to rest. I wanted to know where serotonin is made during life, not where its artifactual motion happens to stop.

For some time, I was stalled by the problem of the initial fixative-induced loss of radioactive serotonin. The solution turned out to be delightfully simple. The aldehydes that I was using did not properly balance the salts and other molecules in the tissues and were causing cells to swell. When I corrected the salt balance, the outflow of radioactive serotonin ceased as soon as the tissue entered the fixative, and after fixation, no solvent could extract it. As an extra bonus, I also found that none of the other radioactive compounds that were present in the tissues of animals injected with radioactive 5-HTP were similarly fixed. After fixation, serotonin was the only radioactive compound left in the tissue, and I had evidence that the process of fixation did not change serotonin's location. Formaldehye had made a molecular PIN code unnecessary. If an ATM has only one user—in this case, serotonin—it does not take a PIN code to figure out its identity.

My next experiments, made possible by iproniazid and aldehyde fixation, were uncomplicated. At various times after the injection of radioactive 5-HTP (serotonin's precursor), I set out to determine the location of radioactive serotonin in the mice. My goal was not just to find out what organ, or even which layer of an organ, contained the labeled serotonin but to find first the cells and then the parts of the cells (called *organelles*) that made it.

To locate the "hot" serotonin at this level of resolution, I used a technique known as *radioautography*. Radioautography is another long word, like so many employed by scientists, but conceptually the technique is simplicity itself, and in contrast to many of the other long words of science, radioautography is a logically constructed name. To locate a radioactive substance by radioautography, radioactive sections of tissue are coated with a photographic emulsion and put away in the dark for several weeks. During this time, the subatomic particles of radioactive decay bombard the emulsion immediately above the tissue sections and a latent image forms. After exposure, the coated sections are developed as if they were film. Silver is precipitated in the region of the latent image, just as if the latent image had been formed by light. In essence, therefore, the radioactive material in the tissue takes its own picture. The picture is thus a "radioactive autograph," which is shortened to radioautograph, and the method of making radioautographs is therefore called radioautography.

The radioautographic result of my experiment, which riveted my attention on the gut, was that every time I injected radioactive 5-HTP into mice, I found radioactive serotonin in their enteric nervous system. Moreover, and just as important, I did *not* find radioactive serotonin in any other nerves outside the brain. This demonstrated that nerves in the bowel have an affinity for serotonin that other peripheral nerves do not share.

While my demonstration that enteric nerves have a unique affinity for serotonin did not entitle me to conclude that serotonin is a neurotransmitter in the gut, that seemed to be the simplest explanation. Cognizant of the dictum that "when you hear the sound of hoofbeats, don't think of zebras," I conducted one more experiment to see if serotonin would indeed behave like an enteric neurotransmitter. This time, I stimulated reflex activity in the gut to make its nerves work. Working nerves release their neurotransmitter. Sure enough, when the nerves of the gut from the mice that I had injected with radioactive 5-HTP were stimulated, they released radioactive serotonin.

The experiments I had conducted to this point gave me a feeling of confidence that my work could withstand anyone's scrutiny, which I assumed (foolishly, it turned out) would be both logical and reasonable. I also thought that my data would be considered to be important by other neuroscientists. I wrote up my results in a series of three articles that appeared in *Science* and the *Journal of Physiology*. My suggestion that serotonin might be an enteric neurotransmitter was based on the following pieces of information: (i) Serotonin is manufactured and stored in the bowel. (ii) Following its biosynthesis from its immediate precursor, serotonin is preferentially located in enteric nerves. (iii) These nerves release serotonin when they are stimulated. (iv) Others had previously shown that serotonin exerts the same effect on the bowel as does the stimulation of enteric nerves. If serotonin was not a neurotransmitter, therefore, it was certainly giving a pretty good imitation of one.

My Mother Never Told Me It Would Be Like This

Since I had not anticipated that my suggestion that serotonin might be a neurotransmitter in the gut would be viewed by the scientific world as outrageous, I was upset by the reaction I actually encountered. My first impulse was to feel empathy with those of my ancestors who faced the Inquisition. Later, after I became numb and ceased to feel pain, I understood the reaction that I had inadvertently caused. According to the scientific gospel that was prevalent at the time, only two transmitters, acetylcholine and norepinephrine, took care of all of the neurotransmission that went on in the peripheral nervous system. The thought that an additional molecule might be a peripheral neurotransmitter was considered not just wrong but perverse and immoral. Scientists, more than most people, admire order, and the order that had been established in the peripheral nervous system left no room for another neurotransmitter.

Disorder is so widespread in nature that when scientists believe that order has been encountered, they immediately think that some great force has been at work to overcome the sinister effects of chance. All fledgling scientists learn in Physics 101, if they have not been taught it earlier in Introductory Chemistry, that disorder in the universe is always increasing. This ever-escalating disorder is called *entropy*. To overcome entropy, the Darth Vader of reality, serious work has to be done. The molecules that assemble to form the human body would never do so on their own if they were simply mixed together. Countless thousands of unlikely chemical reactions have to occur in just the right place and at just the right time. Those with a deeply religious inclination contemplate the sheer improbability of these events and turn to God for an explanation. Scientists, however, have surrendered this option, even if they, like me, believe in God.

When we scientists see order, we tend to think that we have found biological reality. Biological processes exert the kind of work and provide the energy necessary to overcome entropy. They impose order on the otherwise reluctant molecules of life, getting them to react with one another to establish the form that we have come to love. For me to upset the order that people thought had been found in the peripheral nervous system was not to be tolerated lightly. My idea that serotonin might be an enteric neurotransmitter was incompatible with the orderly belief that had been held for a long time and thus was much cherished. Back in 1965, the entire peripheral nervous system could easily be described in a simple chart:

	MOTOR SYSTEMS		
	Skeletal (voluntary)	Autonomic (involuntary)	
		Sympathetic	Parasympathetic
Final transmitter:	acetylcholine	norepinephrine	acetylcholine
Targets:	skeletal muscles	glands, blood vessels, heart and visceral muscles	

Note that the autonomic nervous system only had two subdivisions, because the third was in eclipse. The two acceptable autonomic divisions, the sympathetic and the parasympathetic, were believed always to oppose one another, like the free world and the communist empire of recent history. There was a concept of a beautiful duality, two autonomic systems, two neurotransmitters, ever in battle, ever in opposition. My innocent suggestion that serotonin was an autonomic neurotransmitter thus was unsettling. If I was right, the perceived duality was wrong. I was a heretic, and was being treated accordingly.

I decided to be obstinate. I was young in 1965, as well as combative and idealistic. I had confidence that truth, as I saw it, would inevitably win out. Even the most orthodox of my detractors, I believed, could not prevent facts from emerging to demolish the dogma of the fundamentalists. Besides, in those days the National Institutes of Health was tolerant of ideas that opposed received wisdom. Its peer-review panels provided funding even for experiments that were not guaranteed, in advance, to work. Funds were thus available to me to pursue the issue of serotonin as an enteric neurotransmitter.

I could, and did, proceed to put serotonin through a set of tests (modeled after "Koch's postulates," which establish the cause of an infectious disease), which are accepted by virtually all neuroscientists as the criteria that must be met by any substance if it is to be accepted into the pantheon of established neurotransmitters. To satisfy this biochemical equivalent of the labors of Hercules for enteric serotonin, it was necessary to prove:

1. that serotonin is really present in the nerve endings at the sites where I had proposed that serotonin might be the neurotransmitter;
2. that serotonin exactly mimics the effects of the natural neurotransmitter;
3. that serotonin is actually released when the nerves that contain it are stimulated;
4. that blocking the action of serotonin (or depleting serotonin) abolishes the effects of nerve stimulation; and
5. that there is an effective inactivating mechanism, which can turn off the response of nerve cells to serotonin once neurotransmission has been accomplished.

Oxford

Fortunately, having produced some unsettling data, I had the opportunity to begin to study some of the actions of the enteric nerves that use serotonin as a neurotransmitter. I had landed a postdoctoral fellowship to work at Oxford in 1965 and 1966.

Oxford, in those years, was a spectacular place for a young American to learn to be a scientist. The traditions of scholarship were deeply embedded, and the atmosphere was one of contagious learning. After I located the sites of the biosynthesis of serotonin in the enteric nervous system, I discovered the ongoing work of Edith Bülbring, carried out at Oxford, on the initiation of the peristaltic reflex. This was the same reflex that Bayliss and Starling had investigated in dogs and Trendelenburg had studied in isolated loops of guinea pig intestine. Edith had published evidence that suggested that serotonin was released when pressure was placed upon cells in the lining of the bowel, and that this serotonin stimulated intrinsic enteric sensory nerves to start the peristaltic reflex. When I wrote to her, describing my observations, she was enthusiastic about having me come to work with her to pursue the role of serotonin in the enteric nervous system.

The National Institutes of Health agreed to fund my training in Edith's laboratory. It was my good fortune to be able to get training in the days before overwrought chauvinism interfered with the willingness of our government to enable us to learn from the expertise of foreigners and to benefit from the facilities provided by laboratories in other countries. My wife, Anne, had just finished a year of interning at New York Hospital and she, too, applied for funding by the National Institutes of Health to work in the Sir William Dunn School of Pathology at Oxford. Given that she was to train in the laboratory where penicillin had been discovered, we thought she had a good chance of obtaining grant money. We expected finances to be tight until we heard about her scholarship, which would not occur until the fall, but with the confidence of youth, we gambled on her funding coming through and sailed to England with our three-and-a-half-year-old son.

The laboratory of Edith Bülbring in 1965 was a very advanced place to do research on gastrointestinal smooth muscle. Edith provided atmosphere, direction, and funding. The other postdoctoral fellows provided the hands-on training. Edith's laboratory was run in a Germanic style. She swept into the laboratory every day at ten and we all stood to say "good morning, Dr. Bülbring." She was Edith to everyone, except when you spoke to her; then she became either Professor Bülbring or Dr. Bülbring. Actually, *almost* everyone stood. Graeme Campbell, a very large Australian, stood up for no one. He just sat at his bench with his feet up and a cigarette in his mouth. I am absolutely sure that he timed his morning smoke to correspond with Edith's arrival. She hated smoking in the laboratory, because she was sure that the nicotine was bad for the preparations. She was sure that the smoke made the gut contract. Graeme's performance always had the desired effect. It drove Edith to distraction. After seeing Graeme, Edith would sit in her office and sulk for at least a half hour. When Graeme left in the middle of the year, Edith's whole being seemed visibly brighter. My demeanor fell, however, because Graeme had been my principal teacher.

In Edith's laboratory, I pursued several lines of research, each of which led to major publications. One investigation identified the site of action of sympathetic nerves within the gut. Another characterized the effects of *tetrodotoxin,* a poison made by puffer fish, which are found in coastal waters off Japan, on the bowel and other organs that contain smooth muscle. The meat of the puffer fish (*fugu*) is a delicacy that is extremely popular in Japan. Skill is needed in preparing the fish, however, or a dinner of puffer fish can become a last meal. The organs that contain the toxin must not be allowed to contaminate the meat.

Tetrodotoxin blocks the conduction of signals by nerve cells and skeletal muscles, but it does not affect smooth muscle. Taken internally, therefore, tetrodotoxin will stop breathing, which depends on intact nerves and intact skeletal muscles. Pharmacologically, however, when used to study tissues maintained outside the body in an organ bath, tetrodotoxin, like muscarine, nicotine, and botulinum toxin, is a powerful experimental tool. The mechanism by which a drug brings about a visible response in an organ can be ambiguous. For example, smooth muscle might contract, either because the drug stimulates nerves, which indirectly evoke the observed response, or because the drug affects the muscle directly. Tetrodotoxin distinguishes between these two possibilities. Indirect effects that are mediated by nerves are abolished by tetrodotoxin, while those due to direct actions of drugs on smooth muscle persist unchanged.

The last of my Oxford studies, and the ones that were for me pivotal in shifting my focus from serotonin to the second brain itself, strongly suggested that serotonin is a neurotransmitter that stimulates a set of intrinsic nerve cells in the wall of the guinea pig stomach. These particular cells, which can also be activated by stimulating the vagus nerves, had the ability to cause the stomach to relax. Tetrodotoxin abolished the relaxant response to serotonin, confirming that it was due to an action of serotonin on nerve cells and not a direct response of the smooth muscle of the stomach. These observations suggested to me that serotonin might function not in the direct conversation between nerves and smooth muscle but in the important cross talk that goes on between nerve cells. The nerve cells that specialize in this type of kibitzing are called *interneurons.*

Interneurons are the cells that add the layers of complexity and sophistication that distinguish the central and enteric nervous systems from the banal peripheral ganglia found outside the bowel. The nerves of the gut do not just slavishly pass signals from sensory receptors to muscles, glands, or blood vessels. Because of its interneurons, the enteric nervous system can modulate and process the information it receives. Serotonin, I speculated, as the neurotransmitter of an interneuron, might be one of the molecules that enable the bowel to function as an independent information-processing center. In fact, I suggested, drugs designed to modify the actions of serotonin might be great tools for treating patients whose bowel is malfunctioning. It seemed to me to be far more likely to be beneficial, from a therapeutic point of view, to try to influence the activity of interneurons, such as the serotonin-containing nerve cells of the gut, than to try to affect either the sensory nerve cells that initiate critical reflexes or the motor nerve cells that activate muscle or glands. The nervous system of the bowel must acquire information in order to work at all, and it must be able to act on the basis of the information it receives. Interference with either the ability of the enteric nervous system to receive information or to send signals to its effectors would be likely to paralyze the gut, a therapeutic outcome that I doubted would be perceived by any patient as desirable.

Cincinnati

From the time I returned from Oxford until 1981, I devoted myself to jousting in the scientific literature with a variety of colleagues who seemed dedicated to proving me wrong. My papers were always followed by their papers, which interpreted my data in ways that I never thought possible. In 1981, however, things had crystallized in Cincinnati.

I am a neuroscientist. As such, with most of my colleagues, I was in Cincinnati in November of 1981 to attend the meeting of the Society for Neuroscience. Over twenty thousand scientists attend this meeting today, but even in 1981 it drew a crowd. People go to find out what is new in the field and to make their own work known. Success in science comes not just from excelling but from excelling in such a way that important colleagues know you have excelled. Leonardo da Vinci, for example, made accurate drawings of the human form before Andreas Vesalius, but Vesalius made his anatomical work known to his contemporaries, and Leonardo did not. As a result, modern anatomy is thought to begin with Vesalius, even though he was scooped by Leonardo.

Vesalius had more than style going for him. He also presciently applied a great technique, still in use. He challenged received wisdom. Vesalius refuted the word of Galen (a physician of ancient Rome), whose dogma almost everyone else accepted. It is wonderful to be correct about anything, but it is even better to be right when being so shows that others are wrong. Before Vesalius, the teaching of anatomy involved, by custom, reading from Galen and pointing out the structures Galen discussed, whether or not these structures were actually there; unfortunately, many were not.

Vesalius's approach to anatomy was radically different from that of his contemporaries. He dissected and described only what he could actually see. Although Vesalius's publications outraged his contemporaries, they had to dissect to prove him wrong. Vesalius's antagonists could not defeat him because it is not possible to verify the presence of what does not exist, or to cause what does exist to disappear. Vesalius thus accomplished what any scientist dreams of doing. He proved the current wisdom wrong, drew a great deal of notice to himself in the process, and his work was independently reproduced by others and confirmed (the essence of scientific progress).

I am no Vesalius, but in a way that I never suspected at the time, it turned out to be his footsteps that I would follow in Cincinnati. I too was right, the majority of others in my field were wrong, and many of them had become very nasty in their reactions to what I was saying. The Neuroscience meeting in 1981 was to be, both for myself and my field, the turning point. My antagonists capitulated.

The meetings of the Society for Neuroscience are a place where old concepts go to die and new knowledge gets a chance to fight for its place in the canon. They are a Darwinian arena where theories clash and only the fittest survive. Facts presented for the first time at these meetings can descend on the unwitting without warning, like V2 rockets out of the 1945 London sky. Unfortunately, when a theory is shot down, its theorist may go with it. Neuroscientists thus present their work at the meetings with smiles on their faces and sweat on their palms.

A second reason why the Neuroscience meetings can be unsettling is the issue of priority. Since scientists are observers, they gain respect by being the first to note and report something of interest. In the major leagues of science, as in baseball, coming in second is a losing option. To this day, under the best of circumstances, the Neuroscience meetings make me nervous, a feeling that I perceive, unfortunately, from the bottom up. My anxiety begins in my colon. At the Neuroscience meetings in 1981 I had more reason than usual to be anxious. I had been asked to arrange a "workshop" on the enteric nervous system, the intrinsic network of nerve cells and fibers in the gut. I should have been pleased. The enteric nervous system was then the center of my universe. Only my wife and children were (and still are) more important to me. Given the way the field had been ignored, moreover, the fact that the "workshop" was scheduled at all indicated that my subject had arrived. This meeting could be an epiphany, the final turning of my worm.

I had, however, become controversial. It was *my* theories that were about to fight the Darwinian fight at that meeting. I thought I was right, but as I now hear incessantly in commercials for the New York State lottery: "Hey, you never know." Scientists, including myself, do not create our results. Since Vesalius, that approach has been out of style. "Creative" is not a word we scientists want to have applied to our data. Even "imaginative" is acceptable only if it pertains to our methods. Our actual observations, our real output, are neither creative nor imaginative. They are what they are, factual descriptions of what *is,* unembellished and unvarnished.

Although my theories were under siege, I had decided, nevertheless, not to play it safe. If there was going to be an epiphany, it was going to be a good one. I had agreed to set up the workshop, and I had planned it as a duel: two of us, myself and a colleague, Jackie Wood, to fight for what I thought was right; and two others, Marcello Costa and Alan North, against. Idealistically, it was the right thing to do. Let the ideas collide and the truth would win. Nine months earlier that had seemed obvious to me. Now that the collision was about to occur, however, I was scared and would have been very happy to be somewhere other than in Cincinnati.

I had envisioned the upcoming workshop on the enteric nervous system as a coming-out event, a neuroscientific debutante ball, introducing an appealing new subject to polite society. The major question we were going to debate, however, was not the nature of the system but the idea that serotonin is a neurotransmitter in the bowel.

By 1981, serotonin had acquired a long track record, and virtually everyone assumed that serotonin was a very important neurotransmitter in the brain. In fact, the neuroscientific world was transfixed by the possibilities that it thought the understanding of serotonin would open up for manipulating mood and treating psychiatric illness. Serotonin was then, and still is, a "hot" neurochemical. Even in 1981, the gut was known to manufacture very nearly all of the body's serotonin. Since only 1 percent or so of the body's serotonin is made in the brain, the brain's output can thus be thought of as a minor supplement. If the tiny bit of serotonin in the brain could be a source of tremendous excitement, then it seemed reasonable to me that the massive concentration of serotonin in the bowel might elicit at least a flicker of interest. Furthermore, since serotonin was already regarded as a neurotransmitter in the brain, there seemed to be no reason why it should strike people as inconceivable that serotonin might also be a neurotransmitter in the enteric nervous system. Besides, I always had been careful to note that my experiments were merely suggestive and that further work would be necessary to prove whether or not my suggestion was correct. Having thus hedged my bet while stating what I thought was obvious, I was prepared for trouble.

The Conference

By 1981, the time of the debate in Cincinnati, I had tested serotonin's ability to fulfill each of the five postulates (see page 43) necessary to establish it as an enteric neurotransmitter. In experiment after experiment, serotonin was provided with an opportunity to fail, but it never did. Every requirement had been met. I thus thought that the time was now right to tell the world not that serotonin *might be* an enteric neurotransmitter but that it *is* one. The old myth that there were only two divisions and two final neurotransmitters in the autonomic nervous system was about to vanish from the neurobiological world. I believed strongly in my ideas and in the results of my own experiments. Still, push was now coming to shove, and was doing so before a large and critical audience.

Many people shut or cover their eyes when they are frightened so as not to see whatever it is they dread. I do not have to do this. I can look straight ahead when I am scared and see nothing. That is what I was doing as I approached Cincinnati. I saw nothing as my mind wandered to the *dramatis personae* of the little docudrama I was about to stage.

JACKIE WOOD

I knew Jackie Wood best. Jack is his nickname; the name his parents gave him is actually Jackie. Jack is soft-spoken, gentle, and good-humored. When he speaks in public, everything comes across as simple and heartfelt. Jack seems open, straightforward, and utterly uncomplicated. Although Jack looks like he would be far more at home in a football stadium than in a laboratory, things are not always as they seem. Jack is as driven and ambitious as the rest of us.

Jack had always been interested in the biophysical and electrical properties of enteric nerve cells, while I had been concentrating on their anatomy, chemistry, and pharmacology. We were thus able to reinforce one another without getting in each other's way. We had a common bond between us. For each of us to succeed as scientists, we would have to make the enteric nervous system a household (or, in this case, a laboratory) word. Cooperation and collaboration, rather than competition, was in our interest. We were comrades, enlisted men in the same tiny army, fighting a war for recognition of a discipline that we were creating. It troubled us that most neuroscientists knew almost nothing about what either of us studied, and, of course, it troubled us even more that most of them did not even care. As far as the neuroscience community at that time was concerned, the enteric nervous system did not exist. In fact, at a previous meeting of the Society for Neuroscience, I was appalled to find that my papers were scheduled in a session on invertebrate neurophysiology. Invertebrates are squishy and crunchy beasts that have no backbone (vertebral column) and no enteric nervous system. My research had been carried out in guinea pigs, which, like humans, are vertebrates in good standing. The committee of neuroscientists who had planned the program did not even know what the enteric nervous system was; they thought that the myenteric plexus belonged to an exotic organism without a backbone.

My immediate reaction was to bring a guinea pig to the meeting, but I rejected that idea as unfair to the animal. It was also inappropriate. I may think combatively, but I rarely act that way. I simply swallowed my liver and presented my data. I understood, however, that, like Moses in the land of Midian, I was a stranger in a strange land when I talked to neuroscientists about the enteric nervous system. It would clearly be necessary to tell the neuroscientists not only what the enteric nervous system was but also why they should want to know about it.

Neuroscientists, as a group, often tend to believe that the entire body exists just to support the brain. It is the brain that thinks, emotes, and remembers. The brain is happy or unhappy, content or troubled. Philosophy, poetry, faith, and reason are all products of the brain. Neuroscientists, if they thought of the enteric nervous system at all, therefore considered it a bit player, a supporting actor in a drama in which the brain was, is, and always will be the star. Of course, now that the enteric nervous system has gained recognition as a second brain, increasing numbers of neuroscientists are willing to grant it second billing. Jack Wood and I both knew that the enteric nervous system is not so different from the brain. We understood that the second brain is very similar to the first. In reality, the enteric nervous system can even be thought of as the brain gone south.

Jack is a pioneer, one of a tiny group of intrepid workers who showed that useful data could actually be obtained by recording the electrical activity of individual enteric nerve cells. It was not that anyone doubted the importance of this type of measurement. We just wondered whether this kind of data could routinely be obtained from nerve cells in the gut. Recording the electrical activity of individual enteric nerve cells, until relatively recently, meant piercing the surface membrane of a single cell with a sharp glass pipette (micropipette) filled with a conducting solution, so that one can measure the electrical potential difference across the membrane of the cell. This may sound simple, but individual enteric nerve cells are small targets, they are buried deep in the substance of the wall of the gut, and since the muscle around them contracts and relaxes, the nerve cells tend to move. It is not easy to hit tiny moving targets buried in a piece of tissue. To stick a pipette into one requires that the nerve cells be dug out and visualized in living tissue, the movement of the gut be stopped without altering the activity of the nerve cells, and whoever wields the pipette have a good aim.

Actually, Jack was not the first scientist to impale an enteric nerve cell with a microelectrode and record its electrical activity; nevertheless, he systematized that approach; established its effectiveness; taught many others, including me, how to do it; and made its use routine. Jack is to the electrophysiology of the enteric nervous system what Picasso is to twentieth-century painting. In a simple, kindly sort of way, moreover, Jack gives in to no one. I expected that Jack was going to be a formidable advocate and a major source of support in the coming battle of Cincinnati.

Jack Wood had made critical electrical observations that demonstrated that serotonin was uncannily precise in mimicking the effects of one of the transmitters released from stimulated enteric nerves. Before this work, a troubling paradox had existed. I had been studying the effects of serotonin on large populations of neurons, while Jack and others had been examining actions on individual cells. My approach was wholesale; Jack's was retail. I had seen the effects of serotonin on the whole population that Jack had not seen in studying individual cells. We did not understand how serotonin could activate groups of neurons in my experiments if it did not stimulate individual cells in his. Something had to be wrong. Jack's new data had resolved the paradox. In fact, Jack was so excited by his observations that he made a special trip to my laboratory in New York to share them with me.

Jack resolved the paradox by introducing a new experimental method. As in previous experiments, Jack had impaled enteric nerve cells with glass microelectrodes in isolated preparations of gut. This enabled him to record the electrical responses of the impaled cells to nerve stimulation and to serotonin; however, now instead of simply adding serotonin to the tissue bath, Jack had spritzed serotonin at impaled nerve cells from a second micropipette. This new technique approximated the way an activated nerve releases its neurotransmitter. The rapidly ejected serotonin was able to reach cell surfaces before becoming inactivated by the tissue. Receptors could "see" the serotonin and trigger a response before they became desensitized and unresponsive. Jack's spritzes worked, and nerve cells could now be seen to respond to serotonin as predicted from my pharmacological work.

Much more interesting than simply confirming my predictions, Jack's research also showed that both the transmitter released from stimulated enteric nerves and the application of serotonin evoked an identical *slow excitatory response* in a specific type of enteric nerve cell. Given that I was postulating that serotonin and the neurotransmitter naturally released from certain enteric nerve cells are one and the same, it would have been terribly embarrassing if Jack's observations had been different. As it was, Jack made my case very strong. After hearing his data, arguing against serotonin as a neurotransmitter became a little like the argument (found in an errant high school essay) that "the plays of Shakespeare were not written by Shakespeare, but by another man of the same name." In fact, the slow excitatory response that Jack studied was unusual because the electrical resistance of the nerve cell membrane actually increased during the event. This meant that the channels in the membrane through which salt ions could flow were closing. It is much more common for membrane ion channels to open during the excitatory responses of nerve cells. The fact that both serotonin and the natural transmitter should close the same ion channels in the same cells thus seemed especially significant. Molecular mimicry of this kind is unlikely to happen by chance alone. It is therefore strong evidence that the natural transmitter is really serotonin. It was to take another fifteen years to learn the molecular basis of these changes. We have, however, finally been able to identify the serotonin receptor responsible for them as well as the molecular events the stimulation of this receptor by serotonin sets in motion to close ion channels in responding nerve cells.

Closing these ion channels in the way that serotonin closes them makes the enteric nervous system much more excitable. Nerve impulses are much more likely to travel long distances down the gut. It is possible, therefore, to soothe the irritation of an irritable bowel by interfering with the action of serotonin at this particular receptor. The beauty of identifying the receptor is that drugs can be designed to interfere only with this action of serotonin and not others, which are mediated by different receptors. A second promise of a drug acting at this site is that it would only reduce the irritability of the gut; it would not paralyze the bowel. Transmission mediated by acetylcholine and other neurotransmitters would be left untouched.

Jack's data in 1981 thus provided an elegant complement to my own. I had, by then, demonstrated chemically that serotonin was present and synthesized by nerves in the gut, that these nerves released serotonin when they were stimulated, that the effects of both serotonin and nerve stimulation could be blocked by antagonizing serotonin's action or by depleting serotonin, and that the nerves that released serotonin subsequently inactivated it by taking it back (a process called *reuptake*). Between us, therefore, Jack Wood and I had satisfied all of the neurotransmitter "postulates." I should thus have been confident of the outcome of our forthcoming workshop, but I was not.

MARCELLO COSTA

I knew Marcello Costa far less well than I knew Jack Wood. Marcello worked then, and still does today, in Australia, which means that we do not meet regularly. He is a fascinating human being. Marcello speaks with an accent that I have never heard duplicated by anyone else. Except that Marcello does it, I would have thought it impossible to speak simultaneously with the accents of Australia and Italy. It is as if a common man and an aristocrat had been melded into a single individual. The combination is devastating to an American audience. No one can resist; Marcello is totally believable. In the forthcoming debate, therefore, I expected that he would be a formidable opponent.

Marcello was not about to accept my proposal that serotonin might be an enteric neurotransmitter. Marcello appeared to think that I was proposing scientific polygamy. Two transmitters were sacred; adding a third was a sin. Unfortunately, this difference of opinion was very important to Marcello, bordering, it seemed to me, on obsession. Almost all of his papers appeared to end with the conclusion that I had to be wrong. I appreciated that saying that I *had* to be wrong was not quite as bad as saying that I *was* wrong; a little loophole was left.

Marcello is a pharmacologist who began his career by effectively using drugs as tools to unravel the complex workings of the enteric nervous system. At the time of the 1981 Neuroscience meeting in Cincinnati, however, Marcello was not doing much pharmacology. He and his longtime Australian colleague, John Furness, were busy exploiting what was then a new technique, called *immunocytochemistry,* to classify enteric nerve cells according to their neurotransmitter. Immunocytochemistry involves using antibodies (natural products of the immune system) as chemical detectors to locate molecules in microscopic sections of tissue. As a result, Marcello was no longer a pharmacologist but a neurobiological linguist. For him, immunocytochemistry turned microscopes into dictionaries in which could be read the definition of each of the nerve cells of the gut.

Marcello and John Furness described their goal as the establishment of a "chemical code" that would permit each of the nerve cells of the enteric nervous system to be named and identified. Marcello and John may thus have thought of themselves as taxonomists, but I considered their purpose much more grand. To me, they were modern-day cryptographers, whose work was even more significant than that of the geniuses who had been assembled at Bletchley in 1941–1943 to break the "Enigma Code" used by the German army. Marcello and John were beginning to decipher the "Enigma Code" of the enteric nervous system, which was the only code I really wanted to know. I had the distinct feeling that I may have appreciated Marcello's work quite a bit more than he did mine. In fact, I really admired what Marcello was doing. If I was not so frightened of what he was going to say about me, I would have looked forward to his talk.

ALAN NORTH

Alan North was the last person I had invited to take part in the workshop. Alan was tall, dark, and very handsome. If nothing else, his appearance would compensate for the fact that Jack Wood, Marcello Costa, and I are each small. We needed some height to make an impression. A cabal of little people pushing an exotic part of the nervous system would have a hard time persuading neuroscientists to pay attention. Alan is different. He is noticed even when he says nothing, which is often. Alan does not like to waste words. He chooses what he has to say with great care and shoots his words out like an archer, aiming them with deadly accuracy at targets that they invariably skewer. Alan is a man of real humor, but he hides that well. The words themselves bristle with the trills and deep booms of northern Britain. When Alan makes a declarative statement, it seems to be unassailable. Alan buttresses his words with a unique form of facial expressions that soundlessly communicate agreement or disagreement. Agreement is signaled by the faintest hint of a grin, while disagreement is transmitted mainly with his eyebrows, which he uses as clubs. You might call Alan dour, but that is too jovial a term; stern or Lincolnesque seem more appropriate.

Like Jack Wood, Alan was an electrophysiologist who made a living by impaling intestinal nerve cells with glass micropipettes. His work differed from Jack's in the emphasis Alan placed on exact physical measurements. Alan's work was extremely elegant, very complete, and incredibly appealing to physiology wonks. Alan may not always have been the first to make a particular observation, but once he made it, he reported the discovery with such definitive precision that the advance became his in the minds of his fellow scientists. Alan did not just find facts of nature, he established them as eternal truths.

Alan is unique in that he can disagree with someone—for example, me—without letting the disagreement spoil a good relationship. Alan likes people he respects even if he thinks they are wrong. Alan is an aficionado of good work. In my case, Alan clearly had decided that my research, even if it might have seemed to him to be temporarily misguided, was OK. On the other hand, Alan was as convinced as Marcello that on the central question of the forthcoming workshop—serotonin as an enteric neurotransmitter—I was dead wrong.

At the time of the Cincinnati meeting, Alan was pursuing an agenda that was different from mine. He hated serotonin because he had another molecule, called *substance P,* in mind as a neurotransmitter. Substance P evoked the same slow excitatory response in enteric neurons as serotonin and the natural transmitter. Once again, ion channels closed. Unfortunately, Alan did not entertain the possibility that more than one neurotransmitter might evoke the same response. To Alan it was a zero-sum game. He thus believed that if serotonin was the neurotransmitter that mediated slow excitation, then his candidate, substance P, could not be. He also believed the converse, that if substance P was the transmitter, then serotonin could not be. The thought that two different nerve cells might each evoke a slow excitatory response, one by releasing serotonin and the other by secreting substance P, was not acceptable to Alan. He liked order. If there was to be a slow excitatory response, then there had to be one transmitter responsible for it. As far as I knew, Alan had no evidence against serotonin, he just interpreted those of his observations that were compatible with substance P as being incompatible with serotonin. The denouement, of course, has turned out to be that both substance P and serotonin are neurotransmitters. Serotonin does long distances; substance P works locally.

I believed I could deal with Alan's logic if my own data could stand up to his withering scrutiny. Still, as I neared my hotel in Cincinnati, I thought of Alan's menacing eyebrows and I shivered. Tomorrow could be a bad day.

Omens

My taxi eventually arrived at my hotel. Unfortunately, my room was not ready and would not be for at least four hours. At this point, I once again discovered the kind of effect the brain in the head can exert over the second brain. I had heartburn, and my stomach began to respond and my colon tightened. A helpful bellman suggested that I check my bags and go to the zoo. There was a white Siberian tiger at that zoo. I empathized with that animal because his image clearly moved against the tide of tiger fashion. Wearing white if you are a tiger seemed to be like proposing serotonin if you are talking about peripheral neurotransmitters. The tiger and I thus shared a bond. Each of us, in our own way, stood apart from our colleagues.

I was staring sympathetically at the tiger when I heard a loud and cultured voice ask: "Are you awake yet?" I turned around, but I knew who it was even before I saw him. It had to be Mike Bennett, an eminent neuroscientist from the Albert Einstein College of Medicine in the Bronx. Three years earlier, I had fallen asleep while projecting his slides during a seminar he gave in my department at Columbia. Mike is very nice, and he forgave me my nap; neither, however, did he let me forget the incident. An inquiry into my state of wakefulness had become his standard greeting. Mike was in Cincinnati for the conference and had been out jogging when he found himself lost in the zoo. He had noticed me contemplating the tiger and decided to chat. He just wanted to let me know that he had noticed my workshop on the program and was looking forward to it. As usual, Mike was being nice, but in being so he made me feel worse. If someone whose interests were as far from the bowel as Mike Bennett's would be there, so too would a lot of other people. The workshop was going to have an audience. If I was going to fail, therefore, I was going to fail in public.

Once again, I felt messages coming from the top and bottom of my gut. I may have been ready to give my all for the second brain, but my own enteric nervous system was not giving its all for me. Instead, it seemed to be doing everything it could to me.

4

THE WORKSHOP

THE CONVENTION HALL where the meetings took place was large and unattractive, as such centers usually are. The dollars saved through the frugal omission of structurally unnecessary but aesthetically appealing details were very much in evidence. Glass was flaunted in front, revealing to the street a view of a lobby that appeared to have been designed to illustrate the crushing depersonalization of the industrial age. Behind the lobby, the stark meeting rooms were devoid of windows, had little adornment, and made lavish use of concrete. The room assigned to our workshop was especially large and only partially filled by an array of about two hundred folding chairs. A large open gap was left between the chairs and the doors, which opened to the audience's left. The gap functioned to make latecomers uneasy, because it provided no cover and made them an unwanted center of attention as they crossed the room to get to the seats. All of the late-arriving people appeared to shrink just a little as they tried to look inconspicuous while their timid footsteps reverberated loudly on the hard floor.

In front of the chairs was a large wooden platform with a podium and a cloth-covered folding table, behind which were places for the participants. Wires for the amplification system extended from microphones on the podium and the table, like the tentacles of an underfed octopus. The room lacked charm, but as the chairs filled, it began to acquire the lively atmosphere of an arena. The audience seemed to be more animated than usual. I imagined that it was expecting blood to flow.

I cannot remember how I entered the convention hall. When I now try to recall the event, only the meeting room, the Colosseum-like setting, and my dry mouth spring to mind. It is as if the affair had no beginning but was created from a celestial void in the middle of Cincinnati. Memories do funny things when they register under pressure. When I now attempt to reenvision the workshop, I see myself sitting on the platform waiting nervously for my coparticipants to assemble and the time to start. Prominent in my mental picture is Hirsch Gershenfeld, an eminent physiologist whom I had met and befriended, years ago, while I was a postdoctoral fellow in England. Hirsch was sitting in the front row looking like an oracle. Since we were waiting for the workshop to begin, he motioned to me to come down and chat.

I was very surprised to see Hirsch at our session and told him so. Hirsch worked on the nervous system of *Aplysia,* a repellent marine organism otherwise known as the sea slug. What these beasts lack in beauty, however, they gain in easy access to their nervous systems. *Aplysia* nerve cells are so large, identifiable, and reproducible in their location in different animals that individual cells are actually named. To me, the enteric nervous system and the ganglia of the sea slug seemed so disparate that I had not expected *Aplysia* people to be attending our workshop. When I mentioned this to Hirsch, he responded by telling me that he wasn't sure that the two nervous systems were as unrelated as I thought. The bowel was, after all, pretty primitive. He suspected that the enteric nervous system might be a remnant of our invertebrate predecessors that evolution had not discarded but left within us. The enteric nervous system, he proposed, was a vertebrate mirror of the relatively simple invertebrate nervous system. Besides, he reminded me, he was a "serotoninophile." Hirsch was probably the first person to demonstrate that there was not one receptor on nerve cells for serotonin, which had been the prevailing assumption, but many. "In any case," Hirsch said with a chuckle, "I love a good fight." At this, I said good-bye with a forced grin and returned to the speakers' platform. Hirsch's presence was welcome, but his love of good fights was not.

As I waited to begin, Hirsch's comments stayed with me. His ideas were sophisticated, and had I not known about the long history of research on the enteric nervous system, I would have found them appealing. I *did* know this history, however, and I thus realized that Hirsch could not be correct. The enteric nervous system was not a simplistic vestige of an invertebrate past. If anything, it was a vertebrate brain moved south. The enteric nervous system had far more in common with the brain in the head than with a ganglion in an *Aplysia.*

My conversation with Hirsch demonstrated once more how foreign the enteric nervous system was to even the most accomplished of neuroscientists. This was now the second meeting of the Society for Neuroscience where what I had to say about the enteric nervous system was linked, albeit now with some degree of thought, to invertebrates. I decided to junk the comments that I had prepared for my introductory remarks and instead to tell the story of the discovery of the enteric nervous system. Junking a prepared speech, however, with two minutes to go before I had to give it represented a bit of a problem. Fortunately, I had previously written a chapter for a book honoring my postdoctoral research sponsor, Edith Bülbring, which outlined the history of research on neurotransmitters and the autonomic and enteric nervous systems. That chapter provided the necessary information. I now had only to use it and compose a talk, quickly.

The workshop began. Since I was the organizer, I introduced each of the speakers. I had provided a slot for an "Introduction to the Enteric Nervous System," which I was to give. This was the "Introduction" that I had just decided to change, and it was now time for me to start talking. My gut churned, which I suppose was ironic, considering the subject of the workshop, but the intestinal kibitzing was bad for my morale. My palms also sweated, and I could feel the beating of my heart. In fact, I doubt that there was ever a time in my life when I felt worse. As I began to speak, however, the pages of my chapter in the book for Edith Bülbring came literally into view. On rare occasions, when I come under a great deal of stress, images of written pages come into my head. I actually see them and can read them, verbatim. The rest was easy. I just read the words, not verbatim, but conversationally, so as not to seem stilted. My talk was far better organized than usual and very economical with words. That comes with having a written text, even if the writing exists only in the mind.

As the "law of the intestine," the "local nervous mechanism" of the gut, the peristaltic reflex in an isolated intestinal segment, and the definitions of Langley went by, I could see that the audience was with me. I had hoped to shock them with "old news," and it was apparent that I was doing that. This audience was clearly ready to accept that the enteric nervous system was not at all what they thought it was, and I guessed they would be willing to accept serotonin as well, if I could present a compelling case. As I turned to introduce the speakers, I could feel my colon beginning to calm down. It might be possible to get through this day after all.

Jack's Talk

Jack Wood was our first speaker. Jack gave his talk in his usual self-assured style. He was clear, simple, and direct. To me, Jack sounded very much like a cowboy riding out of the West to announce an enteric rodeo. The various cells of the enteric nervous system could have been bulls with exotic names whose characteristics were electrical in nature. I worried a bit about whether the audience would find Jack condescending. Simple is great, but it can be overdone. This was not a group to whom you could condescend and hope to survive with your credentials intact. The neuroscientists in the hall, however, seemed to like what Jack was saying and, even more, his manner of saying it. I noted affirmative nods and smiles in the audience. The data were strong and Jack's evidence was compelling. He also made everything easy to understand, and the audience appreciated his doing so.

Although the crowd was obviously sophisticated in its knowledge of the nervous system as a whole, the enteric nervous system was as new to them as I had thought it would be. The cells, therefore, had unfamiliar names and properties. Jack made the introduction gentle. Since it is far more common for scientists to make their introductions complex and brutal than simple and gentle, Jack's presentation was perceived as a memorable one. By the time he had finished, the audience had no trouble not only remembering the names of the cells but using the terminology. That was obvious from the questions directed to Jack in the discussion. People who had never before considered that there might be anything of interest outside the skull seemed to be as comfortable with "AH/type II neurons" (the term for one of the nerve cells in the gut) as they might have been with their classmates at a college reunion.

Jack first took the neuroscientists through the arcane details they needed to know to understand how the enteric nervous system works. To do this, he presented a "Who's Who" of the gut, describing the nerve cells and naming them. Soon, however, Jack moved on from the passive portrayal of the players to his own experiments on the nerve cells in action. Jack told the audience how he dissected the bowel, so that he could actually see the nerve cells and ganglia in living tissue. Jack then explained his clever use of tiny "pressure feet," made out of bent platinum wires, to hold the muscle still so that he could impale the nerve cells with sharp glass micropipettes. This was the trick that had enabled Jack to succeed where many others had failed before him. The microelectrodes, in turn, recorded the electrical events that occurred when the impaled nerve cells were perturbed by neurotransmission.

When Jack stimulated a bundle of the nerve fibers that entered the ganglion that contained the nerve cell he had impaled, the microelectrode usually revealed that the cell was excited. The onset and the duration of the excitatory response, however, varied greatly. Since events of the type Jack recorded are the actual changes in electrical potential evoked by the connections (or synapses) of the stimulated nerves with the impaled nerve cells, the responses are called *excitatory postsynaptic potentials* (or EPSPs). Some of the EPSPs, which Jack named "fast EPSPs," began almost immediately after the nerve fibers were stimulated, but they were extremely evanescent. Another type of excitatory response, which Jack named a "slow EPSP," did not even begin until after the fast EPSP was finished. However, the slow EPSP lasted for over a minute, which for a nerve cell is an extremely long time, close to an eternity.

The excitatory nerves in enteric ganglia were thus surprisingly coy in the manner in which they spread messages to their postsynaptic targets. They communicate like politicians speaking to the press. On the other hand, postsynaptic nerve cells are not unlike reporters in action. Both register and disseminate what they are told. The first words that the excitatory nerve cells say—the fast EPSPs—are like the official communiqué a politician is wont to put out, terse and lacking important pieces of information. The later parts of the message—the slow EPSPs—are provided by the politicians in the form of leaks. The information is revealed slowly, keeps on coming out for an excruciatingly long period of time, and makes the recipients of the information, the press, irritable. The slow EPSPs may not tell their targets what to do in terms that are as direct and unmistakable as a fast EPSP, but they are nevertheless a powerful way to communicate.

Jack reported that fast EPSPs were always abolished by any of the antagonists, including curare (and a synthetic compound, called *hexamethonium*), that block the action of acetylcholine at nicotinic receptors. The fast EPSP was also identical to the responses of the same cells to acetylcholine and nicotine. Enteric ganglia, like those of the other divisions of the autonomic nervous system, thus contain the ubiquitous nicotinic receptor.

In contrast to the fast EPSPs, slow EPSPs persisted unchanged, despite the application of curare or any other nicotinic antagonist. Slow EPSPs were faithfully mimicked not by the nicotinic effects of acetylcholine but by serotonin. Jack characterized the response to serotonin in exquisite detail, emphasizing just how precisely it resembled a slow EPSP. His observations led Jack to postulate that fast transmission between enteric nerve cells was the province of acetylcholine and its nicotinic receptors, while slow transmission belonged to serotonin. Like Alan North, Jack was playing a zero-sum game. In his case, however, the winner was serotonin, not substance P. Ironically, as it turned out, everyone was right. It was the zero-sum game that was wrong. All of these substances played a role (and the sympathetic transmitter, norepinephrine, was not involved in those events at all).

What Jack needed to clinch his argument was something that he did not have at that time: a good antagonist to interfere with the actions of serotonin on enteric nerve cells. Such a drug would provide an unambiguous test of the serotonin hypothesis. A drug that abolished the slow response to serotonin would be expected do the same for slow EPSPs. If it did, the hypothesis would be confirmed. If it did not, the hypothesis could be discarded. Unfortunately, none of the then-known antagonists of serotonin's actions at other sites (the classical serotonin antagonists) worked at this one. We were later to find the antagonists Jack needed, and in fact they would confirm his ideas, but that is a subsequent story. At this workshop, Jack concluded that the serotonin receptors on enteric nerve cells must be different from any that were then known to exist. Despite the lack of a suitable antagonist, Jack decided that when his information was added to what I had previously published, the case for serotonin as a neurotransmitter that could evoke slow EPSPs was compelling.

Jack's talk was well received. The questions all showed sympathetic interest. Hirsch Gershenfeld, in particular, commented on how analogous the responses to serotonin of enteric nerve cells were to those of *Aplysia*, especially in regard to the seemingly unique nature of the serotonin receptors. Hirsch pointed out that he too had found that there were multiple forms of serotonin receptors in *Aplysia* and that some of them were also resistant to blockade by classical serotonin antagonists. Current drugs were clearly inadequate to deal with the system. Serotonin research, he suggested, was going to turn out to be far more complicated than any of us suspected. This comment was quite prescient. The fact that there are many subtypes of serotonin receptor is a modern idea that, at that time, was not appreciated (but which now occupies a great deal of my waking moments). In those days, the armamentarium of so-called "serotonin antagonists" was very small and limited mainly to compounds that blocked the serotonin-induced contraction of the rat uterus, which, however exotic, was the serotonin bioassay then in favor. The possibility that the serotonin receptors on enteric nerve cells might be different from those of the rat uterus had not yet entered most people's minds. Clearly, step one in any discovery of something new is to think of it.

My Turn

My scientific talk followed Jack's. I had set up the workshop so that that Jack and I could present a case for serotonin as a neurotransmitter in the gut, after which Marcello and Alan would have a chance to try to rebut this claim. Back in the heady days of conference scheduling, I had the idea that once our evidence was out in the open, there was no way anyone could doubt it. I had the strong conviction that going first would make our case very hard, if not impossible, to attack. Jack had now given me a great opening. I had a receptive audience and, I thought, one that was ready to suspend its disbelief.

I croaked a hoarse "thank-you" to Jack as he turned over the portable microphone and began to help me fasten it onto my lapel. By now my anxiety had given me a dry mouth as well as a miscreant bowel. Fortunately for me, Jack is perceptive. Years of scientific observation have had a favorable effect on his everyday life. He noticed the croak, and while I began to speak he quietly strolled to the back of the room, filled a glass with water, and brought it to me. I cannot remember what I said before the water arrived, but I do recall the ecstasy that accompanied the first sip. I was now able to speak, but the panic attack had not abated and I simply forgot what I had planned to say. I really was speechless. Not knowing what else to do under the circumstances, I called for my first slide. Those were literally the only words that I was capable of saying. Fortunately, they were also enough. I may have lost the ability to think, but at least I could still read. The lights dimmed and the slide came on the screen. It may have seemed silly to read out loud, word for word, the legend that was projected on the screen. Perhaps the audience thought I was doing it to achieve a dramatic effect. Whatever they thought, however, the words sufficed. Gradually, as I read the legend, the panic ebbed and my mind cleared. I regained my capacity for rational thought.

Marcello coughed. He was sitting on the podium, watching me from behind, so that he was well within my field of vision while I read my slide. Marcello was grinning. I could hardly believe it. Jack Wood had just finished implying that what Marcello had been advocating for the past decade was wrong, and I was about to do so again. What was Marcello smiling about? I looked over at Alan North. His expression was suitably sour. Marcello, however, was clearly enjoying himself. Something was up. I was about to go back into panic, when unmistakably, Marcello winked. In retrospect, I now realize that Marcello, who is a very kind human being, had recognized my unease and was sending me a signal to relax. At the time, however, I missed the meaning of the wink. Instead of seeing it as friendly, I read it as a sign of contempt. That made me angry, which helped. From then on, I did not speak so much as I roared. I had become alive. If I could have, I would have concluded each point with: "Take that, damn it!"

The plan of my talk, which my water-loosened tongue was now able to present, was to list the criteria needed to identify a neurotransmitter and show how serotonin in the enteric nervous system had met every one. It was as if I were describing a rite of passage in which an adolescent named Serotonin had to overcome a series of obstacles in order to gain acceptance in the adult world as a mature neurotransmitter. My first slide was a list of criteria, the tests that serotonin had to pass. Subsequent slides showed the labors of serotonin in the form of experiments that demonstrated the successful completion of every test. Serotonin was present in enteric nerves, it was released when these nerves were stimulated, it mimicked the effects of a natural transmitter, and blocking the action of serotonin antagonized the effects of nerve stimulation. Serotonin, moreover, was synthesized in the wall of the bowel, and enteric nerves possessed a means of inactivating serotonin after it stimulated its receptors. As far as I was concerned, *QED*.

My Aunt Ruth, who was a teacher of mathematics, taught me to write *QED* (*quod erat demonstrandum*) after every theorem that I proved to be correct in high school geometry. Since then, the initials *QED* come to mind every time I finish what I think is a successful argument. My argument for serotonin, however, was not exactly the same as the Euclidean proof of a geometric theorem. Biology is never that precise. There are always weak points, and in this case I knew what some of them were. One of the most glaring, which I thought Marcello was going to exploit, was the seemingly simple demonstration that enteric nerve cells actually contain serotonin. Although there was solid biochemical proof that serotonin is present in the layer of the bowel wall that contains the enteric nervous system, the microscopic evidence that this serotonin is actually located in nerve cells was less definitive.

Marcello and John Furness previously had not accepted my radioautographic visualization of radioactive serotonin in enteric neurons as compelling. In order to use radioautography to find the locations where serotonin was made and stored, I had to administer (as described earlier) a radioactive tracer that was converted in the body to radioactive serotonin. Marcello and John did not trust my tracer. As far as they were concerned, there was no way of knowing what cells that tracer was going to find its way into. The fact that my radioactive tracer, 5-HTP, was the immediate precursor of serotonin did not impress them. Marcello and John preferred to believe that I had inadvertently filled sympathetic nerves with radioactive serotonin. Because of their scepticism about serotonin as a neurotransmitter in the peripheral nervous system, Marcello and John were willing to attribute a great many hoofbeats to this particular zebra.

I had several important facts to support my point of view, although Marcello and John chose to overlook them. First, no other sympathetic nerves became radioactive after the injection of radioactive 5-HTP. Second, nerves in the gut still became radioactively labeled even after sympathetic nerves were destroyed with a selective neurotoxin (a person-made poison called *6-hydroxydopamine* that is specifically and suicidally taken up by sympathetic nerves). Third, enteric nerves in developing animals became labeled by radioactive serotonin, even before sympathetic nerves grew into the bowel. Fourth, radioactive serotonin itself was avidly and specifically taken up by nerves in the gut, but not by sympathetic nerves in any other organ. Fifth, the uptake of serotonin into enteric nerves was *not* blocked by drugs that inhibit the uptake of norepinephrine by sympathetic nerves (serotonin, therefore, was not piggybacking its way into nerves by riding on norepinephrine's transporter system). Finally, the uptake of radioactive serotonin by enteric nerves was blocked by the very same drugs (Prozac and its relatives—more about this later) that were known to block the uptake of serotonin by known serotonin-containing nerve terminals in the brain.

Marcello and John would settle for nothing less than the direct demonstration that enteric nerves contain endogenous serotonin that the nerves made all by themselves, unprompted from the outside by me or any other investigator. Prior to 1980, the only way to detect endogenous serotonin in microscopic sections of tissue was to freeze-dry preparations and expose them to formaldehyde vapor. Serotonin reacts with formaldehyde to form a fluorescent product that can be detected microscopically. The method works, but not well. The tissue stores of serotonin have to be increased in order to make the fluorescence bright enough to visualize, and, even then, the fluorescence fades rapidly while you look at it. Still, when the bowel was freeze-dried and exposed to formaldehyde, a yellow fluorescence appeared in the enteric nervous system that had the signature (in the wavelength of the light used to excite the fluorescence and in the wavelength of the emitted light) of serotonin. The fluorescence, however, was transient and diffuse. I *thought* that endogenous serotonin was present in enteric nerves, but my colleagues from Australia had said in print that they did not.

A second weak point was the primitive state of our understanding of serotonin receptors in 1981. We did not then appreciate how many serotonin receptor subtypes there are, nor did we have available adequate antagonists. This weakness was also apparent in Jack Wood's presentation. The best we could do to block the action of serotonin was to expose nerve cells to high concentrations of serotonin. This desensitizes serotonin receptors, while leaving the receptors for other neurotransmitters intact. In particular, both Jack and I had shown that when serotonin receptors were desensitized, responses of enteric nerve cells both to serotonin that we applied and nerve stimulation were lost. This was exactly the effect that one would have anticipated if serotonin was really the neurotransmitter. Desensitization of serotonin receptors, moreover, did not prevent enteric nerve cells from responding to nicotine, showing that the treatment was specific.

In the past, John Furness and Marcello had discounted the effects of serotonin-receptor desensitization, arguing that the high concentration of serotonin we had used to desensitize receptors might also have inhibited the secretion of neurotransmitters. They pointed out, correctly, that neurotransmission will be blocked if stimulated nerves cannot release their neurotransmitter. That is, for example, the effect of botulinum toxin. The association of desensitization of serotonin receptors with the failure of enteric neurotransmission, they argued, might just be a coincidence in which two different effects occur simultaneously. In the face of a high concentration of serotonin, the receptors might desensitize *and* nerves may no longer secrete.

Prior to the conference, there was general agreement that desensitization of serotonin receptors blocked slow EPSPs. John and Marcello, however, would not accept this as evidence that serotonin was the missing transmitter. First, we would have to prove to their satisfaction that the natural transmitter was actually secreted when we stimulated nerves to the bowel. To me this seemed like still another case of looking for zebras when the sound of hoofbeats was heard in the land; nevertheless, a challenge is a challenge and I accepted it.

As I concluded my talk, I discussed the obvious weak points in my story and emphasized the need for a more precise localization of serotonin in the wall of the bowel and better antagonists. When I spoke of needing better tools to locate serotonin, I distinctly heard Marcello chuckle. When I looked over in his direction, I could see that he was now grinning broadly. There was now no question about it. A surprise was coming that I did not think I would like. I wondered whether I had inadvertently set myself up for my own demolition. Nevertheless, I quickly summarized my argument and asked for questions.

The questions indicated that by this time many in the audience no longer doubted that serotonin is an enteric neurotransmitter. The questions were mostly softballs, easily fielded. Several colleagues asked for more details about the kinds of reflexes that would bring the nerves that use serotonin as their neurotransmitter into action. At the time of the workshop, my answer to that question was somewhat speculative, but I was able at this point to bring up the experiments that I had carried out at Oxford in 1965 and 1966. These had clearly suggested that serotonin was important for intestinal motility.

After reviewing my Oxford research and my speculations on its therapeutic implications in the discussion period, I turned to introduce Marcello, who was to be the next speaker. I cannot remember what I said about Marcello in my introduction, but I assume it was favorable because Marcello looked pleased when he walked to the podium.

Surprise!

Marcello thanked me and then started his talk in a way that I will never forget. He began by reminding his audience about our disagreement. "All these years," he said, "we thought Michael was wrong about serotonin being a neurotransmitter. Now, I am here to tell you that he really was right all along, and I am going to show you the definitive evidence that proves it is so."

I had been expecting open warfare, but what I got was Appomattox. T. S. Eliot had maintained that the world would end "not with a bang, but a whimper," and so it had. Marcello was graciously giving in! In an open and perfectly straightforward way, Marcello was happily eating crow in public. I do not think there are many people who can show that kind of style and honor. I remembered the chuckle that I heard during my talk and suddenly understood its real meaning. It was not the laugh of an evil genius but the amusement of a friend about to reveal a happy surprise. I loved the sentiment, but I was a little ashamed of my previous anger.

What Marcello had done was to use the technique of immunocytochemistry to fill in and strengthen the weakest link in my argument. He had located serotonin, right where my theory had predicted that it should be, inside enteric nerve cells. It was not in sympathetic nerve endings but right there in nerve cells that were intrinsic residents of the bowel wall. To understand Marcello's (and his colleague John Furness's) accomplishment, you have to know something about the immune system and how microscopy is done.

Marcello used antibodies the way the customs police use dogs. Dogs find drugs by smelling them in sites where smugglers have hidden them. Marcello employed antibodies to find molecules of interest in sites where nature has hidden them. Antibodies are truly fantastic chemical tools. They are not, however, magic swords, a biological Excalibur provided by a supernatural power. Animals make them as part of an immune response.

Proteins and big sugars are complicated molecules that are collectively called *macromolecules* because of their large size. Macromolecules evoke immune responses when they invade or are injected into an animal. The immune response is basically defensive in nature. When macromolecules that normally are not found in an organism get into that individual, they are recognized as foreign by the individual's immune system. An inventory is kept by the immune system, so that it can readily distinguish whether any given macromolecule is *self* (one of its own) or *not-self* (not one its own). This inventory is not a dictionary or a grocery list but an incredibly diverse population of reacting cells.

A foreign macromolecule, upon entering an animal, binds to a small number of cells of the recipient's immune system. These cells are known as *lymphocytes*. In the process of binding to the surface of a lymphocyte, the foreign macromolecule, in effect, selects that cell. The selected lymphocytes are specialized to make the antibodies that are able to react specifically with particular regions of the foreign macromolecule. These regions of the foreign macromolecule are called *determinants* because they are the ones that select the corresponding lymphocytes. In its lifetime, any animal may encounter billions of unique determinants present in a vast number of foreign macromolecules. To cope with this potential, animals literally have to contain billions of unique lymphocytes, which, in fact, they do.

Any substance that provokes an immune response is called an *antigen*. All foreign antigens entering a vertebrate are met by corresponding lymphocytes. These lymphocytes are ready and waiting to react with the invading molecules even before the animal and the foreign substance have had any contact with one another. The foreign molecules thus do not mold or instruct cells to make antibodies; they simply pick out the cells that already have the necessary capability. The population of lymphocytes acquires this tremendous repertoire through a process of genetic scrambling (apparently random) during fetal life, which provides the population with an extraordinary degree of adaptability.

During this massive scrambling process, some of the determinants recognized by the resulting lymphocytes end up corresponding to determinants found on the body's own molecules. These self-reactive lymphocytes have to be destroyed, or else they will make antibodies that will react with the body itself. This self-destructive reactivity, now called *auto-immunity*, was first recognized by the great microbiologist Paul Ehrlich, who named it "horror autotoxicus." Auto-immunity perverts the major weapons of the body's defense and converts otherwise benign lymphocytes into traitorous and self-destructive moles, subverting from within.

To avoid auto-immunity, the body carefully monitors lymphocytes while they are being generated. The testing of lymphocytes is mercilessly stringent, and any lymphocytes that fail the test, and are capable of reacting with self-molecules, are eliminated. The standards to which juvenile lymphocytes are held are thus very tough indeed. There is no such thing as a permissive or progressive education. No cell is given a second chance. One rule holds: Pass the test or wind up being eaten by an avaricious monitor cell, known as a *macrophage*. The lymphocytes that pass the test, however, wear the antibodies that they make as badges of honor. The antibodies become surface proteins, components of the membranes of the cells, with their reactive, antigen-recognizing surface facing out. In that position, the cell surface antibodies can serve as receptors, which bind to a foreign macromolecule as soon as it is encountered. This is the binding process that enables an antigen to select a particular lymphocyte from the billions in the body's library.

Once a lymphocyte is selected by an antigen, it proliferates and gives rise to many identical successors. A population of cells that are clones of the selected lymphocyte results. Within the clonal population, every individual lymphocyte is devoted to making the same antibody molecule. Some of the clonal lymphocytes also change their form and differentiate into a different kind of cell called a *plasma cell.* This specialized cell no longer just inserts the antibody into its surface membrane, like its lymphocyte predecessor, but instead secretes the antibody molecules, pouring them into the blood at a rate of over two thousand molecules per minute. The foreign proteins on the surface of, let's say, a bacterium invading an immune animal can thus instantly be coated with antibodies that have been produced in advance and held in readiness. Antibodies enable the body to respond immediately to an invading antigen. There is no time-consuming delay while cells are mobilized. The coating of antibodies marks the bacteria for rapid destruction by white blood cells and macrophages. Similarly, a toxin entering an immunized animal can immediately be neutralized by antibodies that recognize it and prevent the toxin from doing harm.

Besides the plasma cells, which die off after a week or so, additional members of the antigen-selected clonal population become long-lived and remain as memory cells. These cells are retained after the initial immune response has come and gone. The memory cells allow the body to react more quickly and with greater strength to a second encounter with the original antigen. The body thus learns from encounters with foreign macromolecules and remembers them. It is thus better prepared for a second round, should there be one. Unlike generals who always fight the last war, the body's lymphocytes learn from their battles and adapt to do better the next time.

Scientists like Marcello Costa have long known that the immune system can be harnessed and utilized as a research tool. For chemical analyses, the specificity of the immune system cannot be equaled by anything artificial. Animals such as rabbits, goats, or horses can easily be employed as chemical factories to churn out huge quantities of antibodies. If one can obtain a small amount of almost any protein in pure form, that protein can be injected into a recipient animal, which will obligingly produce the corresponding antibodies. The antibodies can then be isolated from the animal's blood and used as probes to detect the antigen that they recognize essentially anywhere that antigen is found. Neurotransmitters may not be macromolecules, but even small molecules like serotonin and norepinephrine can be chemically coupled to proteins to generate antibodies that recognize the small molecules as determinants. Marcello Costa and

John Furness did not themselves make antibodies, but the exciting advance they were exploiting in 1981 was a burgeoning availability (from colleagues and commercial sources) of antibodies that recognized the neurotransmitters and associated molecules of the enteric nervous system. These antibodies allowed Marcello and John to precisely locate these neurotransmitters on a cellular and even subcellular level.

By the beginning of the 1980s, antibodies had been generated to many of the neurotransmitters that are found in the enteric nervous system. Marcello Costa and John Furness had assembled a large team of coworkers and were absolutely relentless in their pursuit of antibodies. Whenever an antibody of interest got raised anywhere in the world, you could be sure that a sample would find its way to Australia. Once there, it would be applied to the guinea pig gut. Guinea pigs had long since established themselves as the animal of choice for the study of the motility of the bowel. It just happens that, in contrast to the gut of most other animals, a segment of the small intestine of a guinea pig exhibits very little spontaneous activity when it is removed from the animal and suspended in splendid isolation, in a warm, nourishing, and oxygenated fluid. As a result, it is relatively easy to study the effects of drugs, neurotransmitters, and other compounds on the guinea pig bowel.

Nerve cells, as we have seen, are specialized for signaling and communication. They talk to one another and to muscles, vessels, and glands by means of a chemical language. The words of this language are the neurotransmitters. Neurotransmitters can be small molecules, like acetylcholine, norepinephrine, and serotonin, or larger molecules, called *peptides,* which are assembled by stringing together a chain of small molecules, called *amino acids.* Some amino acids, such as glutamate and glycine, can themselves be used not only as components of peptide chains but as neurotransmitters. Some molecules that have recently been discovered to be neurotransmitters are gases, including nitric oxide and carbon monoxide. These molecules obviously require special handling and will be discussed later.

A nerve cell that responds to a command that it receives in the form of a neurotransmitter can be thought of as a *follower* cell because it responds to a message from on high. Follower cells hear or detect the chemical signal by means of receptors on their surface. When a neurotransmitter combines with its receptor, the binding of these two molecules initiates an operation in the receptive cell that converts the message passed by the neurotransmitter into cellular action. The intracellular events that follow the binding of a neurotransmitter and that convert the binding to a cellular response are known as the transduction mechanism. The transduction machinery either changes the state (open or closed) of watery channels that pass through the surface membrane of the follower cell or it initiates a complex series of chemical steps that generate second messenger molecules to carry the signal inside the cell or to other regions of the cell surface.

The chemical language of the nervous system is quite complicated, because different nerve cells speak with different neurotransmitters or combinations of neurotransmitters. A single neurotransmitter, moreover, can activate many different receptors, making it possible for a single neurotransmitter to induce a variety of responses. For example, there are more than fifteen different receptor subtypes specialized to respond to serotonin alone. Each of these subtypes is a unique molecule coupled to its own transduction pathway. The response of two nerve cells to serotonin, therefore, may be dramatically different, depending on which subtypes of serotonin receptor the two cells express. In fact, even the response of a single nerve cell to serotonin may be complicated if the cell expresses several subtypes of serotonin receptor. Clearly, to understand the function of a region of the nervous system, it is necessary to identify the neurotransmitters and the receptors found there. Even then, one may still not know the language, its grammar or its syntax, but one will at least have begun to learn the words.

Antibodies and Chemical Coding

Since it was the gold standard, the nervous system of the guinea pig small intestine became the first element of the enteric nervous system to which the newly available armamentarium of antibodies was applied by Marcello and John. They had systematized their approach. To some extent, their methods were similar to those used by everyone else. They had to be. To paraphrase Gertrude Stein once again, microscopy is microscopy is microscopy. Certain things one simply has to do, and it makes very little difference who is doing these things, or to what one is doing them. Most biological structures are transparent or translucent; therefore, to make sense of the structure of almost any animal tissue viewed through a microscope, it is necessary to add contrast to its components. Contrast can be added by using a specialized microscope fitted with exotic and very expensive optics or, more simply, by staining elements of the tissue. The use of biological stains, which add color to tissue components with varying degrees of chemical selectivity, was a great advance made in the nineteenth century. It was the introduction of stains that made possible the investigation of the structure of both normal (histology) and abnormal (pathology) tissue. Stains, which come in brilliant colors, are informative because they make sense out of what would otherwise be an incoherent mass of monochromatic sameness. They are also extremely satisfying because they make microscopy an aesthetic experience. Each slide is the biological Jackson Pollock.

Marcello and John used antibodies as stains. Many people, all over the globe, did so as well. This technique, called immunocytochemistry (briefly described in an earlier section in which I outlined the kind of research carried out by Marcello), is relatively simple and was well established by the time Marcello and John made such effective use of it. Antibodies can be employed as stains when they are chemically linked to molecules that fluoresce when excited by light with a suitably short wavelength. Antibodies can also be chemically attached to enzymes that, when the conditions are right, cause colored, insoluble reaction products to be precipitated at sites in the tissue where the antibodies are bound. Antibodies that have been chemically linked to other fluorescent molecules or enzymes are known as labeled antibodies. Immunocytochemistry is wonderful in that it uses the extreme specificity of antibodies to identify and locate individual molecules in cells and tissues. Mistakes in identification are possible, but the accuracy is that of immune recognition, which, especially when accompanied by standard controls, is superb.

In practice, very few antibody molecules bind to the small quantities of neurotransmitter present in tissue sections. It is thus usually necessary to enhance the sensitivity of the technique. To do this, one tries to cause as many labeled antibodies as possible to bind to the structure one would like to detect. The idea is to create sandwiches of antibodies in the tissue, to literally pile them on. The critical antibody, the one that combines with a molecule in the tissue, is added first and thus is called the *primary antibody*. You do not lightly tamper with the primary antibody. You want to keep it intact so that it loses none of its natural avidity for its antigen. The primary antibody is thus not usually directly coupled to a fluorescent or enzymatic label.

To detect the primary antibodies, one takes advantage of the fact that antibodies themselves are macromolecules and thus very good provocateurs of an immune response when they are administered to another species. In this situation, therefore, antibodies become antigens. If you inject rabbit antibodies into a goat, for example, the goat will make its own antibodies that will react against those of the rabbit. Language fails us at this point, because we are talking about antibodies to antibodies, which seems, if not confusingly redundant, then at least clumsy. Despite their linguistic shortcoming, however, antibodies to antibodies are effective tools. To avoid confusion, no one actually calls them antibodies to antibodies, but instead these molecules are called *secondary antibodies* and they are classified by using a kind of combative animal-against-animal terminology. One speaks of "goat antirabbit," "donkey antisheep," and "mouse antihuman" antibodies. Whatever they are called, secondary antibodies naturally bind to antibodies from the other species of animal. Goat antirabbit secondary antibodies thus will, if they are labeled, bind to and locate any rabbit antibodies present in a given piece of tissue.

In practice, the primary antibodies are usually applied first and given enough time (hours to days) to become bound to antigens in the specimens. The loose, unbound antibodies are then washed off and a secondary antibody, which is coupled to a label, is added. The secondary antibodies bind to many regions of the primary antibody, so that many labeled molecules become attached. The sensitivity of the technique is thus amplified greatly, and the structure to which the primary antibodies have bound can be visualized.

Staining, by itself, is not enough to make microscopy work. To make sense of the underlying structure, even if it is stained, it is also usually necessary to examine thin sections of organs, not the whole specimen. You can get a sense of the problem by imagining an apartment house with a glass roof and floors. You might, if a light were sufficiently powerful, shine that light all the way through such a house from its cellar to its roof. Still, even so, if you stood on the roof of the transilluminated house and looked down, you would be hard-pressed to discern order in the scene below. The light would reach you, and the people and furniture on each floor have contrast and thus should be identifiable. Unfortunately, however, the people and objects on the upper floors would be superimposed over those on the lower floors. The order of objects present on each floor would thus be obscured.

A specimen to be examined in a microscope presents an observer with an analogous difficulty. Again, as with the transparent apartment house, the tissue can be transilluminated. The light is sent through magnifying lenses, but you peer down the tube of the microscope and look vertically through the specimen. The many tiny objects of which the tissue is composed thus lie on top of one another in their natural order. If these tiny objects have been stained, then they will be superimposed on one another and obscure the view. The order of structures present in layers of the tissue would be no more evident than that of the people and objects on the various floors of the apartment house. If you could cut out an individual floor of the house and look only at that, then the order of its people and objects would become apparent. Individual items would no longer be obscured by structures lying above or below them.

It is, of course, impractical to cut up apartment houses, but that is exactly what is done with biological tissues. By cutting tissue into sections thin enough to prevent components from lying on top of one another, one can recognize the profiles of individual components as distinct entities. Sectioning, however, makes the reconstruction of the three-dimensional architecture of the tissue difficult. Thinness is also a virtue that can be overdone. As sections get thinner, they encompass less and less. A thin section includes only the objects present in that layer of tissue. Components of tissue that are above or below the section are missed.

Marcello and John managed to obtain preparations of ideal thickness. They dissected the gut wall into incredibly thin layers and then simply examined each layer of the wall of the bowel as a whole mount. The layers were thin enough to avoid obscuring overlapping objects but not so thin that they missed structures or had difficulty in reconstructing the bowel in three dimensions. Marcello and John also used secondary antibodies with fluorescent tags that, when excited, glowed with their own light. For fluorescence microscopy, you look not directly at the light that you shine through a specimen but at the light that emanates from the fluorescent dye-stained structures within the tissue itself. The original light is used to excite the dye molecules and then is removed from the optical path, so that it does not reach the eye of the observer. If nothing emits light within the specimen, the field is dark. The fluorescent structures, which do emit light, thus leap into view, glowing in iridescent brilliance against a background that is black. Since the antibodies determine what is to be stained, relatively few structures within a tissue become fluorescent. The problem of establishing order in thick tissue becomes less severe. Relatively few visible structures are superimposed on top of one another. Marcello and John were making rapid progress.

To enhance what they could discover from immunocytochemistry, Marcello and John had learned to use the dissection of the guinea pig bowel as an experimental procedure. Delicately removing individual layers of the gut wall enabled Marcello and John, in anesthetized animals, to surgically cut through enteric nerves in order to deduce the direction of their projections. Nerve cells tend to be extraordinarily directional or polarized in their shape. One end or pole of the nerve cell, occupied by the cell body, is usually receptive. The nerve cell body and its associated projections, called *dendrites,* receive input from other nerve cells. A single long thin snoutlike extension, which is called the *axon,* extends from the receptive pole of the nerve cell. The axon is conductive and an electrical signal flows rapidly along its surface, away from the cell body. The axon ends in a *terminal,* which is often swollen in appearance because it expands to contain many little packages of neurotransmitter. These tiny neurotransmitter-filled sacs, or vesicles, are called, appropriately enough given their location at synapses, *synaptic vesicles.*

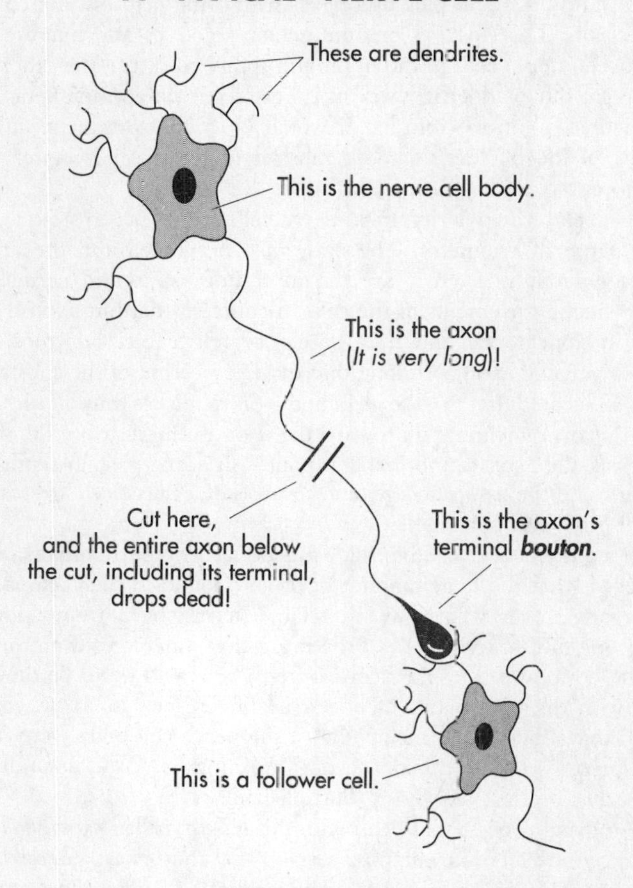

If an axon ends in a single enlargement, the swelling is called a *terminal bouton.* Axons of the intestine often end, not in a single terminal bouton but in a chain of such swellings. The individual swellings are referred to as *varicosities,* like the blue swellings in the veins of the legs of elderly people. Sympathetic and parasympathetic nerve endings in cardiac muscle, smooth muscle, blood vessels, or glands are examples of nerves that have long varicose nerve endings. These nerves spray their transmitter over large areas and cannot be very precise about where they put it. In contrast, the skeletal neuromuscular junction, and most of the nerve endings that communicate with other nerve cells, are of the bouton variety. These endings are usually very precise about where they aim the sprays of their transmitter. This difference seems to reflect function pretty well. We use our hands and fingers to pick up objects with some deftness and considerable precision. This involves neuromuscular junctions and many connections between nerve cells in the brain and spinal cord. On the other hand, when we get our bladder to work, it is best if our parasympathetic nerves get the whole organ to contract. It would be counterproductive to get a select few of the bladder's smooth muscles to twitch. It is better just to force it to empty itself of urine.

In their extreme polarity, most nerve cells are utterly grotesque. They have no sense of symmetry. The body of a nerve cell and the arbor of dendrites emanating from it are the most impressive and certainly the most eye-catching elements of the cell. In contrast, the thin axon is unimpressive in appearance and may barely be noticeable. For most nerve cells, however, the axon is so long that over 95 percent of the substance of the cell is located not in the cell body or dendrites but in the axon. Despite this overwhelming tilt toward the axon in the distribution of cellular material, the axon cannot make protein. All of the protein-synthesizing machinery and the associated genetic controls of a nerve cell are restricted to its cell body and dendrites.

One consequence of the odd functional compartmentalization of nerve cells is that the huge amount of protein found in their axons has to be transported over what may, in relative terms, be a tremendous distance. If one of the nerve cells of the brain that innervates a motor nerve cell in the bottom of the spinal cord were to be scaled up to be the size of a room with the dimensions of fifteen by fifteen feet, its axon would be over 250 miles long. That means that if the nerve cell body were in New York, its axon would terminate beyond Washington, D.C., assuming that it could navigate the New Jersey Turnpike.

The moving process that deals with this length, called *axonal transport,* is managed by an elaborate shipping service, one that even Federal Express should envy. Materials destined for the terminal, including the neurotransmitter, are efficiently packaged in little vesicles that are driven by a motor molecule along specialized tracks (called *microtubules*) down to the terminals. This type of movement is *fast axonal transport.* The structural components of the axon are moved in a far more leisurely manner by an entirely different machinery, *slow axonal transport,* which is much less well understood. There is even an efficient mechanism for return delivery, as dirty old worn-out vesicles are sent back to the cell body by fast transport (now working with a different motor) to be digested in the cell's garbage-disposal system.

A second consequence of the absence of protein-manufacturing machinery in axons is that after an axon is cut, the segment of the axon that is cut off from the cell body drops totally dead. The proximal part, which is still attached to the cell body, tends to survive, because the cut surface membrane seals over in a process that can be likened to the fusion that occurs between colliding soap bubbles. Since axonal transport continues after the axon has been cut, transported material, including vesicles of neurotransmitter, tends to accumulate on the cell body's side of the cut.

Marcello and John first cut enteric axons, then used antibodies to identify and locate neurotransmitters. In this way, Marcello and John were able to establish their "chemical code" for the classification of enteric nerve cells. The neurotransmitters disappeared from the fields where cut axons ended and accumulated above the cuts. Marcello and John could thus readily ascertain whether the axons of specific nerve cells ascended or descended within the bowel, or whether they projected from one layer of the wall of the gut to another.

Marcello and John Furness had obtained antibodies to serotonin from Harry Steinbusch, who raised them in Holland. Harry was one of the first to realize that to produce antibodies that recognize a small molecule, like serotonin, he needed to introduce a trick. Normally, as we have seen, lymphocytes look for big molecules, not small ones. Harry Steinbusch's trick was to use formaldehyde to chemically link serotonin to a big carrier protein. He then injected the modified complex into an animal. Serotonin (or more accurately, the reaction product formed by serotonin and formaldehyde) became a determinant recognized by some of the antibodies against the complex. Antibodies, of course, were also made to other, uninteresting parts of the bigger carrier molecule to which serotonin was coupled, but these were easily removed by absorption, as described earlier. When tissues are fixed with formaldehyde, the serotonin in the tissue becomes linked to neighboring proteins (the mechanism that I had exploited to hold radioactive serotonin in place for me to find it by radioautography). Serotonin linked by formaldehyde to protein was precisely what Harry Steinbusch's antibodies to serotonin were looking for.

Using Harry's antibodies to serotonin (tacked by formaldehyde to tissue protein), Marcello was thus able to confirm that serotonin is present in enteric nerve cells. He also showed that all of these nerve cells are present in the myenteric plexus (the larger of the two enteric plexuses) and that all of their axons descend within the bowel. In the direction of their projections, therefore, the serotonin-containing enteric nerve cells have an unerring eye fixed on the anus.

As Marcello's talk unfolded, I became overwhelmed by a feeling of release. Here was a man who had disagreed with me for years, and now, to hear him talk, it was hard to understand why. Marcello never mentioned his previous objections to my conclusions, and in fact, without saying so he confirmed almost everything I had suggested. We now agreed that serotonin-containing nerve cells are present in the enteric nervous system and that these cells are probably interneurons. Marcello, however, still could not bring himself to call serotonin a neurotransmitter without a qualifying adjective, like "putative." I supposed that it would take more time for Marcello to banish the last zebra; nevertheless, I could live with "putative," especially coming from someone whose work now dovetailed so nicely with my own.

When Marcello was finished with his talk, the expression on Alan North's face was totally black. I imagined steam coming from his ears, but of course there was none. He had been undercut in the worst of all possible ways. Marcello had never been Alan's teammate, so he was under no obligation to divulge his surprise to Alan. His talk, however, was a double cross for Alan, who was not taking it well. Not that I think, under the circumstances, that I would have done any better. In any case, Alan went on with his talk, but his situation was untenable. Three people had just concluded that his thesis, which did not allow for serotonin to be a neurotransmitter in the gut, was wrong. Instead of an even debate, everyone was ganging up on him. Alan's message was in trouble before it had chance to fight.

Ironically, there was nothing wrong with anything Alan had to say, except for his fixed idea that if serotonin was the transmitter responsible for slow excitation, then his candidate, substance P, could not be. Substance P and serotonin are probably both neurotransmitters that evoke the same kind of response. They are released, however, from different nerve cells under different circumstances. In any case, Alan went on to give his talk and get killed in the discussion.

Alan had hoped to take advantage of the fact that substance P is a peptide while serotonin is not. Peptides, like proteins, are digested by the enzymes they encounter in the gut. They are cut up by these enzymes, releasing the amino acids of which they are composed. In contrast, serotonin is itself a derivative of an amino acid (tryptophan) and thus is not affected by the enzymes that chew up peptides. Alan's idea was to apply digestive enzymes (from the pancreas) to isolated enteric ganglia and determine their effect on slow EPSPs. He reasoned that if substance P was the neurotransmitter, then the digestive enzymes would destroy it and slow EPSPs would therefore not occur. On the other hand, he expected the enzymes to have no effect on the slow EPSPs if the neurotransmitter was serotonin. Sure enough, the digestive enzymes inhibited slow EPSPs. Alan thus concluded that substance P and not serotonin was responsible for slow EPSPs.

Unfortunately for Alan, the audience was unmoved by his data. Hirsch Gershenfeld immediately pointed out that digestive enzymes are not specific agents, like the poisons that had revealed so much about neurotransmission. They not only digested substance P and other peptide signaling molecules, but they could be expected to digest the surfaces of the nerve cells in the enteric ganglia as well. There was no way to know how many potentially important molecules were digested when the enzymes were in contact with tissue. The slow EPSP may have been inhibited by the enzymes for many reasons, some of which could have had nothing to do with the digestion of the neurotransmitter. Even worse, from Alan's point of view, I had shown, years earlier, that serotonin, which is positively charged, becomes tightly bound by digestive enzymes of the pancreas, which are negatively charged. The digestive enzymes, therefore, could easily have inhibited slow EPSPs not only by digesting substance P but by binding (and thus inactivating) serotonin.

Alan's experiments were beautifully presented, but his demeanor was sullen and angry, and he did not handle the questions well. By the time the workshop ended, it was obvious that Alan had not had a good day. I, however, had never done better. The enteric nervous system was on the neurobiological map, and my suggestion that serotonin is an enteric neurotransmitter had suddenly moved from the underground to the mainstream. I was no longer a dissenter.

As I left the platform, with a strong urge to take care of what I then believed was a severe alcohol deficiency in my bloodstream, a young woman came over to talk to me. She introduced herself as Theresa Branchek but immediately told me to call her Terri. She said she was about to receive her doctorate from the University of Oregon. Terri was amazingly direct in her speech. Whatever was on her mind was soon on her tongue. More important, she asked good questions and soon revealed a very extensive knowledge of my previous work. Clearly, Terri had done some homework and had come to the workshop prepared for this discussion. Our meeting was no accident. I decided to forget—at least for now—the drink that I thought I needed and sat down to enjoy our conversation.

It turned out that Terri was looking for a laboratory in which to do postdoctoral work. The workshop had not only convinced her that serotonin is an enteric neurotransmitter, but it also had convinced her to come to work with me. I was delighted by the prospect and told her so. Terri was not one to temporize, so she agreed immediately. After a few more minutes of small talk, we traded addresses and telephone numbers and we each turned to leave the hall, which was rapidly emptying of people. I found some colleagues, who were exultantly upbeat about the workshop, and resumed my pursuit of a drink.

Curiously, although everything had gone about as right as I could ever have hoped, I was not as enthused as my colleagues, or as I should have been. In fact, I felt a little let down. After all, in bringing the enteric nervous system to public consciousness, I had, in a sense, rediscovered the wheel. I thought my accomplishments were worthwhile, but Bayliss and Starling, Trendelenburg, and Langley had visited the critical ground before me. With respect to serotonin, I felt vindicated, but I had made the essential discoveries years earlier. Now, as I left the darkening hall, my mind turned to what remained to be done. We had established that serotonin is a player in the drama of the gut, but I had yet to learn what it was that serotonin did for the bowel and how serotonin did it. More important, while I may have convinced the neuroscientific skeptics in the workshop audience that there is a second brain, I had not even begun to deal with the big question of why a second brain is necessary. After all, the first brain is a pretty impressive instrument. You might think that if it can deal with things like politics and quantum mechanics, the brain in the head should be able to handle something as simple as a gut. Of course, the fact alone that a second brain evolved simultaneously suggests that running the bowel is not a trivial activity. To understand why a second brain is needed in the gut, it is important to consider first just what does go on in the bowel.

A short time after I left the conference center, I sat with my colleagues in a dark, wood-lined bar. My thoughts turned inward, and I paid little attention to the revelry that surrounded me. My focus turned to the question of what to do next with my scientific life. An image of Dwight Eisenhower came to mind. Eisenhower was, of course, the highly successful general who succeeded masterfully in keeping a large number of fractious individuals with conflicting self-agendas focused on defeating Nazi Germany in World War II. There came a day, however, when Hitler was done in, Nazi Germany was disposed of, and Eisenhower was faced with the question of what to do next. One always likes to go on to bigger and better things, but winning World War II is hard to beat. Establishing that serotonin is a neurotransmitter in the enteric nervous system is not an accomplishment that can be compared to winning a world war, but there *had* been a war over serotonin, and I had won it. The sudden declaration of peace now meant that I needed a new cause.

Clearly, Eisenhower found productive ways to use his time. He went on to become president, first of Columbia University (my school), and then of the United States. Perhaps, in my own less exalted fashion, I too could continue to have some success in my postwar period. Terri Branchek provided me with a good way to retool. She had never thought previously about the bowel and needed orientation. I decided to draw up an outline of the things that happen in the gut, which parts of the bowel accomplish them, and how these activities are influenced by the enteric nervous system. This outline, once completed, improved my own understanding of why we need a second brain.

Part II

THE TRAVELOGUE

5

BEYOND THE TEETH: THE DOMAIN STALKED BY HEARTBURN AND ULCER

A MAJOR BENEFIT of taking a comprehensive look at the gut is that the neat way in which the structure and function of the bowel explain one another becomes obvious. The anatomy of the gut evolved so as to permit the organ to do its job; therefore, form and function are best considered and investigated together. This idea seems obvious, perhaps even simplistic, when stated directly; nevertheless, I have found that in practice, the concept is not widely appreciated by workers in the field. Some, with a well-developed aesthetic sense, become bemused with the incredible beauty of biological structures and want to capture their loveliness in words and pictures. These people hate to be distracted by what they regard as the "mere utility" of their subjects. Others, who consider themselves to be "hard" scientists, are rationalists who find anatomy disturbingly complex and resistant to compression into a logical framework. The structure of living things cannot easily be described by mathematical formulae and must be documented by illustrations rather than by numbers. These individuals, who find pictorial evidence to be "soft" and devoid of the precision they associate with numerical data, turn with relief to the physical and chemical phenomena that underlie biological activity. Such processes can be more comfortably conceptualized with graphs and, in the best of circumstances, by equations. In truth, any attitude toward research that tries to separate form from function, is artificial and impedes progress.

The structure of the bowel was not created by the Eternal for the sole purpose of providing visual pleasure to those who discover it. Neither was it provided by the devil to torture rational scientists with illogical detail. Instead, the gut is enabled by its structure to do what it has to do. Knowledge of structure can thus be used to predict and explain function, and information about function can be employed to predict and explain structure. The extensive folding of the lining of the inside of the intestine, for example, can be understood as a structural modification with a critical purpose: It increases the surface area available for the completion of digestion and the absorption of nutrients.

To extract and be able to internalize fuel and materials to run and sustain the body from something as complex as a steak is really a very difficult thing to do. You cannot simply grind up the meat and take it intravenously. First, a great deal of highly sophisticated chemistry has to take place to liberate what we need from what we eat, and then we have to transport those critical nutrients into the body. To get the necessary chemical reactions to proceed, the environment within the bowel has to be regulated, the contents of the gut have to be mixed, and the enzymes that attack foods have to be present in precisely the right concentrations. To get everything right, it is necessary to have a system of sensors in place that can detect the progress of digestion and evaluate conditions in the bowel on a moment-to-moment basis. The information derived from these sensors then has to be coordinated to assure that the internal environment within the gut will favor digestion and absorption. Beyond just nutrition, the gut also has to defend itself—and, by extension, the rest of the body—from invasion by an army of hostile germs that is forever poised to attack should the bowel ever let down its guard. Only the kind of militaristic control that a brain can exert over an organ system can assure that every element in the bowel's apparatus for digestion, absorption, and defense works well and is there when it is needed. So much nervous horsepower is involved in getting the gut to operate properly that it makes good sense for evolution to have put the requisite brain right in the organ itself. So many nerve cells have to be involved that if they were all to be controlled centrally in the head, the thickness of the interconnecting nerves would be intolerable. These cables would also be a source of danger; cut them and the digestive lifeline of the body would be severed. It is thus both safer and more convenient to let the gut look after itself. The brain in the head is also liberated to pursue things that are far more interesting that liquefying a steak.

We All Are Hollow

The design of the body can be understood by paraphrasing T. S. Eliot. We are indeed hollow men and, although Eliot did not say it (he was a sexist pig), also hollow women. The space enclosed within the wall of the bowel, its *lumen,* is part of the outside world. The open tube that begins at the mouth ends at the anus. Paradoxical as it may seem, the gut is a tunnel that permits the exterior to run right through us. Whatever is in the lumen of the gut is thus actually outside of our bodies, no matter how counterintuitive that seems. The body proper stops at the wall of the gut. Nothing is truly in us until it crosses that boundary and is absorbed; moreover, anything that moves across the intestinal wall in the reverse direction, into the lumen, is gone. If we bleed into the lumen of the bowel, the blood is as lost from us as it is when it drips on the floor. When an alcoholic bleeds from varicose (swollen and bulging) veins into the lumen of the esophagus, he or she can easily bleed to death. A fatal hemorrhage can thus occur without a drop of blood being externally visible. Even water moving from the body proper into the lumen of the gut can be as lethally dehydrating as a trip through the Sahara without a canteen. Diarrhea kills babies by drying them out in just this way, and cholera is a fatal disease for the same reason.

Since we are hollow, we have two surfaces that separate our inside from our outside. One boundary, which is visible, and therefore obvious, is the skin. This surface is straightforwardly impermeable. Water does not pass through it very easily. We thus do not evaporate in dry weather, nor do we dissolve in the bathtub. The skin also is tough and protective, resisting abrasion despite constant abuse and preventing invasion by bacteria, which, if not kept out by the skin, would find us a superb broth in which to live. The other boundary, which separates us from an exterior environment that is no less real than that which confronts the skin, is the inner lining of the bowel. This surface, however, is modified to solve problems that are far more complicated than those faced by the skin.

The lining of the gut, like the skin, must protect us by preventing excessive loss of water (into the lumen of the bowel) and invasion by hostile microbes. We swallow germs with our food, and in addition, some are permanent residents of our mouths and our colon. On the other hand, unlike the skin, the intestinal lining has also to participate in the critical processes of *digestion* and *absorption.* Digestion is the term for the variety of means by which the complex and often very large molecules in food are converted to simpler and smaller molecules that can then be moved from the lumen of the gut into the body. Absorption is the term for the transport of the products of digestion across the lining of the bowel to reach blood and lymph vessels in the wall of the intestine. These vessels carry away the absorbed nutrients and distribute them to all of the cells of the body. Digestion and absorption are thus essential for life, just as essential, in fact, as the beating of the heart and the drawing of breath. When either digestion or absorption fails, starvation looms. The intestinal lining, therefore, cannot be constructed like the skin as a tough and impenetrable barrier. Instead, the intestinal lining has to permit nutrients to go right through it. Many of these nutrients actually have to be helped across the intestinal surface by the cells that line the gut, which work very hard and overcome serious obstacles to get these substances absorbed. The bowel therefore displays a cleverness in carrying out its basic functions that is not shown by the skin.

Although the gut's lining separates in from out, this separation is carried out selectively. Water and other molecules move in both directions across the intestinal lining, yet when the gut functions normally, what needs to stay in remains "in," and what is best kept out stays "out." On the one hand, neither water nor anything else that we need is lost excessively in feces. On the other hand, the bacteria we swallow all the time do not invade us. Clearly, the gut is an extraordinary organ, which, in fact, is probably why it has evolved a mind of its own.

Swallowing

Food enters the bowel through the mouth. We chew our food, sometimes with insufficient degrees of thoroughness. As we chew, the ingested material is moistened with saliva. If food is not chewed adequately, or if it remains dry, it can stick in the "gullet" (*esophagus*) and cause agonizing discomfort. Although the windpipe (*trachea*) may not be blocked, the phenomenon is nonetheless excruciatingly painful and accompanied by massive salivation (sometimes with blood), nausea, and a frustrating inability to throw up. A finger down the throat will not help; neither will a drink of water. One has to wait for slow reflexes, triggered by the second brain, to dislodge the food and push it down. Sometimes a tiny bite of bread will help to jog the sluggish reflexes into action. Although this can happen to anyone, and is almost never fatal, it is wise to remember to chew before swallowing.

Saliva does contain some digestive enzymes, but food does not linger in the mouth long enough to be digested by these enzymes, and they are inactivated in the stomach. A little of what we eat dissolves in our mouths and some of it even vaporizes. The dissolved components can be detected as taste by special sensory receptors known as taste buds, which are present on the rough surface of the tongue. The vaporized molecules can reach the nose, where they can be smelled, because the cavities of the mouth and the nose communicate in the back of the throat, or *pharynx* (where "postnasal drip" is experienced). Actually, the nose and the mouth are separated by a movable partition that regulates their communication. The sensation that we perceive as taste is really a combination of taste and smell. In fact, since the sensory receptors that detect odor are much more discriminating than those that detect taste, the nose makes a greater contribution to a gourmet's palate than does the tongue.

Taste buds are most concentrated on the back of the tongue near the site where the oral and nasal cavities communicate. As a result, food is best experienced in the back of the mouth, especially when it is rolled around and massaged so as to allow it to give off vapors. Wine connoisseurs understand these requirements very well. To taste a wine and judge its quality, they take a sip, hold the wine in their mouths, suck some air in through the teeth, and then tilt their heads back just a bit to let the wine flow to the critical region where they gurgle it. This procedure presents the wine to the maximum number of taste buds while simultaneously allowing it to vaporize and stimulate odor receptors in the nose. In contrast, bad wine (or fast food for that matter) is best dealt with by the tip of the tongue and swallowed as fast as possible.

Once a bolus of food is pushed far enough back in the throat, control of it is lost. Further handling switches to autopilot. The details of swallowing are not something that is plotted but a reflex that is triggered, like a knee jerk in response to the tap of a doctor's hammer. The bolus acts as the tap, and the contractions of the muscles of the larynx (voice box) and esophagus act as the jerk. The nose is cut off from the mouth, the *epiglottis* swings into place, shutting off the trachea, and the food is driven "down the hatch," which, in this case, is the esophagus.

Although the swallowing reflex defies conscious control, it is nevertheless an activity that is run by the brain. There is little or no involvement here of a peripheral control of activity. Epicurean experiences and other sorts of cerebral activities are called "cerebral" because they are appreciated at the highest level of the brain, the *cerebral cortex*. Automatic, visceral kinds of things, like swallowing and breathing, which may not be consciously experienced at all, are the responsibility of the lowest level of the brain, the *brain stem* (or *medulla*). A collection of nerve cells has been set aside in the brain stem to deal with swallowing. Cranial nerves carry the sensory signals from receptors in the throat to these nerve cells, which process the information and stimulate nearby motor nerve cells, also in the brain stem, which control the muscles of the larynx and esophagus. These muscles are skeletal in type, and in keeping with the design of the peripheral nervous system (discussed earlier), they are innervated directly by the nerve cells of the brain stem, without benefit of an intervening synapse.

Of course, higher brain centers influence lower brain centers. As a result, the brain stem may not be given the unfettered opportunity to deal with swallowing. Emotions are able to intervene. Choking or "choking up" are familiar sensations. If the brain acts, it can always, under detrimental circumstances in some people, act neurotically. The brain may also dry up the mouth by its ability to affect the autonomic nervous system. This type of cerebral misadventure is, fortunately, recognizable by physicians and, in contrast to problems of the gut that do not involve the brain, susceptible to psychiatric or psychopharmacological treatment.

It is important to realize that food does not just fall down the gullet to reach the stomach. The gullet also has to participate. Food is driven in an oral-to-anal direction by the movements of the muscles in the esophageal wall. If a stroke destroys the little group of brain stem cells that direct swallowing, the muscles of the pharynx, upper, and middle esophagus become paralyzed and no amount of effort can restore the ability of the stricken individual to swallow. A surgeon will have to create an opening in the abdominal wall so that feeding can be accomplished by manually inserting meals directly into the stomach. Once the food is in the stomach, however, its further movement can occur even in individuals who are brain dead. The central nervous system is thus needed for swallowing, but from the time food is swallowed to the moment its remains are expelled from the anus, the gut can regulate events all by itself. Defecation, like swallowing, requires central nervous system participation to be accomplished normally.

Although nothing very spectacular happens to food between the mouth and the stomach, it is important to realize that the region in between performs a critical function for us. The pharynx and the esophagus are conducting elements of the digestive system. They are essential. Cut them out, or weaken them, and you die. The linings of these portions of the bowel are thick and impermeable. The surface cells are stratified, sitting on top of one another in multiple layers, very much like those of the skin. Fortunately, these linings are modified to prevent permeation and to resist abrasion, a role made ever more important by the speed of modern life, in which a lunch ill-prepared in haste may be gulped down with lightning speed. Chewing may, for some people, be an afterthought.

Stomach Juice

Digestion essentially begins in the stomach. To initiate the process of digestion, the stomach produces its own digestive enzymes and acid. The most important digestive enzyme synthesized by the stomach is *pepsin,* which catalyzes the breakdown of proteins in food to smaller components called *peptides.* Pepsin needs an acidic environment in which to work, and the stomach obliges by producing *hydrochloric acid,* which is a remarkable product given its noxious quality. In fact, the stomach does not just dabble in acid production. When it gets up a good head of steam, the stomach can churn out hydrochloric acid in industrial strength. Metallic iron will dissolve nicely in gastric juice, which can also burn a hole in your shirt or blind you if it gets in your eyes. The contents of the stomach are thus a witch's brew, which can turn steak into soup and kill most of the germs that we eat along with our food. Germ killing, of course, is beneficial, but coexistence with a deadly antiseptic would seem to pose a problem for the lining of any organ that contains it.

The caustic nature of gastric juice highlights an extraordinary property of the lining of the stomach. The lining cells secrete both pepsin and hydrochloric acid, yet, in contrast to swallowed hamburger or iron filings, neither pepsin nor hydrochloric acid dissolve the cells that make them. Resistance to digestion by hydrochloric acid and pepsin is a unique property of the gastric lining that is not shared by the skin or the lining of the esophagus. The ability of the gastric lining to avoid being consumed by its own juice is due to another of its secretory products, an alkaline mucous gel that clings tenaciously to the luminal surface of the lining cells. This gel neutralizes stomach acid and pepsin and prevents them from gaining access to cell surfaces, which thus are exposed only to an environment that remains happily neutral, even while extreme acidity and digestive enzymes swirl among the contents of the gastric lumen.

The stomach utilizes an additional mechanism to keep it safe. The stomach does not actually secrete pepsin itself but an inactive precursor, *pepsinogen.* Acid begins the process of converting pepsinogen to pepsin, but as soon as some pepsin is produced, it exerts a positive feedback on the process by catalyzing the further conversion of the remaining pepsinogen to pepsin. This process cannot occur on cell surfaces, which, because they are clothed in the alkaline mucous gel, are not acidic. Pepsin thus arises only in the lumen of the stomach where food awaits.

Everything Has Its Place, Especially Gastric Juice

The lining of the esophagus, as much as that of the stomach itself, has to be protected from gastric juice. There is, however, no alkaline mucous gel in the esophagus. To its credit, the esophageal lining is tough and resistant to abrasion. It also has a healthy ability to repair itself when it is damaged. Gastric acid and pepsin, however, are another matter. Putting them on the lining of the esophagus is like taking a bath in the drainings of automobile batteries. It burns. When gastric acid affects the esophagus, we experience a feeling we call *heartburn* (or *dyspepsia* if we are trying to sound medical).

Small amounts of heartburn can be tolerated, but as the profits of a major industry demonstrate, our ability to put up with heartburn is very limited. The number of products sold to combat this sensation is large and growing. These products either neutralize stomach acid (Tums, Maalox, Rolaids) or inhibit its secretion (Tagamet [cimetidine], Zantac [ranitidine], Pepsid AC [famotidine], Prilosec [omeprazole]). Gastric juice is fine when it stays in the stomach, which is modified to live with it. When *reflux* occurs, however, and gastric contents flush up into the esophagus, the stuff is both bothersome and dangerous. Unfortunately, the esophagus is not very sophisticated in the repertoire of distress signals that it has available to send. A geyser of acid backing up into the esophagus may thus feel no different from a relatively trivial "Maalox moment." A single brief episode of heartburn is not likely to be serious. Repeated episodes that tend to persist for long periods of time are something else.

It is not only our comfort that is put in jeopardy by gastric acid in the esophagus. The esophageal lining can be eroded and even destroyed by gastric acid. Chronic reflux disease (known in the trade as *G*astro-*E*sophageal *R*eflux *D*isease, or *GERD*) is thus often an extremely serious problem. GERD, which is disturbingly common, is most frequently treated by sacrificing the ability of the stomach to produce acid. Administration of a potent modern drug, such as Prilosec (omeprazole), which inhibits the action of a critical pump necessary for the generation of hydrochloric acid, can cut acid production to zero. This usually brings about rapid relief and healing of burns and erosions in the esophagus, but the treated patient has to make do without the benefit of acid in the stomach.

Lack of stomach acid is an unnatural state of affairs that is not good. In experimental trials in rats, the long-term consequences of having no acid in the stomach have proven to be very bad indeed. Tumors (called *carcinoids*), which arise from one of the cells of the stomach's lining, tend to occur in these animals when acid production is suppressed for long periods of time. Humans are not rats and have not been observed to develop these tumors. Still, physicians are hesitant to keep patients for extended periods on a drug that drives stomach acid to zero. Conventional wisdom thus is to administer Prilosec only for a limited time, during which the esophagus gets a chance to heal, and then to stop the treatment and hope that GERD does not return. Unfortunately, however, GERD usually does return. Patients who suffer from GERD and experience the relief afforded by Prilosec, therefore, tend to take the drug for periods that are becoming longer and longer.

Nothing has yet been reported that would indicate that Prilosec is unsafe when used in humans; nevertheless, blocking the production of a natural product, hydrochloric acid, that is otherwise useful seems to be a superficial way to treat a disease, even if it works. A better method to treat GERD would seem to be to restore the natural mechanisms that prevent reflux of gastric acid into the esophagus. The marvels of Prilosec have thus not yet put researchers out of business.

Normally, reflux of gastric acid into the esophagus is prevented by a structure found at the junction between the two organs, called the *lower esophageal sphincter* (known to its friends as the *LES*). The LES acts as a movable diaphragm that opens and shuts. As you might imagine, the LES remains shut most of the time. When it is shut there is no communication between the lumens of the esophagus and the stomach. The contents of the stomach thus remain in the stomach, even if pressure rises inside that organ. If you stand on your head, for example, the LES still holds back the acid and protects your esophagus. The LES opens only transiently to transfer swallowed food from the esophagus to the stomach. Reflux of stomach acid tends not to occur on these occasions when the LES is briefly open because the openings are associated with movements of the esophageal muscles that drive food toward the stomach. These esophageal muscle contractions cause the pressure inside the esophagus to be greater than that in the stomach, which prevents backflow. If a little acid does leak into the esophagus while the LES is open, it is neutralized by the alkaline secretions of specialized glands in the area. Heartburn thus is not a problem when the LES, esophageal motility, and backup glands are all coordinated and working together. Why the complex nervous control that keeps all three elements synchronized evolved is thus easily understood. When they fail, life becomes intolerable and sometimes shorter. If evolution is driven by survival of the fittest, then certainly prevention of heartburn must be a potent factor. The pharmacological means of dealing with heartburn that does occur is a relatively recent development.

The stomach is more than a digestive organ. It has an extraordinary ability to expand and act as a reservoir. Food is not consumed continuously, nor is it necessarily consumed in reasonable quantities. Amazingly, the stomach can accommodate not only the delivery of three square meals a day but, when it has to, even the most egregious kinds of gluttony. The wall of the stomach is *compliant*; that is, the stomach can simply make itself larger without increasing the pressure inside its lumen. The stomach is not a balloon with an elastic wall that squeezes back when its contents push at it. Since internal pressure does not increase when the stomach expands, the ingestion of an enormous amount of food does not force the gastric contents either up into the esophagus or down into the small intestine. The loss of this reservoir function is one of the most disabling consequences of the surgical removal of the stomach. Three meals a day are no longer possible. People must eat six or more meals a day to stay alive. There is no equivalent reservoir anywhere else in the bowel.

The Gut Must Know What It Is Doing

It should be apparent at this point that the digestive process is not simple. Many events take place and dangerous substances are utilized. Regulation of each step is thus critical; furthermore, regulation has to be rapid and specific. We have seen, for example, that the LES is opened momentarily to let food through but otherwise stays shut. This is easily described but hard to pull off in practice. The timing of the open phase is exquisite and has to be tightly controlled. Food approaches, the esophageal muscles contract, the LES opens, food goes through, and the LES slams shut. Boom-boom, snap-pop, coordination is everything, and the timing has to be perfect. The complex actions of the stomach, too, must be regulated with the same kind of precision. The stomach does not, for example, make hydrochloric acid all of the time. Acid is produced only when it is needed. The effort of making acid (which, incidentally, is prodigious) is avoided when it is not necessary. Not making acid perpetually also eliminates the risk of living with it every second of every day. Similarly, the secretion of pepsin and the production of the alkaline mucous gel also occur on an as-needed basis.

The stomach does not simply store food. It also kneads large clumps of food into minuscule particles, which are the only kind of material that can be tolerated by the next segment of bowel, the small intestine. Nothing escapes the stomach unless it passes a rigorous test of size. The stomach has to decide when to store and when to knead, when to churn and when to propel. The activities of the bowel thus change frequently, and one pattern of activity may even oppose the one that precedes or follows it.

If an activity occurs only when it is required, then something has to sense the need and communicate that need to the effector cells. Regulation and speed are the specialties of the nervous system, and it is the nervous system that coordinates the complex acts of digestion. In fact, this regulation by the gut's own nerves is just as necessary for digestion as the enzymes that liberate nutrients from a complex meal. Digestion is not a mechanism by which food is unceremoniously dumped into a waiting solution that dissolves it. The presence of food within the bowel is detected by the enteric nervous system, which then evokes the secretion of digestive materials and induces the gut to exhibit patterns of motility that are appropriate to whatever digestive events are occurring at any given moment. The finely tuned synchrony of a perfectly ordinary enteric nervous system is surely every bit as marvelous as the creative output of the brain. Certainly, unless this synchrony is delivered by the second brain, the creative capacity of the first brain will be lost.

Like Politics, All Digestion Is Local

The several activities of the stomach are relegated by the organ to different regions and cells. The stomach that is seen without the aid of a microscope (see the accompanying diagram) can be divided into four zones, called the *cardiac* region, the *fundus,* the *corpus* (body), and the *pyloric antrum.*

The structure of the cardiac region implies that this zone has evolved in order to provide aid and succor to the esophagus. The cardiac stomach produces neither acid nor pepsin. Instead, its lining is modified to form glands that secrete an alkaline mucous, which tends to neutralize stomach acid and prevent its reflux into the esophagus. The esophageal-gastric junction is thus a specialized acid-free zone. It is a firebreak, which prevents the need for "instant relief." The cardiac region of the stomach, therefore, can be considered as a kind of natural Maalox dispenser.

The corpus and fundus of the stomach can be thought of as a common region. Their glandular structure is virtually identical and they function together as a unit. The fundus is that part of the stomach that rises above the opening of the esophagus. The air that is swallowed along with food tends to rise and accumulate in the fundus because of its elevated position. Burps bring this air, redolent of gastric perfume, to consciousness and often also to public attention. The gas in the fundus is clearly visible on flat X rays of the abdomen and is a major landmark that enables radi-

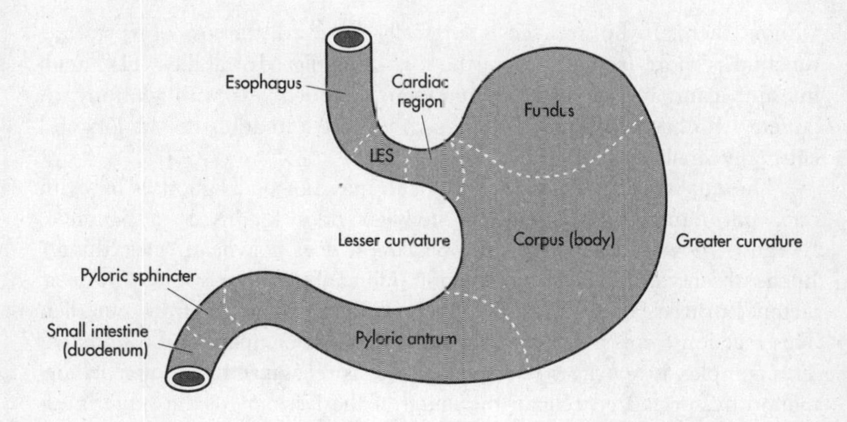

ologists to discern the position and size of the stomach. The lining of the fundus and corpus is modified to form glands that secrete all of the pepsinogen (converted in the lumen to pepsin) and the acid of the stomach. Neither pepsinogen nor hydrochloric acid is produced anywhere else. Of course, the glands of the fundus and corpus also secrete the alkaline mucous gel that enables the lining of the stomach (these regions included) to survive.

In a sense, the stomach has been balkanized by evolution. Its functions have been parceled out to individual cell types, and each kind of cell inhabits its own turf. For every function there is a region. The control of the Senate or the House of Representatives may pass from one party to the other, not because of pressing national issues that the country decides in a referendum but because of the cumulative effects of many races decided on the basis of parochial concerns. The stomach is like that. The fate of the organ as a whole is determined by the success or failure of each of its regions.

All You Really Need Is Intrinsic Factor

In addition to acid, pepsin, and the alkaline mucous gel, the fundus and corpus secrete yet another molecule, *intrinsic factor*. Of all the substances produced by the stomach, intrinsic factor is the only one that is absolutely essential for life. Incredible as it may seem, life can go on without acid and pepsin. The small intestine and its associated glands can make do without them. If the stomach is surgically removed, the loss of its storage function is more troubling than the loss of its digestive ability. The small intestine cannot cope with a square meal, much less with gluttony. A patient who has lost his or her stomach will have to adjust to the loss and eat many small meals every day.

The quality of life without a stomach may not be as good as life with one, but almost everything the stomach does for us is dispensable. "Almost," however, is a key word, because it does not mean "everything." In the absence of medical intervention, life cannot go on without intrinsic factor. Intrinsic factor binds vitamin B_{12}, which we obtain from our diet. This reaction forms a vitamin B_{12}/intrinsic factor complex in the stomach. This complex is not digested in the gut but is recognized long after its formation by specialized cells in the lining of the last part of the small intestine, the *ileum.* When these cells recognize the B_{12}/intrinsic factor complex, they can absorb the vitamin. In the absence of intrinsic factor, no matter how much vitamin B_{12} is eaten, it cannot be absorbed. Since vitamin B_{12} is required for the maintenance of many nerve cells of the brain and spinal cord and also for the formation of red blood cells, a deficiency of vitamin B_{12} is debilitating and ultimately fatal. The disease resulting from B_{12} deficiency is called *pernicious anemia,* an appropriate name if ever there was one. Anemia (an inadequate number of red blood cells) starves tissues of oxygen, leaving its victims in a perpetual state of exhaustion. As if that were not enough, the degeneration of nerve cells, which cannot survive without B_{12}, leads to a loss of sensation and paralysis.

Interestingly, the same cells (*parietal* cells) that make hydrochloric acid also produce intrinsic factor. These two molecules are very strange bedfellows to come out of a single cell. Intrinsic factor is a protein. The organelles required to make proteins tend to dominate the innards of almost every other kind of cell that is similarly modified to make proteins for export. Proteins have to be coded by DNA, transcribed to RNA, synthesized as assembled chains of amino acids, and transferred into membrane-enclosed packages for secretion. In contrast, hydrochloric acid is a small molecule that in stomach-type quantities is poisonous to any cell with which it comes into contact. Worse, hydrochloric acid is able to diffuse right through cell membranes. Storing it in membrane-enclosed packages until it is secreted is thus completely out of the question. Clearly, no cell could put hydrochloric acid together in its final concentration and live. Parietal cells, however, do not die. They work their magic of making hydrochloric acid and survive to do so again day after day. How?

The Pump and the Package

Intrinsic factor is assembled inside of parietal cells, stored in little membrane-enclosed packages, and secreted, but hydrochloric acid arises externally. A single molecule of hydrochloric acid contains one hydrogen and one chloride ion. To produce the hydrochloric acid of gastric juice, the parietal cells pump hydrogen ions from the blood into the lumen of the stomach. Chloride ions follow the movement of hydrogen, resulting in the formation of hydrochloric acid outside of cells where the two ions meet.

The trick is to be able to pump the hydrogen ions. This is not easy. Hydrogen ions carry a positive charge. Moving charged particles is difficult because they affect one another. Particles with the same charge repel, and particles with an opposite charge attract each other. A cell thus cannot just gather up a bunch of positively charged hydrogen ions and move them from one place to another. To successfully transfer a large number of positively charged hydrogen ions from one side of a cell to the other, some other particles with the same charge have to be moved back the other way to replace the hydrogen. Without this countermovement, an excess of positive charge would accumulate when the hydrogen ions arrive at their new destination, and a deficit of positive charge would be left in their wake. This kind of separation of charge is not really possible, because the attractive and repelling forces exerted by accumulations or deficits of charged particles are too strong for biological systems to manage. Nature is said to abhor a vacuum, but it is even more opposed to charge separation. It is forbidden by the physical laws that govern biology.

Parietal cells manage to avoid charge separation by making the pumping of hydrogen ions a simple transfer operation. The cells exchange hydrogen ions for potassium ions, which are similarly positively charged. As hydrogen ions move from the blood to the gastric lumen, a compensatory movement of potassium ions flows back the other way, from the gastric lumen to the blood. As a result, there is no net separation of charge. This hydrogen-potassium exchange is the process that is blocked by omeprazole (Prilosec). Once it stops, acid production comes to a screeching halt.

Since the concentration of hydrogen ions in blood is far less than the concentration required in the gastric lumen, the parietal cell pumps against staggeringly unfavorable electrical and chemical gradients. In terms of the amount of work involved, the pumping of hydrogen ions is not unlike going *up* Niagara Falls in a barrel. The effort is vast and requires the consumption of immense quantities of oxygen, the utilization of megacalories, and the production of an amazing amount of the high-energy molecule ATP. ATP is the currency that the cells spend to get the work done.

The magic exercised by parietal cells is to secrete hydrochloric acid without ever having to contain it. What they contain instead is an extraordinary apparatus that allows them to generate enough ATP, and a massive amount of membrane to hold the necessary ion pumps. The ATP comes from a very large number of oversized *mitochondria,* the organelles that are responsible for *respiration,* the process by which oxygen is utilized to "burn" (without actual flames) small carbon-containing molecules to produce ATP. The ion pumps are actually integral membrane proteins, which function as enzymes (*ATPases*) that break down ATP and use the energy liberated by this breakdown to transport hydrogen and potassium ions in otherwise improbable directions.

It is quite surprising that two such dissimilar apparatuses, one for protein secretion (to produce intrinsic factor) and one for hydrogen ion transport (to support hydrochloric acid production), should be found in a single cell. Cells are more commonly designed by evolution to be able to carry out only one such chore. If they are assigned a difficult job to do, cells usually focus on that duty. Since they also have to keep themselves alive and functioning by performing housekeeping jobs, most cells lack the capacity to perform widely different kinds of luxury tasks for the benefit of the body as a whole. For example, the cells that produce pepsinogen (known as *chief cells*) and mucous-producing cells make proteins for export, and their appearance indicates that is what they do. Neither chief cells nor mucous-producing cells do much else that is altruistic. Cells in the ducts of salivary glands transport ions, and their inner construction reflects that role and that role alone. Overall, the appearance of parietal cells, with their vast number of mitochondria and massive amount of membrane, is appropriate for their ability to do the equivalent of cellular hard labor and to transport ions, but it does not suggest that these cells produce intrinsic factor. Although parietal cells clearly contain the requisite protein biosynthetic machinery, that machinery is not obvious. The fact that parietal cells make intrinsic factor thus went undiscovered for many years, because the cellular apparatus for protein synthesis and secretion was masked by the more flamboyant modifications of the cell's interior that are devoted to acid production. Scientists looking at the fine structure of the parietal cells of the gastric glands were understandably deceived. One function, acid production, was evident, but the other, intrinsic factor secretion, was clandestine.

Gefilte Fish

One of the first clues that the same cell is responsible both for hydrochloric acid production and intrinsic factor secretion was the clinical association of pernicious anemia with the absence of acid in the stomach (*achlorhydria*). Many patients who lack the ability to make stomach acid acquire pernicious anemia, and many patients with pernicious anemia have no acid in their stomachs. At first, the association of vitamin B_{12} deficiency and achlorhydria was mysterious. The mystery was eventually solved when it became clear that patients with pernicious anemia and achlorhydria had no parietal cells. For reasons that are as yet unknown, auto-antibodies arise in some people that attack the surface membranes of their own parietal cells. Ultimately, the attacked cells die off, and when they do, achlorhydria and pernicious anemia result. Since the parietal cells make both intrinsic factor and hydrochloric acid, neither is produced after parietal cells have been knocked out.

Fortunately, these patients can be treated. They are able to survive without stomach acid, and vitamin B_{12} can be administered. The vitamin, of course, has to be injected, because in the absence of intrinsic factor, which the patients do not secrete, they cannot absorb vitamin B_{12}. The patients thus learn to inject themselves with vitamin B_{12}, just as diabetics learn to inject themselves with insulin.

Pernicious anemia turns my thoughts to my childhood and a person I dearly loved. My grandmother, who was a surrogate mother for me, suffered from pernicious anemia. She took care of me while my mother worked. Politics was my grandmother's avocation, but cooking was her business. She baked a mighty cheesecake, made a formidable matzo ball, and no one could resist her gefilte fish. The fresh fish were critical, and there was hell to pay if they were not perfect. No one knew what properties the fish had to have to satisfy my grandmother because no one else tasted the raw fish. My grandmother ground and mixed the fish, tasting the mixture as she went to get the proportions right. In retrospect, I realize that this habit may have done her in. Raw fish carry an unseen menace, *Diphylobothrium latum*, the broad tapeworm of fish (and of people who eat the fish).

After many years of distributing gefilte fish to family and friends, my grandmother's health took a turn for the worse. She complained of weakness and an accelerating feeling of fatigue. One day she was struck with an agonizing pain in the abdomen. She vomited, collapsed in a chair, and asked feebly for help. The symptoms and physical findings led to the diagnosis of intestinal obstruction. Surgical intervention was mandatory. As an incidental finding, the preoperative work-up also revealed that my grandmother had pernicious anemia, which had not previously been suspected, although the anemia certainly provided an adequate explanation of why she had been feeling tired and weak.

Unfortunately, my grandmother's physicians did not see the relationship between the pernicious anemia and intestinal obstruction. They also were too busy, too focused, or perhaps too worried to take a complete history, which might have revealed my grandmother's habit of eating raw fish. *Diphylobothrium latum* was in fact the obstruction in my grandmother's gut. She had, over the course of years, made many batches of gefilte fish. She suffered not from one worm but from many. Some people collect stamps, but she collected *Diphylobothrium latum.* Grandmother's gut was put back together by a surgeon who was, I imagine, quite unhappy as he sewed. The proper treatment was medical (worm-killing pills), not surgical.

None of the rest of us, the consumers of my grandmother's finished gefilte fish, ever became ill. We, of course, ate cooked fish. She, however, in her zealous mixing of raw ingredients, sampled the tapeworm eggs along with the fish. *Diphylobothrium latum* competes with its human host for vitamin B_{12}, and when a person is massively infected, the worms can win the competition. My grandmother, it seemed, acquired pernicious anemia because of her tapeworms. This phenomenon used to be relatively common in Jewish housewives who made gefilte fish and in Norwegian housewives who made an analogous delicacy called lüdefisk. Disease caused by the broad tapeworm of fish is less common in modern American society than it used to be, no doubt because gefilte fish more commonly comes out of jars bought at a supermarket than out of a kitchen. Cookbooks and measurements are also replacing the tasting of raw fish as a means of getting the proportions right.

Even after my grandmother was cured of her tapeworms, she still had no acid in her stomach, and her pernicious anemia persisted. It thus seems plausible that her worms did not, in the end, give her pernicious anemia by eating her vitamin B_{12}. She may have had the auto-immune disease that kills parietal cells, in addition to the worms. My grandmother died of stomach cancer, a disease that occurs frequently in people with pernicious anemia and achlorhydria. Perhaps, in her case, *Diphylobothrium latum* got a bum rap.

Glands

Parietal cells are found not on the visible surface of the stomach but together with the chief cells, which make pepsinogen, in deep pits in the stomach's lining called *gastric glands*. Although these glands (see the accompanying figure) are formed by infoldings of the stomach's lining, the cells within the glands themselves are not identical to those on the surface of the stomach.

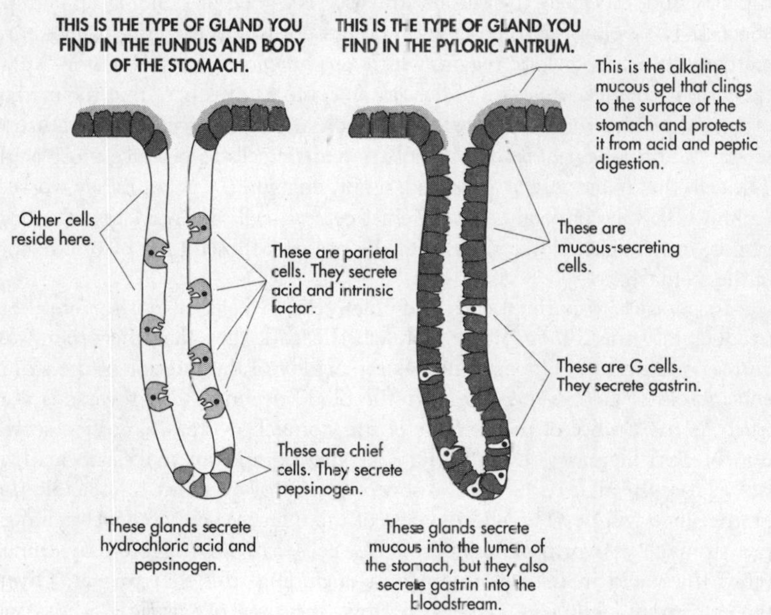

The stomach's surface cells, in all of its regions, are concerned with survival and are mucous cells, pouring out the protective alkaline mucous gel. The cells responsible for the production of acid and pepsin are found deep within the glands, and only within the gastric glands of the fundus and corpus. Parietal cells and chief cells occur in different regions of the same glands. Parietal cells are in residence throughout the necks of the glands, while the chief cells are restricted to their base.

Chief cells are more conventional than parietal cells in their structure. The primary role of chief cells is to secrete a protein, and they look the part. What parietal cells have in mitochondria and membranes with ion pumps, chief cells have in RNA and membranes that do protein packaging. Chief cells are shaped like pyramids. The base of the pyramid is filled with the apparatus to make and package proteins, and the apex is loaded with membrane-enclosed packages of pepsinogen. In the absence of food, chief cells sit like loaded guns waiting for underlying nerves to provide a signal to fire their pepsinogen-filled bullets.

Curiously, although the fundus and corpus produce all of the hydrochloric acid and pepsin of the stomach, relatively little digestion occurs in these regions. The fundus and corpus account for most of the stomach's storage capacity and, especially the corpus, are also very good at grinding up clumps of food. The bulk of gastric digestion occurs not where the critical juices are secreted but in the pyloric region, which also functions as a "laboratory" that carries out chemical analyses of the gastric contents. It is here that the acidity of gastric juice is monitored. When stomach acid is low, cells in the antrum secrete a hormone, *gastrin,* that stimulates parietal cells to produce more acid. The cells that make gastrin are called, simply enough, *G* cells, as if they worked for the FBI. Like parietal cells and chief cells, G cells are residents of gastric glands, but the glands that G cells inhabit are only those of the pyloric region of the stomach.

In contrast to parietal cells and chief cells, G cells do not secrete their product into the lumen of the stomach. Instead, they face the other way and secrete into the bloodstream. As the acidity of the interior of the stomach increases, gastrin secretion into the blood declines. The reverse is also true: As the acidity of the interior of the stomach decreases, gastrin secretion of acid increases. Even though the pyloric antrum makes no acid, it still exerts, through its G cells, a powerful control over just how acidic the gastric juice will be. The effectiveness of the internal acid-control system of the stomach is worth remembering. G cells can be triggered by things other than acid in the pyloric antrum, including stuff that we eat. Given the enormous ability of G cells to stoke the fires of gastric acidity, one should probably devote some thought to what one puts into one's own pyloric stomach. Everyone has a learning curve. The fewer trials it takes to learn something, the steeper and more effective the learning curve. Most people learn very quickly what is going to hit their pyloric stomach like a keg of gasoline. This particular learning curve is very steep because the stakes are high. Mexican red pepper does it for me. One slice too many of pizza does it for my wife. Neither of us needed many repetitions before we knew what to avoid. The damn things are full of fun on the way down, but the esophageal grief is simply not worth that particular candle.

The ability of the stomach to regulate its own acidity can be frustrating to people who take commercial antacids, like Maalox or Tums, to counteract heartburn. When the alkaline chemicals in these preparations neutralize stomach acid, a person's own G cells can turn traitorous and relight the fires of heartburn. The turncoat G cells sense the decline in stomach acid and secrete gastrin. The secretion of gastrin causes still more acid to be secreted, which defeats the antacid. A vicious cycle can thus occur. We may take another Tums or more Maalox. Once again, the antacids begin to neutralize the acid in our stomachs, and they provide the instant, but transient, relief. G cells sense the drop in acidity and secrete gastrin, our stomachs put out still more acid, and on . . . and on . . . and on. Each cycle is worth another few coins in the banks of the Tums and Maalox people.

Fortunately, evolution has been amazingly resourceful in putting us together. Our ability to make hydrochloric acid means that we are perpetually sitting astride a volcano. We have our defenses against the depredations of acid, and they are strong. The alkaline mucous gel is a formidable barrier, and our backup system of glands acts as an array of fire extinguishers that neutralize the acid that leaks up into the esophagus. Still, even under the best of circumstances, something can go wrong, and inevitably, therefore, it does. Life is tough for the cells that line the stomach. Every now and again one of them is bound to be killed, despite all of the precautions the body has taken on their behalf. A salvage mechanism thus has to be in place to deal with the death of these good soldiers. It is not that the casualties need a proper burial or anything like that. The body operates without sentiment. Dead cells just fall into the lumen of the stomach and quickly vanish into the pool of acid that awaits them there. What is needed is a means of stepping into the breach. Each dead cell is a potential gap in the stomach's line of defense. On the one side of the gastric lining, there is a hellish soup of acid and pepsin, while on the other there is the sweet, nurturing broth that provides sustenance to the cells of the body. The acid cannot be allowed to flow freely through the space created by the death of a lining cell. Potential catastrophes do not occur because the cells on either side of their dead comrade are able to quickly plug holes. These cells sense the onset of trouble and rapidly step up, hurling their bodies into the gap created by the loss of a cell, and by spreading across the potential space, they make sure that the lining of the stomach holds.

The agility of the stomach's lining cells, which provides them with their breach-filling ability, is stimulated by distress signals that are manufactured by unhappy cells. These signals belong to a class of molecules called *prostaglandins*. There are many different prostaglandins, and they do a wide variety of things in different organs. For example, they contract the pregnant uterus and can evoke an abortion. Prostaglandins also cause pain to be perceived in many different locations, such as the head and the joints. In the stomach, however, prostaglandins protect cells and thus are called *cytoprotective* agents when they are used as drugs. If one prevents prostaglandins from getting made and doing their thing, one can obtain relief from pain, but only at the price of putting the stomach in peril. Ibuprofen (Motrin) and similar compounds, called as a group *n*on*s*teroidal *a*nti-*i*nflammatory *d*rugs (NSAIDs), inhibit the biosynthesis of prostaglandins and are thus capable of wreaking gastric havoc. Especially when used chronically, they can punch a hole in the best of stomachs. Some people, such as individuals who suffer from rheumatoid arthritis, simply have to take NSAIDs in order to live without pain that would otherwise cripple them. The oral administration of a cytoprotective agent can be a nectar of the gods for such people. They can have their cake (pain relief) and eat it too (freedom from ulcers). Unfortunately, the cytoprotective prostaglandins can also end a pregnancy, and the kind of person who might need a cytoprotective drug—for example, a young arthritic woman living on NSAIDs—is the kind of person who might become pregnant.

I vividly remember a meeting of the Advisory Panel on Gastrointestinal Drugs of the Food and Drug Administration (FDA), of which I was a member. We were asked to consider a cytoprotective prostaglandin to be used by patients being treated with NSAIDs. Our meeting opened with an address by a Right to Life spokesperson, who pleaded with us to ban the drug. To him, the benefit of allowing a young woman to be treated, so as to exist and function in the absence of crippling pain, was outweighed by the risk of abortion. He envisioned the cytoprotective drug being abused by thousands of young women seeking abortions, and thousands more aborting unwittingly as they became pregnant while being treated with the cytoprotectant. I have seen few advocates self-destruct as completely as that spokesperson. He did everything anyone could have expected of a true ideologue. He cajoled us, lied to us, and, in the end, threatened us. His concluding remarks were quietly delivered, but they hit us like thunder. He said, simply, "We know where you live."

Despite his admonition, with its implied threat of personal harm, our panel chose not to follow his lead. We recommended instead that the cytoprotective drug be issued, but we called for the issuance to be accompanied by stern controls and an appropriate warning on the label. Cruelty to women and banning drugs that are effective and safe are not popular with committees of the Food and Drug Administration. In fact, the cytoprotective agents have not been abused. The dose that may cause an abortion is far higher than that which is cytoprotective for the stomach. Furthermore, there are far better means of inducing an abortion, if that is what you want to do, than taking a huge dose of a cytoprotective drug. Moreover, the women whose quality of life depends on NSAIDs and cytoprotection are careful about becoming pregnant. The Hobson's choice between the denial of relief to women in need in order to protect fetuses that are not only unborn but unconceived proved, happily, to be one that need not be made.

Histamine

Acid secretion by parietal cells is not just the result of a dialogue between G cells and the acid in the pyloric antrum. G cells are a major player in the game of acid regulation, but there are others as well. In addition to G cells, specialized cells of the stomach's connective tissue, called *enterochromassin-like cells,* and nerve cells also participate.

The second signal that is sent to parietal cells comes from mast cells. Mast cells send their signal to parietal cells by way of *histamine,* a molecule that has gained more notoriety for the sneezing, wheezing, and nasal stuffiness that it causes than for its role in acid production. The histamine receptors on parietal cells, however, are not the same as those that are responsible for the grief caused by the release of histamine during an allergy attack or a cold. There is only one histamine, but there are three types of histamine receptor. The misery that accompanies allergies and colds is due to a histamine receptor that is called *H1*. Parietal cells do not express H1 and they do not participate in colds. Instead, parietal cells express another histamine receptor called *H2,* which has nothing to do with sneezing. The antihistamines in cold remedies and allergy pills block H1 histamine receptors and thus do not alter acid production. An entirely different class of drugs is needed to block H2 receptors. These H2 blockers are useless in the treatment of hay fever, but they are dynamite for reducing the acidity of gastric juice. There are even H3 histamine receptors, but they do not influence allergies or acid secretion. H3 receptors are found on nerve cells and may be important in the communication that occurs between the nervous and immune systems of the gut.

By secreting histamine, mast cells provide a constant drive that favors the production of acid. Parietal cells react to this kind of stimulation like teenagers react to popular music. It must always be present and, while it does not make them do anything special, they cannot function if it is taken away. Parietal cells need to have histamine playing in the background to keep them awake. This requirement means that H2 blockers, such as Tagamet (cimetidine), Zantac (ranitidine), and Pepsid AC (famotidine), have a wonderful ability to turn off parietal cells. By inhibiting the drive provided by histamine, the H2 blockers blunt the action of signals, such as gastrin, that would otherwise induce parietal cells to secrete.

The H2 blockers revolutionized the treatment of ulcer disease when they were introduced because they provided an alternative to surgery (more about this in a moment). The H2 blockers were revolutionary because they not only reduce the acidity of gastric juice but, in contrast to antacids, keep it down. Ulcers are thus given a chance to heal, and their owners, for the first time, obtained a means of avoiding the surgical knife.

The H2 blockers have recently been made available for sale over the counter. The nonprescription form of these compounds has turned into a popular alternative to antacids. Of course, the antacid makers have not rolled over and played dead, so that a new class of drugs could make off with their market. The various pills sold to counter heartburn are now engaged in a vicious commercial war with one another. Both types of drug, antacids and antihistamines, are relatively safe and relatively effective. The antacid people push the rapidity with which their products provide relief, but they say little about how long-lasting that relief will be. The antihistamine folks push the duration of relief that their products will give you, but they fail to tell you how long it will take before you notice that things have improved. Watching the commercials on television for each type of compound is itself a form of relief. This is the same kind of outlet that is provided by a good cockfight. The heartburn wars are truly a blood sport, waged over the agony of millions who happily provide billions to both the winning and the losing companies.

The Nervous System and Acid Production

The third signal that is sent to parietal cells is acetylcholine, which is delivered by the nervous system. This allows parietal cells to be nimble, agile responders that turn on or off almost immediately as circumstances change. Histamine keeps parietal cells aroused, gastrin provides a slow thermostat-like action (which tends to hold the acidity of gastric juice relatively constant), but nerves provide flexibility. Nerves also innervate mucous-secreting cells, enabling them to adapt at least as quickly as parietal cells; therefore, when the acid pours out, the critical defense of the stomach's lining can be instantly assembled. The alkaline mucous gel is thus not like a ponderous Maginot Line.

The nervous input to parietal cells is provided in part by the brain and in part by the enteric nervous system. Just the contemplation of steak can make the lumen of your stomach turn acid. This is the *cephalic*, or head phase, of digestion, and is evidence of the brain in action. Consciousness of food, or even of the imminence of food, is enough to cause the parietal cells to hitch up and ride. Think "beef" (or, if you are a vegetarian, "tofu") and your parietal cells secrete. Put away some meat (and/or tofu) and your stomach is prepared to receive it, even before you swallow.

Alternatively, if you swallow something without thinking, or even without knowing about it, your stomach will sense the presence of food in the belly and your parietal cells will produce acid. That is your enteric nervous system at work. You can respond to food in your stomach even when the nerves from the brain (the vagus nerves) have been cut. The ability of the stomach to know what is in it and to respond accordingly, of course, explains why it was possible for surgeons to cut the vagus nerves with impunity (for the treatment of ulcers) in the bad old days before H2 blockers were introduced.

Cutting the vagus nerves was not a foolish thing for surgeons to have done. Clearly anxiety—not simply the contemplation of sauce béarnaise—can make the stomach more acidic. The head phase of digestion can be co-opted by a neurotic brain and get the hydrogen ions to flow. It can also make the stomach, and the intestines as well, churn away in high dudgeon. The twists and turns of the excited bowel send return messages back to the brain that result in perceptions that range from butterflies and queasiness in the belly to cramping and frank abdominal pain. A judicious slice through the vagus nerves lets the brain play its neurotic games all it wants without involving the gut. Nerves act directly on parietal cells, which express muscarinic receptors for acetylcholine. (These are the very same muscarinic receptors that are stimulated by the mushroom toxin, muscarine, and helped to earn Sir Henry Dale his Nobel Prize.) The link of gastric acidity to anxiety led people to believe that ulcer disease was a psychosomatic illness. Ulcers were thought to be holes bored into the lining of the stomach or intestine by excessive hydrochloric acid and pepsin, released as a result of aberrant signals sent down the vagus nerves by a misguided brain. In the presence of acid, pepsin digests protein, which is what the lining of the stomach and intestine is made of. It stood to reason that ulcers came from neurotic thoughts. The logic was impeccable, even if the conclusion was wrong.

Many observations, both clinical and experimental, seemed to support the psychosomatic view of ulcer disease. For example, patients undergoing arduous surgical procedures or suffering from stressful illnesses often develop ulcers. These "stress ulcers" frequently arise right in a hospital and are a dreaded complication of treatment. In animal studies, ulcers were found to develop in the bowel of rats that were simply tied down overnight. These lesions are known as "restraint ulcers." One of the most intriguing animal experiments was one that came to be known as the study of the "executive monkeys." Two monkeys were seated in experimental chairs where they could see one another and communicate. One animal had access to a lever that, when pushed the right way, *prevented* the other animal from receiving electrical shocks. The shocks were bothersome, but they were not strong enough to cause outright pain or suffering. In any case, no shocks were delivered to the monkey with the lever, no matter what he/she did. The experiment was designed to cause anxiety, and it succeeded in doing so. Ulcers appeared in the gut of one of the monkeys, but not, as you might expect, in the bowel of the monkey who received the shocks. Ulcers appeared in the gut of the monkey with the lever. The animal charged with the unwanted responsibility of preventing his/her colleague from receiving shocks got the ulcers.

Experiments carried out on restrained animals (rats, monkeys, or other species) that cause bad things to happen are difficult to justify. I do not do research of this sort, in part because the scientific questions I ask are answered by other means, but also because I do not like to be an agent of pain. My own studies are designed to cause no animal or human suffering, beyond that which I go through personally in attempting to interpret results. I am, however, not quick to condemn research done by others that does induce some discomfort in animals, as long as the discomfort is minimal, brief, and can be stopped by the animals as soon as it occurs. Modern animal experimentation fulfills these criteria, and no journal will publish the results if it does not. (The reasons why scientists do experiments with animals are explained in the endnote on page 314.)

It is clear that in both humans and in animals, severe overt stress can induce ulcers to form in the gut. Since severe mental duress can demonstrably cause an ulcer to arise, people concluded (incorrectly, it turned out) that the lesser degree of stress associated with psychoneurotic anxiety must do so as well. Vacations, psychotherapy, and tranquilizers became important adjuncts to ulcer therapy. In practice, it was easy to demonstrate that these therapies worked wonders on mood, but it was never convincingly demonstrated that they healed ulcers. Vacations, psychotherapy, and tranquilizers are still in vogue as therapy, but they are now used to treat anxiety and even angst rather than ulcers.

A famous study of a patient named Tom, carried out many years ago, seemed to add direct support to the hypothesis that ulcers are mental or psychosomatic in origin. Tom had inadvertently swallowed lye, and the resultant damage and scarring had permanently plugged his esophagus. At the time of his injury, it was impossible to repair or replace his esophagus surgically. If nothing was done, however, Tom would have starved to death. The solution was to attach Tom's stomach to his abdominal wall and to create a permanent opening through which Tom could insert his food. The permanently visible inside of Tom's stomach also provided investigators with an opportunity to watch the stomach at, or preparing to go to, work. A great deal of important information about the digestive process was learned.

When Tom was subjected to psychological stress, his stomach became more acidic, but of even greater interest, from the point of view of the theory of the psychosomatic causation of ulcers, the lining of his stomach seemed to become more fragile. Mucous secretion decreased and, when a probe was placed on the stomach lining while Tom was under mental stress, the probe had a greater than normal tendency to cause bleeding. The investigators, for obvious reasons of decency, never pushed the psychological stress they caused Tom to the point of actually provoking an ulcer. In fact, even though Tom had given his informed consent, I do not think that these particular experiments were ethical. To deliberately provoke Tom and make him feel stress, and particularly to engage in an activity that the investigators had reason to believe might even induce an ulcer, is not, I believe, an act that is covered by the Hippocratic oath. Doctors should not harm their patients. To be sure, therapy always carries a risk, but to incur a risk while seeking a cure is one thing, to incur it just for the sake of gaining information is another. In any case, the investigators clearly thought they had discovered a smoking gun. Stress led to ulcers, case closed—or so it was thought.

Helicobacter pylori: Infectious Ulcers

In the light of evidence, reason, and tradition, few people were willing to challenge the idea that ulcers are always due to excessive gastric acid induced by stress. Generations of medical students and thousands of unhappy businessmen were told that ulcers of the stomach or duodenum were the prototype of a psychosomatic illness. Nevertheless, although almost everyone thought that ulcers came from anxiety, ironclad proof that they did so was not easy to obtain. This particular "psychosomatic" disease, for example, was one that was extremely resistant to cure by psychotherapy. Tranquilizers were found to be far more likely to put patients to sleep than to cure their ulcers. Epidemiological confirmation that anxiety caused ulcers, furthermore, was not forthcoming. If anything, after controlled investigations replaced the telling of anecdotes, the epidemiology of ulcer disease suggested the opposite: that anxiety might be a *consequence,* rather than a *cause,* of the majority of ulcers. Patients with ulcers were often anxious, but the psychosomatic theory of the derivation of ulcers required that the anxiety *precede* the ulcer. Clearly, an ulcer might well make a person anxious, but that does not establish that anxious thinking gave rise to the ulcer.

Meanwhile, pathologists kept finding bacteria in people's ulcer lesions that they tended to ignore. Bacteria, after all, would surely find a hole eroded in the lining of the bowel an opportunity too good to pass up. The lining of the gut is itself a defense against infection; therefore, an ulcer is a breach in the body's defensive system. Once the bacteria were studied carefully, however (first by B. J. Marshall in 1984), they turned out not to be transient opportunistic invaders that happened to be in the stomach when an ulcer formed. If this had been the case, a wide variety of organisms would have been anticipated. Any germ that happened to be present would be expected to jump right in. Nevertheless, a wide variety of organisms was not what was found. The bacteria in stomachs and intestines with ulcers almost always seemed to be *Helicobacter pylori*. Once that relationship was understood, it did not take long to establish that *Helicobacter* was not a bystander taking advantage of a psychosomatic illness but the major culprit.

We now know that a great many ulcers, perhaps most, do not arise simply because the brain has made the stomach too acid. Ulcers are a symptom of an infectious disease caused by *Helicobacter pylori*, a peculiar organism that manages to avoid getting killed by the stomach's hydrochloric acid. The modern treatment of ulcers includes the eradication of *Helicobacter*. This can be accomplished by combining an appropriate antibiotic with a drug that eliminates the acid in gastric juice. For reasons that are not entirely clear, *Helicobacter* has proven to be more vulnerable to antibiotics when the stomach lacks acid. The antibiotic, Biaxin (clarithromycin), for example, wipes out *Helicobacter* over 90 percent of the time when it is combined with Prilosec (omeprazole), which is even more effective than the H2 blockers at preventing acid secretion. Many fruitless trips to psychoanalysts' couches could now be replaced by a single fruitful trip to a drugstore for a supply of Biaxin and Prilosec.

In the light of our current concern over the high and escalating costs of medical care, it is interesting to consider how these costs have been affected by the successful application of science to the simple question of what controls the secretion of acid by the parietal cells of the stomach. Ulcer disease is a serious problem. Ulcers cause discomfort and pain that can be intolerable. Even worse, they can bleed into the gut, causing severe but often unrecognized blood loss, and they can actually burrow right through the bowel and perforate it. A perforated gut is a medical emergency and, if not repaired quickly and successfully, is likely to be a lethal event. In fact, because of infection, a perforation can be fatal even when it is repaired quickly and successfully. Ulcers thus cannot be ignored and, once diagnosed, have to be treated. Ulcers are also widespread, occurring literally in millions of people every year. That means that ulcers attract and are responsible for a great deal of utilization of the health-care system.

Early drug treatments for ulcer disease were not notably successful. Since it has been clear for a long time that hydrochloric acid and pepsin are involved in digesting the wall of the bowel to produce an ulcer, the first effective treatments for ulcers were designed to try to stop the secretion of acid (without which pepsin would not work). The ability of the nervous system to induce acid secretion via the release of acetylcholine was the first regulatory mechanism to be discovered, but drugs that blocked the action of acetylcholine on muscarinic receptors were of no real use in treating ulcers. These drugs cannot be targeted to block just the muscarinic receptors on parietal cells. They act all over the body. When given in doses adequate to prevent nerves from stimulating the secretion of gastric acid, therefore, muscarinic antagonists virtually paralyze the entire parasympathetic division of the autonomic nervous system. This kind of cure is worse than the disease. As we have already seen, acetylcholine, of course, is also not the only excitatory signal that parietal cells receive; besides, it is the histamine, and not the acetylcholine, that provides the kind of chronic stimulation that maintains high levels of acid production. Acetylcholine is more involved in producing peaks of acidity than high plateaus. Since drugs did not work very well in the days before the advent of H2 blockers, nothing remained except to send in the surgeons.

A variety of surgical treatments were devised to alleviate ulcer disease. One, mentioned earlier, was to cut the vagus nerves to disconnect the bowel from the brain. Once the vagus nerves are cut, a patient's brain and his/her stomach should be completely independent of one another. After surgery, therefore, a psychoneurotic brain would, at least in theory, be free to indulge itself in an orgy of anxiety without forcing its owner to pay a price in stomach acid. Fortunately, the enteric nervous system perseveres after the brain is cut off. The vagus nerves can thus be cut without killing people.

Another surgical approach was to remove the pyloric antrum. This operation leaves the acid-producing part of the stomach intact, but it cuts out the chemical laboratory that analyzes the acidity of gastric juice, thereby eliminating the feedback secretion of gastrin. When all else failed, surgeons removed either the acid-producing regions of the stomach (the fundus and corpus) or the whole organ. Patients then survived with daily vitamin B_{12} injections and multiple small meals.

None of the surgical procedures was pleasant. All left the patient with at least some degree of handicap, and all were associated with serious complications, like postoperative bleeding and infection, which occurred in occasional patients even when the surgery was performed with great competence. In addition, surgery was always very expensive, and it has not become cheaper with the passage of time. The more one thinks of surgery and the gross rearrangements of innards that used to be commonplace, the more one appreciates the modern age.

The introduction of H2 blockers began the retreat from surgery, the development of hydrogen pump inhibitors such as Prilosec furthered the march, and the discovery of the role of *Helicobacter pylori* has made surgery for ulcer disease rare and almost obsolete. Certainly, the folks who pay health-care bills should thank the Eternal for these developments before they go to sleep at night. The expenditure of small amounts of money on the research, which has given us H2 blockers, hydrogen pump inhibitors, and antibiotics that kill *Helicobacter pylori,* has saved society billions of dollars in health-care costs and an incalculable amount of pain and suffering. Throw in the restoration of self-esteem to patients who discovered that they may not be neurotic after all, and this advance in medicine can only be called revolutionary.

Mom: A Case History of the Revolution in Ulcer Treatment

Unfortunately, my poor mother lived and died before the revolution reached her physicians. Heartburn and upper abdominal pain were constants in her life. They were always there, a part of her, like her love of good music, fine art, and her family. She rarely complained about her dyspepsia, but you could easily judge how it was doing from the rate at which she took antacids. Mom never fit the picture of a typical ulcer victim. She was anything but a hard-driving executive, and, in fact, until late in her life, X rays never revealed enough of an ulcer to quicken the pulse of a surgeon. Still, Mom's stomach had a pronounced tendency to bleed. Fortunately, I happened to be nearby on most of the occasions when it did. She would go pale and collapse. I would collect her as best I could and rush her to New York Hospital, where the bleeding would be diagnosed and a transfusion of a suitable amount of blood would restore her to health. After the bleeding stopped, she would be discharged on a regimen that included an H2 blocker. Once home again, she would take the pills faithfully until there were none left in the bottle; then, because she would be feeling perfectly well, she would forget about the bleeding and her pills until, of course, she began to bleed once more.

The last and final time Mom's stomach began to bleed I was not nearby. I was giving a lecture at a meeting in Florida. By then she had developed Alzheimer's disease and was no longer at the top of her intellectual game. Nevertheless, despite the decline in her mental agility, she was coping reasonably well and was still preparing meals for herself and my father. My wife and I saw my parents frequently, and my children made sure to include their grandmother in their own agendas. The quality of Mom's life was relatively well maintained, despite the Alzheimer's disease, when my father found her unconscious and lying on the floor of her kitchen. Not knowing what else to do, he called a family friend who happened to be a physician. He understood what was happening and did the same thing that I had been doing on similar occasions, except this time he brought my mother to a different hospital, St. Luke's–Roosevelt.

As on the earlier occasions when I had taken my mother to New York Hospital, a correct diagnosis of gastric bleeding was made at St. Luke's–Roosevelt, an appropriate blood transfusion was given, and an H2 blocker was administered. My mother recovered consciousness and her condition stabilized, although she was quite naturally confused and frightened by the situation. This time, however, despite the administration of the H2 blocker, the bleeding in her stomach did not stop. Repeated transfusions were required to keep her alive.

I returned home the next day and conferred with a surgeon who was anxious to operate. My mother, however, had been terrified of surgery ever since my grandmother's death, which she attributed to surgery rather than to the stomach cancer that the surgery had been unable to cure. My mother's objections to surgery, however, were deemed irrational by her physicians, and, in any case, because of her underlying Alzheimer's disease, she was not expected to be able to provide informed consent. My father was asked to provide this, but he had decided to hold out until he could talk to me.

As it happened, my tenure of service on the FDA's Advisory Panel on Gastrointestinal Drugs coincided with the request by Astra-Merck to approve omeprazole (the generic name of Prilosec) for the treatment of ulcers. In particular, I had been asked by the committee to review the data on omeprazole's safety. There had never been any question about the efficacy of omeprazole. In fact, it was a pleasure to look at the efficacy data. We often reviewed information about compounds that either eked out a narrow win in competition with the placebo or actually lost. In contrast, omeprazole not only beat the placebo but wiped out the H2 blockers in head-to-head tests as well. Omeprazole does more than merely inhibit the production of gastric acid; it stops it cold. Once exposed to an adequate concentration of omeprazole a parietal cell cannot put out acid, no matter how strong a stimulus the cell receives. Questions had been raised to the FDA about omeprazole's safety, but it was absolutely clear that there had never previously been anything like omeprazole to stop acid secretion. Once our advisory panel decided that omeprazole was safe for short-term use, we were more than happy to recommend it.

As a result of my work for the FDA, I was thinking of omeprazole when I spoke to the surgeon. I thus suggested that he might consider a therapeutic trial of omeprazole before resorting to what I believed would be a risky and trying operation on a patient who happened to be terrified of surgery. I was not prepared for his response. He told me that omeprazole was a new and expensive drug and that it could not be used unless a hospital committee, which had been formed to contain costs, approved. Naturally, I pointed out that the cost of omeprazole was trivial when compared to that of surgery. The surgeon then advanced the opinion that as a basic scientist and not a practicing physician I had a lot of chutzpah to question his clinical judgment. He concluded by asking what I could possibly know about omeprazole. So I told him. Given that I had recently reviewed the entire world's literature on the drug, it took me some time to do so. When I finished, he seemed less certain about my chutzpah and promised to get the committee's assent to try omeprazole. In return, I promised to consent to surgery if a fair trial of omeprazole failed to stop my mother's bleeding.

In the end, omeprazole was never used. When I visited my mother that night, I found the same old H2 blocker dripping uselessly into one vein while a blood transfusion dripped into another. My father, moreover, had been told that his son was going to kill his wife by refusing to permit essential surgery to be done. My father, frightened and intimidated by the show of medical power, had signed the consent form. The juggernaut was rolling and surgery was scheduled.

Mom never recovered from anesthesia. Her blood pressure fell during the operation and she probably suffered a stroke on the table. She lived at home for another six months, but the result was not pretty. They call this kind of living a vegetative state. Her grandchildren and family still visited, but she never knew it. My father lovingly caressed her every night, but she could get no pleasure from it. Periodically, she would scream as if in pain, but she never said another word. She died at home, sooner I think than she needed to, and under circumstances that were less good than they needed to have been. It really is a pity that Mom missed the revolution in ulcer treatment. Today, even managed-care organizations know that omeprazole is cheaper than surgery.

6

ONWARD AND DOWNWARD

IN EVERY BODY, the brain is king. Its writ is law. At the top of the bowel, the rule of the king is acknowledged, but as one descends deeper and deeper into the depths of the gut, the rule of the king weakens. A new order emerges: that of the second brain. From the mouth to the middle of the esophagus, virtually nothing moves unless the brain decrees that it should. The first tentative signs of the lower will become manifest is the peristaltic movements of the lower esophagus, which require the participation of the enteric nervous system to be anything like normal. The central authority of the king is restored at the lower esophageal sphincter, but only temporarily.

In the stomach the central order is still important, and in the form of orders transmitted by the vagus nerves, the will of the brain looms large. The second brain, however, is now also a potent factor, and should the word of the brain be lost, the enteric nervous system is ready and able to take over the show. That is, the enteric nervous system can take over all aspects of the show except for the running of the pyloric sphincter, which curiously enough is left to the brain to operate by way of the vagus nerves. To descend below the *pyloric sphincter* (the exit of the stomach), however, is to move almost beyond the reach of the king. This is the turf of the enteric nervous system, where the brain can exert only quantitative effects and not make the basic decisions of what to do and when. It is an autonomous region that cares little for the brain and is happy to do without it altogether. The central authority of the brain is not reexerted until one emerges from the colon at the rectum and anus.

The Pyloric Sphincter

Once food has been sufficiently pulverized, sterilized, and partly digested in the stomach, it is dribbled out into the small intestine in quantities that are not too hard for that organ to handle. Emptying is thus an act that the stomach performs with considerable finesse. The stomach feeds the *duodenum* (the region of the small intestine that adjoins the stomach) like a mother feeds an infant. The food is turned into pabulum and delivered in tiny baby bites. The sophistication of the act of gastric emptying involves an interesting collaboration between the enteric and central nervous systems. The delivery process is orchestrated by the enteric nervous system, but the gate through which everything moves is controlled by the brain.

The opening and closing of the portal through which food passes from the stomach to the duodenum is governed by specialized muscle cells that encircle the gut. The structure formed by these cells, the pyloric sphincter, does for the stomach's exit what the LES does for its entrance. When the pyloric sphincter is closed, it keeps the contents of the stomach out of the duodenum. To do so is important, because the lining of the duodenum is no better able to resist digestion by gastric acid and pepsin than is the lining of the esophagus. In fact, the duodenal lining is probably even less resistant because it is much thinner than that of the esophagus. The corrosive nature of stomach acid thus makes it necessary for the pyloric sphincter to keep the gastric juice bottled up in the stomach where it belongs, except for the brief and unavoidable moments when the sphincter has to open to let the food move on.

The nervous signals that regulate the muscle of the pyloric sphincter are carried to and from the brain by the vagus nerves. These nerves bring the information that the brain needs in order to decide when it is sphincter-opening time in the stomach. The brain deciphers the messages from the stomach, acts on them, and sends the appropriate command to the sphincter to open up. Once the food passes through, the brain is again informed and tells the sphincter to close. Because of the critical nature of the role played by the brain in regulating what the pyloric sphincter does, the stomach is in big trouble when the vagus nerves are cut. Essentially, the sphincter becomes paralyzed in the closed position and the poor stomach is unable to empty itself. Clearly, a stomach that does not empty is not only distressing to its owner but incompatible with life.

Fortunately, the problem of pyloric sphincter paralysis associated with cutting the vagus nerves can be fixed by a surgical procedure known as a *pyloroplasty*, which is a euphemism for the virtual destruction of the pyloric sphincter. Miraculously, the stomach adapts nicely to this insult. After drainage to the small intestine is assured, the enteric nervous system regulates the rate of gastric emptying all by itself. Even though it is totally bereft of brain-supplied assistance, therefore, the stomach perseveres. Acid is still secreted when it should be, the stomach goes right on storing food when good solid meals are consumed, and food particles continue to be turned into the kind of pabulum that the duodenum can tolerate. Still more remarkable, even with a seemingly open drainage hole, the stomach pretty much keeps its contents at home and lets them leave only at a rate that remains tolerable to the small intestine.

Even under normal circumstances, when the enteric nervous system, the brain, and the pyloric sphincter are all working to perfection, acid poses a problem for the duodenum. This problem is similar to that faced by the esophagus, but is more severe and requires a more elaborate solution. If the LES does its job, very little gastric juice refluxes up into the esophagus. On the other hand, not even the optimum performance of the pyloric sphincter can prevent gastric acid from accompanying deliveries of food to the duodenum. The difference between the two sphincters is simply a matter of direction. Food moves from the esophagus to the stomach; thus, all a well-intentioned LES has to do to prevent heartburn and esophageal damage is keep gastric juice from going backward. The duodenum, however, is the natural sink into which the stomach drains, and thus it has to receive whatever the stomach has to deliver. This means that the duodenum must take the bad (acid) with the good (stomach-processed food). Job number one for the duodenum, consequently, is to neutralize the acid that the stomach dumps into it, and to do that fast!

The New World of the Small Bowel

The solution to the presence of acid that is favored by the stomach, a tenacious alkaline gel coating its lining, is not an option for the small intestine. Digestion is finished in the small intestine and absorption occurs there. Both of these tasks are the responsibility of cells that line the lumen of the small intestine and in fact require the active participation of their surface membrane. The intestinal lining thus cannot be covered by an adhesive coating of slime, no matter how protective that slime might be. The surface of the small intestine is its critical work space and has to be kept clear and clean. The alkaline mucous gel that prevents digestion of the stomach's lining thus ends at the pyloric sphincter. Once through that portal, a whole new set of cells is found, and the structure of the bowel changes abruptly.

The majority of the cells that line the small intestine are specialized to deal only with digestion and absorption and thus are helpless in the face of stomach acid. These cells, and the duodenum as a whole, need help. In fact, although the small intestine is the site where the passing of the digestive buck ends, the organ cannot produce everything it needs to accomplish the job. The necessary help to rapidly neutralize the gastric acid that arrives with food from the stomach is provided by accessory glands. These glands also provide the enzymes that handle the bulk of digestion. The small intestinal cells themselves have only to apply the small but necessary finishing touches. Given the right environment, which is critical, these cells are extraordinarily good at doing so.

The Pancreas

The most prominent of the accessory glands are the pancreas and the liver. The pancreas is at once a protector of the intestine, a provider of the kind of environment digestive enzymes need in which to function, and a manufacturer of digestive enzymes. In its spare time, the pancreas also secretes hormones (*insulin* and *glucagon*) into the bloodstream. These hormones regulate the blood sugar level. The enzyme-manufacturing capacity of the pancreas is impressive. There is very little that we eat that cannot be digested by at least one of the enzymes the pancreas makes.

The pancreatic cells that produce and secrete digestive enzymes form little grapelike clusters at the ends of complex, highly branched tunnels, or *ducts*. These ducts carry the newly secreted pancreatic enzymes to the duodenum. Aficionados call glands with ducts *exocrine* to distinguish them from glands without ducts, which are called *endocrine*. Exocrine glands always secrete onto body surfaces, such as those of the gut, the respiratory passages, or the skin. Endocrine glands always secrete into the bloodstream, which is why they get away without having ducts. The pancreas thus has a dual personality in that it is able to go both ways, exocrine and endocrine. The endocrine glands of the pancreas are called, colorfully, the *islets of Langerhans*. If you lose your exocrine pancreas, you starve unless you eat pancreatic enzymes with your food. If you lose your islets, you get diabetes.

The ducts of the pancreas do more than just drain the enzymes that exocrine cells of the gland choose to secrete. The ducts are themselves secretory; however, instead of enzymes, the duct cells put out a watery fluid that is alkaline and able to neutralize the acid that accompanies food into the duodenum. The pancreas thus divides its functions among its different cells. The cells at the ends of the tunnels make enzymes, and the cells that line the tunnels make the juice that provides an environment in which these enzymes can work. Both the enzymes and their working fluid are dumped into the lumen of the duodenum, where they find the partially digested meal that the stomach has delivered.

To back up the pancreas in dealing with stomach acid, the wall of the duodenum is uniquely equipped with another set of glands (known as *Brunner's* glands) that also put out an alkaline juice. Similar glands are not found in any other region of the small intestine, but then, nowhere else is there a comparable threat. The plan of the bowel appears to have incorporated a fail-safe redundancy of protective mechanisms whenever the costs of failure are intolerable. Brunner's glands illustrate this principle. The ingenuity of the gut's designer is very impressive.

Orchestration of the Duodenal Defense

To be effective in eliminating acid, it is not enough to just pour a base (aka alkali), which is the antithesis of acid, into the intestine. The duodenum has to be more subtle than that. In fact, it is, and it approaches the problem presented to it by the emptying of the stomach the way Hegel approached philosophy. First, there is thesis (acid), then there is antithesis (base), and finally there is synthesis (neutrality). Clearly, to get to neutrality it is necessary to know how much acid is present in the first place, so that just the right amount of base can be secreted. A solution that is too alkaline is just as corrosive to tissue as one that is too acid. Gastric acid could be eliminated, for example, by secreting the biological equivalent of Drano or Liquid-Plumbr, but if that were to be done the wall of the duodenum would turn into soup. The duodenum has thus evolved a system of sensors that measure the degree of acidity or alkalinity (a measure called the *pH*) of the intestinal lumen. These sensors communicate with both endocrine cells and nerves (intrinsic and extrinsic), which, in turn, regulate the secretion of acid and alkali. When the inside of the duodenum is made acidic, alkali is pumped in, and further emptying of the stomach is inhibited. When the lumen becomes neutral or alkaline, secretion of alkali stops and the inhibition of gastric emptying is reversed. The duodenum will now accept another delivery from the stomach, and the pyloric sphincter will again open briefly to let a little more acidic gastric juice go through. The gastric drainage cycle thus repeats in short bursts until the stomach has nothing more to deliver.

The tight regulation of pH is critical, not only to protect the duodenal wall but also to set the stage for pancreatic enzymes to operate. Unlike pepsin, which is used by the stomach and demands a highly acidic environment in which to work, pancreatic enzymes work only when the pH is nearly neutral or slightly alkaline. They are actually irreversibly done in by exposure to acid.

The details of how the duodenum senses the pH inside its lumen and orchestrates the secretion of base are not entirely clear. The process involves, in part, the secretion of a hormone, *secretin,* which was, incidentally, the very first hormone to be discovered. Secretin is produced by cells in the duodenal lining, which like their gastrin-secreting counterparts in the pyloric antrum, secrete into the blood. The secretin-producing cells appear to be directly sensitive to acid that they detect in the lumen. Pancreatic duct cells are targets of the secretin in the bloodstream; thus, alkali flows from the pancreatic ducts to the duodenum very soon after the level of secretin in the blood rises.

Although the role played by secretin in regulating the pH of the duodenal lumen is undoubtedly important, and secretin is justifiably famous for its activity, secretin is not the only musician in this particular orchestra. The complexity of the job of getting right the coordination of acid secretion up in the stomach, opening the pyloric sphincter, delivering gastric contents, and getting the pancreas (and Brunner's glands) to secrete just enough base to bring the pH of the duodenal contents to near neutrality is far too much for any single cell or hormone to handle. This kind of work requires the intervention of the nervous system. It is apparent that the nervous system intervenes, but unfortunately the modus operandi of the nervous system is not apparent at the present time. The brain participates in opening the pyloric sphincter, but once that sphincter has been circumvented by clever surgery, the brain is no longer needed. Regulation, even after the brain has been cut out of the game, is accomplished smoothly by the enteric nervous system. The thoughtful bowel perseveres, but how the nerves of the gut operate is a subject of ongoing research. Still, it is good to be at the stage of knowing that the enteric nervous system is the player crying out to be understood. Only recently has even this much come to be recognized.

Bile

The liver is the second major accessory gland. It plays many roles in running the body, most of which are not directly related to digestion. One of the most apparent functions of the liver, however, is to secrete *bile,* which leaves the liver through a system of ducts. Making bile is only one of the necessary things the liver does for us, but when it comes to making bile, the liver never rests. Bile pours continuously out of the organ. The constant production of bile is required because the secretion of bile is, in part, an excretory activity, a form of garbage removal. Waste materials that are soluble in water can be filtered by the kidneys and eliminated in urine. Those that are not water soluble cannot be filtered by the kidneys and have to be gotten rid of differently. The liver transports many of these waste products to the bile and in the process converts them into molecules that more or less stay in solution. The distinctive green color of bile, for example, comes from molecular detritus resulting from the destruction of dirty old worn-out red blood cells (a job the liver shares with the spleen). Since the bile flows into the gut, the nasty molecular refuse, if it is not reabsorbed, is eventually evacuated along with feces, which get their brown color from the bile.

The digestive function of the liver, which it carries out simultaneously with its role in waste disposal, is to provide the bowel with the detergents it needs to digest fats. These detergents, or *bile salts,* are made in the liver and added to the bile. Bile salts emulsify fats in food so that a pancreatic enzyme, called *lipase,* which does the actual fat digestion, can get at the fat molecules and attack them. The digestion of fat can thus be blocked either by depriving the gut of bile or by denying it pancreatic lipase.

The constant production of bile is not very efficient from a digestive point of view. The bile salts are needed in the gut only when food containing fats to be emulsified is present. The pancreas, for example, does not secrete enzymes all the time but only when nerves and hormones tell it to. One might similarly expect the liver to secrete in the same way. Sadly, however, it cannot. Since bile contains molecular trash as well as bile salts, an intermittent pattern of secretion would cause the trash that the liver excretes to accumulate in the blood during nonsecretory periods. Excretion would be linked to the presence of fat in the duodenum. This would seem to be counterproductive. Just think of what happens to communities when the regular schedule of garbage pickup is interrupted. Neither God nor evolution would be expected to favor a scheme that leaves the removal of potentially toxic waste to the capricious vagaries of food presentation. Animals can go a long time between meals.

The Gall Bladder

The alternative, which appears to have evolved in order to let the liver continuously get rid of the garbage without being profligate with bile salts, is to store the bile in the *gall bladder* until it is needed. The gall bladder is a blind sac at the end of one of the arms of a Y-shaped duct. The other arm of the Y is the duct that drains bile out of the liver, while the common stem empties into the duodenum.

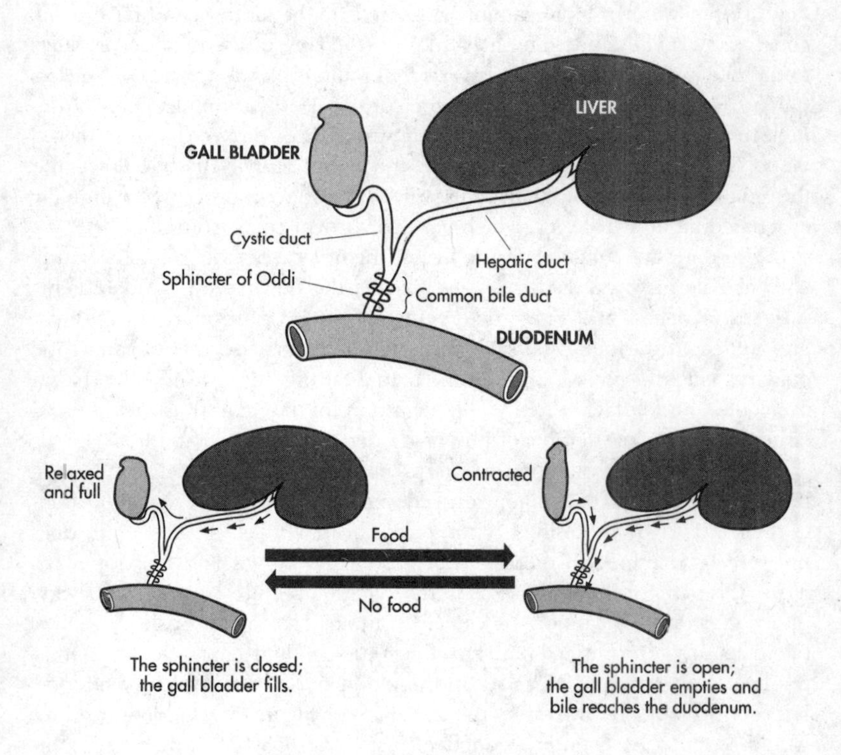

The gall bladder's arm is called the *cystic duct.* The liver's arm is called the *hepatic duct,* and the joint stem is called the *common bile duct.* The main pancreatic duct joins the common bile duct, and the two ducts empty into the duodenum together through a shared hole. The opening of this hole is regulated by a muscle, the *sphincter of Oddi* (not the odious sphincter, as it is known to medical students who have trouble with anatomy and thus hate this sphincter). The sphincter of Oddi does for the bile ducts what the pyloric sphincter does for the stomach.

The anatomy of the biliary system is really not very complicated. In essence, a distensible bag, the gall bladder, is hooked up to a reservoir, the liver, and a hose, the duodenum, by means of a three-way valve, the hepatic, cystic, and common bile ducts. Bile flows out of the liver or the liver is in trouble. Where it goes is determined by what is happening at the sphincter of Oddi. When the sphincter of Oddi is closed, bile cannot enter the duodenum and flows only from one arm of the Y-shaped duct system to the other, that is, from the hepatic duct up the cystic duct and into the gall bladder. The gall bladder fills with bile, distends, and holds the bile until it is needed in the gut. That need is transmitted by hormonal and nervous signals to both the sphincter of Oddi, which opens, and the gall bladder, which contracts. When the sphincter of Oddi opens, bile can flow down the common bile duct and enter the duodenum. The gall bladder also pumps it along, ensuring that the bile is delivered to the gut. The gall bladder, moreover, is not just a passive bag, a device designed to hold bile until it is required. Bile also becomes concentrated during the intervals between meals while it is stored in the gall bladder. Concentration of bile is accomplished by transporting water and salts across the wall of the gall bladder, moving them from the bile to the blood. The presence of the gall bladder thus enables bile to be produced constantly but discharged intermittently and, when it finally reaches the gut, the bile is concentrated and potent.

Stones

The tubes that drain the liver and gall bladder are frequently the site of major medical distress. The list of the characteristics of the patients in whom these problems typically occur is a quadruple alliteration that generations of medical students have loved because it is easily remembered. The patients tend to be fair, fat, female, and forty. The disease is thus sexist, race-conscious, and discriminatory on the basis of age. It has no sense of decency. Patients get into biliary trouble when the process of concentrating their stored bile causes constituents of the bile to precipitate and form stones (*gallstones*) in the gall bladder or bile ducts. These stones, which are quite literally solid rocks, can be nasty things to carry around, even when they stay in the gall bladder. The stones are irritating to the gall bladder's lining, and they may interfere with the drainage of the organ. As a result, stones frequently give rise to some pretty awful gall bladder infections (*cholecystitis,* in medical parlance).

Curiously, large rocks are less likely to make big trouble than tiny pebbles. Small stones are much more liable than big ones to roll down the duct system and become stuck. A stone in the common bile duct can act as a plug. Since bile is constantly secreted by the liver, a plugged common bile duct causes pressure to rise in the whole biliary duct system, including the fragile little ducts inside the liver itself. There is no escape valve. The pressure damages the liver and small bile ducts, leading, essentially, to the leakage of components of the bile into the blood, coloring patients yellow (*jaundice*). The difficulty posed for patients by being yellow with jaundice is not simply cosmetic. Their image problem is negligible compared to their constant nausea, malaise, loss of appetite, and incessant itching. The general symptoms arise because the damaged liver can no longer do what it needs to do in the metabolism of sugar, fat, and protein. The itching is caused by molecules of the bile reaching the skin. When a stone is stuck in the common duct, surgeons have got to go in and get it out.

Gallstones are dangerous. The litany of woes associated with them are like the trials of Job revisited on poor unsuspecting women and, less frequently, men. The stones may initially lie in wait, insidiously holed up in the gall bladder, for long periods of time without giving rise to any symptoms. Then, seemingly for no reason, catastrophe erupts, often at times when a patient is under stress and is unable to deal with it.

Initial symptoms are frequently insidious. The stones may begin by interfering with the flow of bile just enough to deprive the bowel of the bile salts it needs to emulsify fats properly, but not enough to cause jaundice. Fat digestion becomes incomplete. Undigested fats thus move through the gut and are fermented by bacteria in the colon, giving rise to gas. The patients become bloated and frankly gaseous. Afflicted individuals can be driven almost mad because they may uncontrollably blow foul-smelling gas, under the worst of social circumstances, out of both ends of the bowel. Friends, associates, and anyone else nearby may take offense and, not understanding the patient's problem, exhibit overt signs of disapproval. Embarrassment and psychological trauma are thus heaped on top of the physical distress caused by gall bladder disease.

As the blockage of bile drainage worsens, and bile pigments no longer reach the bowel, the stool changes color, becoming pale and clay-colored instead of its customary and reassuring brown. The stool is also full of millions of bacteria happily and malodorously fermenting the abundance of fat the feces now contain. As a result of the fat and the mass of bacteria, the stool become bulky, greasy, foul-smelling, and lighter than water. It tends to float in the toilet bowl and is hard to flush away. Even the familiar act of defecation thus becomes a source of terror and adds to the psychological unrest suffered by the patient. At this point, almost as if the disease has caused all this trouble to soften up its victim, further blockage or infection can occur, and the main events begin.

The signal that the gall bladder is sick beyond redemption is severe pain in the upper right quadrant of the abdomen, radiating through to the back and right shoulder. Nerves from the gall bladder and the right shoulder converge in the central nervous system, and what should be gall bladder pain is mistakenly felt in the right shoulder. Chills and fever announce that the gall bladder and/or bile duct system are infected. The arrival of jaundice, with its associated nausea, vomiting, itching, and malaise, punctuate the severity of the condition. Then, because the biliary and pancreatic ducts share a common opening into the duodenum, disease of the biliary system frequently leads to disease of the pancreas (*pancreatitis*).

Pancreatitis

While the problems of the gall bladder and liver are bad, they pale beside the potential of pancreatitis. The exocrine pancreas is designed to make enzymes that digest almost everything we eat. The trouble is, what we eat is not very different from what we are. A steak, for example, is simply muscle from a cow. Pancreatic enzymes can very rapidly turn a steak into juice. The protein and fat that gives a steak its substance are converted into the kind of small soluble molecules that the gut can absorb. What pancreatic enzymes do to a steak they can obviously also do to you. When all goes well in a healthy individual, the enzymes confine their destructive activity to food in the lumen of the gut. The enzymes are prevented by a variety of mechanisms from digesting their owners.

First, the cells that make the enzymes manufacture most of them in an inactive form. These precursors of enzymes (*proenzymes*) are not themselves dangerous, but even so, they are transferred as an integral part of the process of making them across a protective shield of membrane. Digestive enzymes are thus never exposed to the innards of the cell that produces them. While they are stored in the pancreatic exocrine cells, the proenzymes sit inside membrane-enclosed sacs or packages waiting for a nerve or hormone to tell the cell to spit them out. When that happens, and the proenzymes are secreted, they are delivered to a duct system that is sealed tight. The cells that line the ducts are joined together in *tight junctions,* which completely occlude the space between adjacent lining cells. The proenzymes have nowhere to go but into the duodenum. Once in the duodenum, the proenzymes are activated by intestinal enzymes (called *enterokinases*) that are integral components of the surface membranes of the intestinal lining cells. The potentially deadly pancreatic enzymes thus become active only within the lumen of the bowel, where they are turned loose in a space that is sealed off from the rest of the body and that itself has a special lining that is not digested by them.

Consider the consequences of a stone blocking the exit of the pancreatic duct at a time that coincides with a signal to the pancreas to secrete. Pressure mounts within the ducts and the enzymes leak out into the surrounding tissue. Tight junctions may break or other things may happen, but the bottom line is that pancreatic enzymes wind up where they should not be. The protective shield of membrane and duct system is breached. In contrast to the lumen of the ducts of the pancreas and the gut, ordinary tissue is no more prepared than a chewed-up steak to act as a container for pancreatic enzymes.

One particular pancreatic enzyme, which is not produced as a proenzyme but is immediately active, is the enzyme that digests fat. This enzyme is especially devastating when it is placed in tissue where it does not belong because it can actually digest the membranes of the living cells of the body itself. As cells around the misplaced pancreatic proenzymes die, the dying cells release products that help to activate the proenzymes. Once the conversion of proenzymes to active enzymes begins, the process rapidly accelerates, because some pancreatic enzymes are able to activate others. What happens normally to a slice of steak that has been eaten now happens to a patient. Pancreatitis involves the horror of autodigestion or self-cannibalism. Pain, which is usually sharp and excruciating, is felt in the middle of the belly and characteristically radiates through to the back. A patient may quickly go into shock, and the condition, even with treatment, can rapidly be fatal.

Integration and Regulation of the Accessory Glands

It is clear, given what they do, that the accessory glands are vital, but they have to be handled with care. We cannot live without them, but when things go wrong, we cannot live with them; the accessory glands can kill us. The problem of controlling these glands and keeping them on a tight leash is thus one of considerable importance. Not surprisingly, a job of this significance and complexity has been allocated to the nervous system, which is the only component of the body that possesses the necessary ability to integrate information and use it to coordinate the activity of the glands.

Back in 1981 when Jackie Wood, Marcello Costa, Alan North, and I were debating about serotonin, almost nothing was known about the ability of the enteric nervous system to influence the accessory glands. We simply looked at the pancreas and gall bladder with the requisite awe and assumed that God would not have created them without a suitable headquarters to control that amount of firepower. Nevertheless, despite the reigning ignorance, the accessory glands caught my imagination. Since I realized that living with the pancreas and gall bladder was a little like living with a couple of bombs in one's belly, my intuition told me that there was golden information in those organs that some day I was going to mine.

Gary and the Gall Bladder

Four years after the Neuroscience meeting, I found myself discussing the enteric way of life with a prospective postdoctoral fellow, Gary Mawe. Gary and I could not have been more different. When we met, I thought Gary might be more at home in Galway Bay than Manhattan. Gary is Irish, and, true to the stereotype of the Irish that I carry, he radiated charm, earthiness, and good humor. My humor owes far more to the Yiddish theater and the borscht belt than Sean O'Casey. Still, it took less than five minutes of conversation to establish a friendship that has endured for years. Gary accepted the postdoctoral position and came to work with me.

During our initial conversation, I mentioned the gall bladder, but only briefly. I do not remember exactly why the gall bladder came to mind. Perhaps it was because Gary's Irish bearing turned my thoughts to green objects. Nothing, however, came of the gall bladder discussion at that time. It went by so fast that I doubt that Gary even remembers it today. I had several ongoing projects that I thought might tweak Gary's interest, and these took up most of the part of our conversation that dealt with science. The National Institutes of Health had provided me with money to investigate the enteric nervous system, and that was what I hoped Gary would help me study. I was free to muse about other issues like the gall bladder, but only in my spare time. Fortunately, Gary was enthusiastic about the enteric nervous system and wanted to learn more about it. The gall bladder was thus put off, like the Fenians, to rise again another day.

Gary turned out to be great in the laboratory, and when he worked with me, my productivity was as good as it has ever been. Gary had a rare ability to teach while he learned. When he finally left my research group, I knew I was smarter because he had been with me. Many good scientists, Gary included, are exceptionally bright. What makes Gary extraordinary, however, is his ability to acquire and use many different techniques simultaneously to solve biological problems. No problem is too daunting for Gary to tackle, and no method is too arcane for him to master.

After a few years of our working together successfully and happily, the moment came when Gary was fully trained and ready to set up his own research program. It was time for him to go. I thought of hanging up black crêpe, but it seemed more useful for the two of us to sit down and think of a project Gary could take with him. The right kind of project would be one that was close enough to my interests so that he could begin the work in my laboratory. Gary needed to get started while he was still with me because, to be successful on his own, he would have to be able to show solid preliminary results to a peer review panel of the National Institutes of Health (NIH) to convince the panel that he was serious.

In the United States, the NIH is the primary source of funding for biomedical research. Private donors, foundations, and industry also provide money, but these funds are small change in comparison to those provided by the NIH. Money from the NIH, however, has been hard to come by for a long time, and peer review panels award priority only to projects that they deem likely to succeed. I like to joke that the way to get a grant from the NIH is to propose to prove something that everyone feels is true but to show it by state-of-the-art means. That is, of course, an exaggeration, but only a small one. Bright young stars presenting great ideas go down in flames unless these ideas are accompanied by enough data to make it obvious to reviewers that they can be trusted with the scarce dollars. Innovation is great, but not if it comes with a risk of failure.

Funding from the NIH is not only required to do anything significant, it is also necessary to maintain one's respect in the academic community. A nonfunded scientist is a nonperson. We pass peer review or we pass out of sight. Since Gary was about to become one of my prominent alumni, I had an interest in ensuring his future success. On the other hand, whatever research Gary started in my laboratory would have to be at least a little different from my own work, so that I would not miss the project after Gary left. Nothing can kill a good friendship more effectively than competition.

I thought again about the gall bladder and brought it to Gary's attention. I would like to remember our conversation as something momentous because it set a path that Gary has followed, quite brilliantly, ever since. It should have occurred while we sipped port, or at least beer, next to a raging blaze in an old stone fireplace. Violins in the background would have been nice. Instead, we began our discussion at a table in my laboratory while I ate my customary lunch of cottage cheese (with pineapple), a box of raisins, and an apple. We finished our talk at the computer in my office, where we turned our thoughts into a coherent experimental design.

There were, we knew, nerves and nerve cell bodies within the gall bladder, but we were not sure what they did. I realized that the gall bladder, like the liver and the pancreas, developed during fetal life from an outpouching of the primitive gut. Could the nerves and ganglia of the gall bladder thus be part of the enteric nervous system? Why not? Bayliss, Starling, and Trendelenburg had shown that the enteric nervous system can work on its own. Once one accepts that the enteric nervous system can run the gut without help from the central nervous system, it is only a small additional leap of faith to believe that it can also run a neighboring organ, especially if that organ is a derivative of the primordial bowel. The enteric nervous system of the colon, furthermore, had already been shown to send axons out of the gut. This discovery, startling at the time, was made by Joe Szurszewski, the postdoctoral fellow who succeeded me in Edith Bülbring's laboratory at Oxford and actually rented my old apartment (or rather, "flat," since it was England). Joe, who is now at the Mayo University, had shown as early as 1971 that nerve cells in the myenteric plexus of the colon innervate the ganglion (the *inferior mesenteric*) that provides the end of the colon and the rectum with a sympathetic innervation. The implication of Joe's observation was that the gut, through the enteric nervous system, can effectively hang up the phone and cancel messages from the brain that it does not want sympathetic nerves to deliver. Joe's initial work had been confirmed many times since his original publication, and by 1988, when Gary and I were enmeshed in our gall bladder conversations, was well accepted. Gary liked my gall bladder logic, and he set out to test it.

Our hypothesis was that nerve cells in ganglia of the duodenum connect to nerve cells in the wall of the gall bladder. Gary soon proved that this concept is correct. He demonstrated that tracer molecules injected into a guinea pig's gall bladder appeared soon afterwards in duodenal nerve cells. This phenomenon, known as *retrograde axonal transport,* demonstrated that nerve cells in duodenal ganglia do indeed send their axons into the gall bladder, just as we had postulated they would. The tracer had traveled backwards, up these axons to reach the nerve cell bodies in the bowel.

Gary went on to study the properties of gall bladder ganglia. This seemingly banal investigation demonstrated that the structural and chemical properties of the ganglia of the gall bladder are, remarkably, very *similar* to those of the enteric nervous system and *dissimilar* to the characteristics of ganglia found elsewhere in the peripheral nervous system. The unique structure that the central nervous system shares with the enteric nervous system is also found in the ganglia of the gall bladder. The extraordinary nature of this fact is that this structure is not observed in any part of the peripheral nervous system that lies outside of the bowel. For example, connective tissue fibers, which are present in almost all types of peripheral nerve and hold it together, are missing from enteric and gall bladder ganglia. Like the brain, enteric and gall bladder ganglia are held together not by *collagen,* the biological rope of connective tissue, but by specialized cells called *neuroglia* (nerve glue). These findings thus provided powerful support for the conclusion we drew—that gall bladder ganglia really are an extension of the enteric nervous system.

The studies Gary began in my laboratory provided him with the preliminary data he needed. His grant was funded, and he landed a nice job at the University of Vermont, far from the wild streets of New York. Since his initial study, moreover, Gary has become the world's expert on the nerves of the gall bladder. Put a scientist in this field on a couch and say "gall bladder" and the odds are good that he or she will free-associate and respond by saying "Gary Mawe." As for me, the gall bladder has become a source only of vicarious pleasure and cultural enrichment. I read about the gall bladder and take pride in Gary's accomplishments, but I do not, myself, investigate it.

Annette and the Pancreas

After Gary and I began to get data that suggested that nerve cells in the gut innervate ganglia in the gall bladder, my thoughts turned to the pancreas. Although the regulation of gall bladder emptying and bile concentration are complicated phenomena, control of the secretion of digestive enzymes and sugar-regulating hormones (insulin and glucagon) by the pancreas is even more complex. Not only is there more going on in the pancreas than in the gall bladder, but the need to coordinate what occurs in the pancreas with what is happening in the bowel is also more critical. Once I learned that enteric nerve cells talk to gall bladder ganglia, I felt sure that they were also going to make themselves heard in the pancreas.

Gary Mawe was leaving for Vermont, but I was then collaborating with yet another great young scientist. Through a lucky turn of events, Annette Kirchgessner had come to work in my laboratory. Annette had originally been recruited to Columbia not by me but by a neurologist, Gaj Nilaver. Gaj, however, was himself recruited away and had departed for the West before Annette could do anything substantial with him. Since Annette was not able to leave with Gaj, he asked me if I would be willing to assume responsibility for Annette's postdoctoral training. As far as I am concerned, Gaj Nilaver is Santa Claus. I said yes, and the result has been a lasting collaboration with Annette that has thus far resulted in twenty-seven scientific manuscripts. Annette is still at Columbia, where she now has her own laboratory and is the director of a highly successful research program.

Although Annette had received her Ph.D. in experimental psychology, she had no trouble understanding the similarity between the enteric and central nervous systems. For her, the transition from one brain to the other required very little effort. Upstairs or downstairs, it made no difference. In either case there was a nervous system that controlled behavior, and that was what she wanted to investigate. The brain in the gut, moreover, promised faster results. As complicated as the behavior of the bowel may be, its complexity pales in comparison to that of a whole animal.

Gary's interesting data on the gall bladder prompted Annette and me to test the idea that the enteric nervous system innervates the pancreas. By then, Annette and I had already collaborated on several other important studies, and Annette was an experienced investigator. Annette agreed that the pancreas might be interesting, and, furthermore, she thought that it might do for her what the gall bladder did for Gary Mawe. If she could obtain evidence that the gut innervates the pancreas, she too would have solid preliminary data in hand, and she would be well on her way to her first NIH research award.

Annette started to investigate the pancreas by using the same approach Gary had successfully employed to look at the gut's innervation of the gall bladder. The experimental technique worked just as well for Annette as it had for Gary. Annette put a retrograde tracer into the pancreas of guinea pigs and rats, and, sure enough, the tracer lit up nerve cells in the myenteric plexus of the duodenum and stomach of both species. The tracer had been transported from the pancreas to the bowel by riding up the axons of the enteric nerve cells. Once she obtained that result, Annette reversed course and put an *anterograde* tracer into ganglia of the duodenum. Anterograde tracers move from a nerve cell body down axons to their terminals. This time, Annette found that the anterograde tracer was transported from the gut to the pancreas, where it labeled axons and their terminals. Annette had thus clearly defined, for the first time, a system of entero-pancreatic nerves. It was clear that nerve cells in the myenteric plexus of a limited, defined region of the duodenum and the stomach innervate the pancreas. The targets of these nerve cells, moreover, were revealed by the anterograde tracing experiments, and they turned out *not* to be the exocrine pancreatic glands or the endocrine islets of Langerhans but pancreatic ganglia.

When Annette showed me her data and we analyzed the results together, I wondered why no one had ever described the entero-pancreatic innervation before us. After all, if nerve cells in the bowel send their axons out of the gut and into the pancreas, they should be visible. Someone should have noticed these nerve fibers as they emerge from the gut. I could understand why nerves from the duodenum to the gall bladder could have been missed. They run alongside the common bile and cystic ducts in the stalk of the gall bladder, and one cannot tell where these nerve fibers are coming from or going to just by looking at them. Enteric nerves on their way to the gall bladder cannot be distinguished from any other sort of nerve unless one uses a tracer to see where they are going. The pancreas, however, is a much bigger organ than the gall bladder. I thus did not think that we could be dealing with a small number of nondescript axons in a little bundle next to the pancreatic duct. It also occurred to me that the labeling of nerves by retrograde and retrograde tracers in Annette's experiments had been accomplished too easily. Labeling nerves is like prospecting for oil. If you strike oil everyplace you drill, there must be a lot of it beneath the surface. The number of entero-pancreatic nerves, therefore, could not be small. I thought we had better check out the literature and see if anyone had found these nerves before us.

When we searched through past publications on the pancreas we found what I suspected. Someone had, in fact, seen entero-pancreatic nerves previously. The Ecclesiastes principle had struck again. Once more there was nothing new under the sun. A rather obscure article had appeared in 1977 in the *American Journal of Gastroenterology* called "The Neural Control of Exocrine and Endocrine Pancreas," written by O. Tiscornia, an author whom I had never encountered before. In that paper, Tiscornia described many nerve fibers running between the bowel and the pancreas, which he found by simply by gross dissection; moreover, he correctly speculated on the nature of these nerves. I feel rather sorry for Tiscornia, a person I have never met. We all stand on the shoulders of those who have come before us, but when it is our shoulders that bear the weight, we like to have them noticed. Tiscornia was right about the entero-pancreatic innervation, but his correctness was a matter of good fortune and not skill. He guessed at the function of the nerves he found running between the gut and the pancreas, but he did not even know the direction of signal traffic within these nerve fibers. For all Tiscornia knew from his own data, the nerve fibers coursing between the bowel and the pancreas could as easily have carried information from the pancreas to the bowel as from the gut to the pancreas. It is nice to be right, but it is better to be right for reasons that are established by your own work.

The anatomical position of the pancreas, the stomach, and the duodenum make nerve fibers connecting the organs hard to find. The pancreas and the duodenum do not actually sit in the cavity of the abdomen (the *peritoneal cavity*). Instead, they are buried together in the connective tissue that separates the peritoneal cavity from the back. As a result, the entero-pancreatic nerve fibers are not obvious and had escaped detection by dissectors from the time of Vesalius to the time of Tiscornia. Tiscornia, however, was unable to do more than speculate on the significance of the nerves that he discovered. Unfortunately, in 1977, he lacked the retrograde and anterograde tracers that were available to Annette in 1989. These molecules and their use had not yet been discovered. Tiscornia was therefore unable to demonstrate that the nerves he saw running between the gut and the pancreas are actually entero-pancreatic in nature. As far as he was concerned, they could equally well have been pancreatico-enteric as entero-pancreatic. Tiscornia's publication, therefore, did not play to rave notices and attracted relatively little attention. I, for one, had not noticed it when it came out, and I discovered it only belatedly when it became relevant and I understood its significance.

Gut to Pancreatic Messages

Finding that entero-pancreatic nerves are present is just the beginning of the game. If the nerves exist, the strong presumption, of course, is that they are functional. Presumption, however, is the sin that has done in scores of investigators. Presumption is cheap, but the scientific world runs on data. Demonstrating that entero-pancreatic nerves are functional was therefore high on our list of priorities.

Annette and I decided that we could not easily determine whether nerve cells in the bowel actually influence the pancreas if we were to investigate the issue in whole animals. The analysis of one little part of an animal is very complicated when all the other parts are also present and functioning. If you perturb the gut in an intact animal, for example, there are many different ways that such a perturbation could make itself known to the pancreas. Sensory nerves travel from the bowel to the brain and spinal cord, which, in turn, send nerves to the pancreas. An event in the bowel might thus affect the pancreas indirectly, by way of the central nervous system. Nerves would indeed be responsible for whatever happens in the pancreas, but in this case, the nerves would not be the entero-pancreatic nerves we wished to study.

The bloodstream represents another potential means of sending signals from the gut to the pancreas. The hormones secretin and cholecystokinin are made in the duodenum by endocrine cells in its lining. These hormones are known to stimulate pancreatic secretion. Any stimulus applied to the bowel in an intact animal, therefore, might cause these or other hormones to be secreted into the blood. The blood would then deliver these hormones to the pancreas, where they would do their thing and turn on whatever pancreatic cells respond to them. Events could thus occur in the pancreas as a result of stimulating the bowel that have nothing to do with the nerves that connect the two organs. Annette and I thus decided that we needed to eliminate all variables, such as the brain, the bloodstream, and indeed the bulk of the animal, that might confound our ability to interpret our results. Our strategy, therefore, was to reduce the system to its essentials and isolate the gut and the pancreas.

Annette neatly dissected the pancreas and the duodenum from a series of guinea pigs without cutting the nerve fibers running between the two organs. The resulting preparation of duodenum with an attached segment of pancreas was then mounted in an organ bath, where it was kept alive by bathing it in a suitable nutrient solution. Oxygen bubbled constantly through the fluid, and the temperature was maintained at 37° C (98.6° F), the normal core temperature of the body. After Annette had ascertained that the tissues were stable and happy in their isolated environment, she stimulated nerves in the duodenum and watched what happened in the adjoining pancreas. To her delight, she found that when nerves were stimulated in the duodenum, nerve cells in ganglia within the attached segment of pancreas became excited.

Annette was able to visualize the excited pancreatic nerve cells by monitoring the activity of a gene (*c-fos*) that turns on when nerve cells become active. Every gene that we have is present in every cell in the body. We have only one genome, and it is present in all of its glory in the nucleus of each of our cells. Differences between cells—between those of the immune and nervous systems, for example—are thus determined not by a different set of genes in each but by a different complement of genes that are turned on or off. In a given cell, some genes are perpetually on (*constitutively active*), while others are irreversibly off (*inactivated*). In essence, the genome in each cell's nucleus is a book of instructions, with pages to be read and pages to be overlooked.

Circumstances change, however, and as they do, a cell is driven to particular pages of its book of instructions to look something up. This phenomenon is manifest as the turning on of particular genes in association with some forms of cellular activity. Genes are written in bases, the language of DNA. Proteins, however, which comprise the machinery that carries out the commands encoded in the DNA, are written in the language of amino acids. To get action out of the cell's instruction book, therefore, the cell first transcribes DNA-talk into RNA-talk, which is a complementary language, still written in bases, but which, unlike DNA, can be translated into the language of amino acids and thus cause proteins to be made. The gene, *c-fos,* that Annette was following is called an immediate-early gene because it turns on (is transcribed) as soon as a cell is driven to its genome for a new set of instructions. The protein, Fos, encoded by the *c-fos* gene is a page opener, enabling the cell to read on through the how-to-do-it messages coded in its genome.

When Annette cut the entero-pancreatic nerves, paralyzed them with a neurotoxin, or blocked synaptic transmission, then stimulation of duodenal nerves did nothing to the pancreatic nerve cells. A massive nervous commotion could be elicited in the neighboring duodenum, and if nervous conduction or synaptic transmission was prevented, *c-fos* in the pancreatic nerve cells slept happily on. That is, the gene was not transcribed or translated and no Fos protein was made in pancreatic nerve cells.

Since there was no brain, spinal cord, or any other organ in the bath with the pancreas and the duodenum, the messages that the duodenum passed on to the pancreas had clearly traveled on a direct link connecting the two organs. The effects of neurotoxins and synaptic blockade, furthermore, confirmed that the linkage had, in this case, involved nerves and not hormones. Hormones may have been secreted in the stimulated gut, but they did not, in Annette's experiments, affect the pancreas. This result is understandable because hormones require a functional blood circulation to deliver them to their targets. In the absence of the heart or a mechanical pump, neither of which was present in the bath with the duodenum and pancreas, blood does not circulate.

Having established to everyone's satisfaction that entero-pancreatic nerves are real and that they work, Annette went on to try to determine what kind of stimuli normally activate entero-pancreatic nerves and to attempt to identify the duodenal nerve cells that send axons to pancreatic targets. This work is still in progress, and now that Annette has left my laboratory, it constitutes a significant proportion of what she does for a living. Thus far, Annette has established that putting glucose into the duodenal lumen, or increasing the pressure inside the organ, turns on entero-pancreatic nerve cells. Some, if not all, of the sensory nerve cells that detect these stimuli are located in the submucosal plexus of the duodenum, while the cell bodies of the nerve cells that actually send their axons into the pancreas are all in the myenteric plexus. The functional targets of the entero-pancreatic nerves, as the data obtained with tracers suggested, are nerve cells in pancreatic ganglia. That is why blockade of synaptic transmission destroys the ability of the bowel to influence the pancreas. Step one in exerting this influence is for entero-pancreatic nerves to activate nerve cells in the pancreas. Step two is for the pancreatic nerve cells to stimulate pancreatic exocrine cells to secrete digestive enzymes, or islet cells to secrete insulin. All this activation can be accomplished in about a second. Nerves work quickly.

The experiments with *c-fos* told us that the gut does indeed talk to the pancreas, but they told us nothing about what the bowel had to say or how it said it. In a sense, we were like nineteenth-century settlers, spotting smoke signals while making their way through the Wild West. The smoke signals would have informed the settlers that nearby native peoples were sending messages to one another, but that fact alone would not have told the settlers what they needed to know. The settlers would surely have been interested in knowing what instructions were encoded in the puffs of smoke, such as, "let them through" or "blow them away." To decipher the messages and get to the bottom line, the settlers needed to know the smoke-signal code. It is thus likely that guides who made a living escorting settlers across the West devoted considerable energy to discovering and learning the material encoded in smoke signals. In our case too, Annette and I needed to get beyond the discovery of *c-fos* transcription (our smoke signals) and learn the identity of the neurotransmitters and the outcome of entero-pancreatic signaling.

To decipher the gut-to-pancreas signal code and to learn its bottom line, Annette continued to use the isolated duodenum with an attached segment of pancreas as her experimental preparation of choice. She soon discovered that when she stimulated the duodenum, not only did nerve cells become excited in the attached pancreas, but digestive enzymes also came out of the organ. The signals from the bowel thus are meant to get the pancreas to put out, in a hurry, the enzymes that the bowel perceives it needs. Annette found that the secretion of these enzymes could most conveniently be monitored by following the release of *amylase,* the pancreatic enzyme that digests starch. The activity of amylase is easy to measure. An analysis of the secretion of amylase in response to duodenal stimulation soon revealed the chemical identity of the nerve cells responsible for the phenomenon. Scopolamine, the plant toxin that poisons muscarinic receptors, and hexamethonium, the synthetic drug that acts like curare and blocks nicotinic receptors, each prevented the enteric nerve-induced secretion of amylase. Hexamethonium but not scopolamine also prevented the excitation of *c-fos* in pancreatic nerve cells in response to duodenal stimulation. These observations showed that the entero-pancreatic nerves that excite pancreatic nerve cells use acetylcholine as their neurotransmitter and that the relevant excitatory receptors are nicotinic. The pancreatic nerve cells also use acetylcholine as their neurotransmitter, but the receptors they excite on pancreatic exocrine cells are muscarinic.

Annette had now defined the signals that permit the gut to get what it needs out of the pancreas, but she knew that she had only half deciphered the entero-pancreatic code. There was still one more smoke signal that remained mysterious. She knew which neurotransmitters meant "on," but she still did not know if there were also neurotransmitters that meant "off."

The studies that Annette had carried out with a retrograde tracer placed in the pancreas had revealed that some of the nerve cells in the bowel that became labeled by the tracer were nerve cells that contained serotonin. These observations meant that, besides the entero-pancreatic nerves that use acetylcholine to excite nerve cells in the pancreas, there are also entero-pancreatic nerve fibers that contain serotonin. At first, the function of these serotonin-containing nerves was enigmatic. When Annette added serotonin to the organ bath, the secretion of amylase was not provoked, and indeed nothing obvious was induced to happen in the pancreas. That result led me to suggest to Annette that she should stimulate nerves in the duodenum and then see what happens with serotonin present in the bath. My thought was that serotonin might be a modifier of the response of pancreatic nerve cells to acetylcholine. We decided, of course, to try to find out.

Knowing the full scope of the entero-pancreatic signal code is important. Diabetes, the disease of insulin deficiency, is very common and very serious. In contrast to juvenile diabetics, whose islet cells have usually been destroyed by an auto-immune process, islet cells are present in most adult diabetics, but they do not secrete enough insulin to cope with the number of fat cells these patients have accumulated over the years. A good understanding of how the gut stimulates islet cells to secrete insulin might thus be exploited, with suitable drugs in adult diabetics, to coax adequate amounts of insulin out of their recalcitrant islet cells. Insight might also be obtained that will help explain why the response of the adult diabetic islet cells to loads of glucose is insufficient.

Pancreatitis is also a pressing issue. In fact, pancreatitis can be so deadly, so fast, that I think of the organ as the P-bomb. When its enzymes explode out beyond their membranous confines, a person can be turned into history as readily as Hiroshima was in 1945. Pancreatitis can be a devastating sequel to the formation of gallstones, a deadly occurrence in an alcoholic, or a catastrophic complication of surgery. Pancreatitis can also strike without warning and without obvious cause in seemingly normal individuals. Knowledge of what the gut says to the pancreas may help us understand why pancreatitis occurs when it does, and may also provide a novel way to treat it. Certainly, if the bowel can "just say no" to the pancreas, that message, mimicked by a drug, might be very helpful in the therapy of pancreatitis.

Sure enough, when Annette stimulated the duodenum in the presence of serotonin, the normal secretion of amylase was inhibited. Furthermore, when Annette stimulated the duodenum in the presence of a drug that prevents serotonin from acting on its receptors (a serotonin antagonist), the secretion of amylase was actually much greater than it had been in the absence of the antagonist. The serotonin antagonist had actually potentiated the secretion of amylase. What these data suggested was that the serotonin-containing nerves are inhibitory. When Annette stimulated all of the nerves in the duodenum, she must have simultaneously excited some nerves that promote secretion and some nerves that inhibit it. The net effect, in the absence of any drugs, is secretion, but not as much secretion as would have occurred if the inhibitory nerves had been silent. The addition of the serotonin antagonist blocked the negative effects of the serotonin-containing nerves. Acetylcholine, therefore, had no restraint, and the full force of its excitation became apparent. The serotonin-containing entero-pancreatic nerves thus exist not to *turn on* pancreatic nerve cells but to *oppose* their activation by acetylcholine. In fact, Annette found that the serotonin receptors are located right on the acetylcholine-containing nerve endings themselves. When stimulated, the serotonin receptors *decrease* the release of acetylcholine and thus diminish the effect of the excitatory nerves. Serotonin, in the entero-pancreatic universe, thus plays yang to acetylcholine's yin. The system is highly sophisticated.

Multiple Controls over the P-Bomb

When I first thought about the complexity of the entero-pancreatic innervation, the effect of serotonin seemed a bit like overkill. I could easily understand why an excitatory system of nerve fibers might have evolved. Although the duodenum had hormonal signals that it could send to the pancreas to summon the enzymes and alkaline juice it needed to cope with food and stomach acid, the action of hormones is relatively slow. They have to find their way into the blood circulation and then get pumped to the pancreas, where they have to exit from blood vessels and reach the cells they stimulate. Hormones percolate slowly through tissue fluids with nothing but concentration gradients to guide them. In contrast, nerves are fast. They deliver a signal to a target promptly and directly. By using nerves instead of hormones, or to supplement the action of hormones, the gut can obtain what it needs exactly when it needs it.

Anyone who has spent any time in front of a computer screen waiting for the machine to boot up can easily understand this logic. You turn the computer on, but between the time you press the on switch and the time the computer gives you a meaningful response an eternity seems to go by. Nerves assure that there is essentially no downtime before anything happens. The hormones, in their leisurely way, can provide a backup for the nerves to be sure that the secretory response occurs and that it is sustained. With both nervous and endocrine systems in place, cooperating over different time courses, the pancreatic digestive enzymes would seem more likely to come out right. Since some of the entero-pancreatic nerve cells are in the stomach, it is even possible that the gut can anticipate its needs before they occur. The pancreas may be turned on by the stomach *before* it delivers food and gastric acid to the duodenum. The duodenum would therefore be primed and ready to handle whatever the stomach is about to dump into it. Excitation by entero-pancreatic nerves thus seemed eminently reasonable to me. The inhibition by serotonin, however, was unexpected.

The ability of the enteric nervous system to inhibit pancreatic secretion made more sense to me when I thought about the multitude of ways that secretion by the pancreas is controlled. The gut and the pancreas in a living animal or human are not isolated the way they are in an organ bath. In an intact animal, the brain, as well as the bowel, has an opportunity to excite the pancreas. The vagus nerves carry parasympathetic axons from the brain to the same nerve cells in pancreatic ganglia that receive a nervous input from the gut. In fact, before Annette and I published our paper describing the entero-pancreatic innervation, the vagus nerves were thought to be the only nerves that excite the pancreas. The brain is thus quite capable of inducing pancreatic secretion, and it does so.

In part, the brain stimulates pancreatic secretion on the basis of information it receives from sensory nerves coming from the bowel, but the brain also has its own agenda. It takes in the whole environment of an individual—the capture of prey, the smells and sounds of incoming food, and the tastes associated with eating. These inputs can all be processed by the brain and used to affect what it tells the pancreas to do. Learning and memory can be harnessed by the brain to facilitate digestion. Most of what the brain does is appropriate. There are also, however, complex, often unconscious and ill-understood events that go on inside the head.

Not everything the brain does is beneficial. Some thoughts would be better left unthought, and some of the brain's output to other organs would be better left unsent. Conceivably, therefore, the gut may find occasions when it cannot cope with the juice that the brain tells the pancreas to secrete. The inhibitory component of the entero-pancreatic innervation provides the bowel with the means to deal with these situations. The enteric nervous system can use its serotonin-containing inhibitory nerves to turn off pancreatic secretion when it is stimulated either by the brain, by hormones, or even by its own acetylcholine-containing excitatory nerves. The enteric nervous system is thus the master not only of a fast-acting on switch but also of an equally fast-acting off switch. Once more, as with the innervation of the prevertebral sympathetic ganglia by the enteric nervous system, the gut has a way of hanging up the phone when the incoming message (this time delivered to the pancreas by way of the vagus nerves) from the brain is not to its liking.

Digestion, as you can see, is not a simple process. It involves a very complicated series of organic chemical reactions that would be extremely unlikely to occur unless the right enzymes are present in precisely the correct amounts, and unless the conditions for these enzymes to act are also exactly perfect. Regulation of enzyme secretion and the maintenance of the right set of conditions for them to function thus involves a highly sophisticated array of sensors, nerves, hormones, and glands. When we find that the controls of digestion are complicated, the only surprise should be that we are surprised. Things are always more complicated than we expect them to be. When it comes to the enteric nervous system and its interactions, it is our preconceived notions that are faulty. Eventually we will learn that simplicity is never what we are going to find in the bowel. Even though I know this is true, I continue to be amazed by each layer of complexity of the enteric nervous system that we uncover.

Culmination, Absorption

Back in 1981, at the time I was defending a role for serotonin as an enteric neurotransmitter at the Neuroscience meeting in Cincinnati, I knew nothing about the nerves connecting the bowel to the gall bladder or about the entero-pancreatic innervation. At that time, if I thought about the control of these organs at all, I dismissed them with the piety that someday someone would learn how they operate. While it is true that enteric nervous control of the accessory glands is not yet fully understood, what we now know about the ability of the enteric nervous system to influence other organs represents a quantum leap past the state of the field that I was trying to introduce in Cincinnati. That is one of the troubles with modern science: blink and you find yourself hopelessly out of date.

Up to this point, most of what the gut has been doing can be considered foreplay. The food that has been consumed has been ground, kneaded, soaked in acid, emulsified, and digested. This sets the stage for the culminating event that the whole process is designed to make possible, absorption: the entry of the products of digestion into the body itself. After the bile salts and digestive enzymes secreted by the accessory glands have completed their jobs, digestion is still not finished, but the next events are fast and linked to absorption. There is, as yet, no evidence that the enteric nervous system is directly involved in either the final steps of digestion or absorption, although there has been a fair amount of speculation that the enteric nervous system might affect absorption. Salts and the ions, especially sodium that salts provide, have a lot to do with absorption. The enteric nervous system exerts a profound influence over the transport of salts from body fluids into the lumen of the bowel; therefore, the enteric nervous system might indirectly influence absorption by altering the distribution of salts across the intestinal lining. Whether the enteric nervous system does so, how it does it, and under what circumstances are all still unknown.

After food has been digested by pancreatic enzymes, with an assist from bile salts to deal with fats, the large, complex molecules that were originally present in a meal have mostly been broken down into smaller ones; nevertheless, many of these molecules are still too big for the small intestine to absorb. Big molecules simply do not go through cell membranes easily, and the small intestine is totally lined by cells. These cells separate our inside, represented by the fluid in the wall of the bowel, from the outside, represented by the intestinal lumen. The intestinal lining cells form the critical boundary that prevents us from flowing out into our guts. Clearly, this barrier cannot be breached to let molecules in, no matter how much we might need them. To get into the body, therefore, molecules in the lumen of the gut have to cross the *apical* (lumen-facing) membranes of intestinal lining cells, go through the cytoplasm of these cells, and come out the membranes (*baso-lateral*) at the other end. Nothing passes between adjoining intestinal lining cells because the space between them is obliterated by tight junctions that seal it off. Proteins in adjacent cell membranes line up and touch one another across the gap that would otherwise exist between cells. These proteins form sealing strands, just beneath the intestinal lumen. Absorption thus is a mechanism to take in what we need while at the same time letting nothing critical flow out. Since the tight junctions that link the individual cells of the lining of the bowel to one another obliterate the space between them, most of what is absorbed passes through the lining cells themselves. Absorption is thus a process that is very much under cellular control.

Once eaten, proteins are reduced, first by pepsin and then by pancreatic enzymes, to small peptides. Starches and other *polysaccharides* (long chains of sugar molecules) are cut down to *disaccharides* (molecules that contain only two sugars). To be absorbed, the small peptides and the disaccharides must first be attacked and split by intestinal enzymes, which are not free in the lumen of the gut but are integral components of the apical membranes of intestinal lining cells. These cells, called *villus absorptive cells* or *enterocytes,* are the same ones that finally are called upon to absorb the end products of digestion.

Anchoring critical digestive enzymes is not all that the apical membranes of villus absorptive cells are called upon to do for digestion. These membranes also contain the enterokinases that convert the inactive precursors of pancreatic enzymes to the active enzymes, which accomplish the bulk of digestion. This spatial arrangement of enzymes and absorptive apparatus on the surfaces of absorptive cells makes great sense. As soon as the final products of digestion emerge, they are located right at the membrane that captures them for absorption. The small molecules are not given a chance to fly away into the lumen of the gut where they might escape. The intestinal lining is thus the site where pancreatic enzymes are activated, digestion is completed, and absorption occurs. This is an impressive set of jobs, and as you might imagine, a great deal of surface area is required to get them done. A massive surface area also facilitates the absorption of beneficial molecules, like vitamin C, that are small, go though cell membranes, and simply diffuse into the body.

The surface of the small intestine is huge. Its area is magnified by a structure in which the surface folds, folds again, and then folds once more. The surface layer of the intestine, the mucosa, forms coarse folds that have as their cores the dense connective tissue of the submucosal layer of the gut. These big "valves" form the first kind of fold.

The second type of fold is located within the mucosa and is much finer. The mucosa folds on itself to form fingerlike projections called *villi* (singular *villus*) that extend into the intestinal lumen. The cores of the villi are the loose internal connective tissue of the mucosa.

The third degree of folding is of the luminal membranes of the villus absorptive cells. These membranes fold into many long extensions that project from the apex of the cells. The extensions are called *microvilli* because they look like miniature villi. Both are fingerlike in appearance, but villi are composed of many cells, while microvilli are surface modifications of single cells. The cores of microvilli are made up of internal cell *cytoplasm.* Actually, this cytoplasm is filled with bundles of thin filaments (*microfilaments*), which are part of the cell's skeleton. The microfilaments inside the microvilli provide them with a certain amount of structural rigidity.

The intestinal surface area is amplified by the structural folding to such an extent that the surface membrane of one centimeter of human intestine is enough to cover a doubles tennis court. Project that coverage

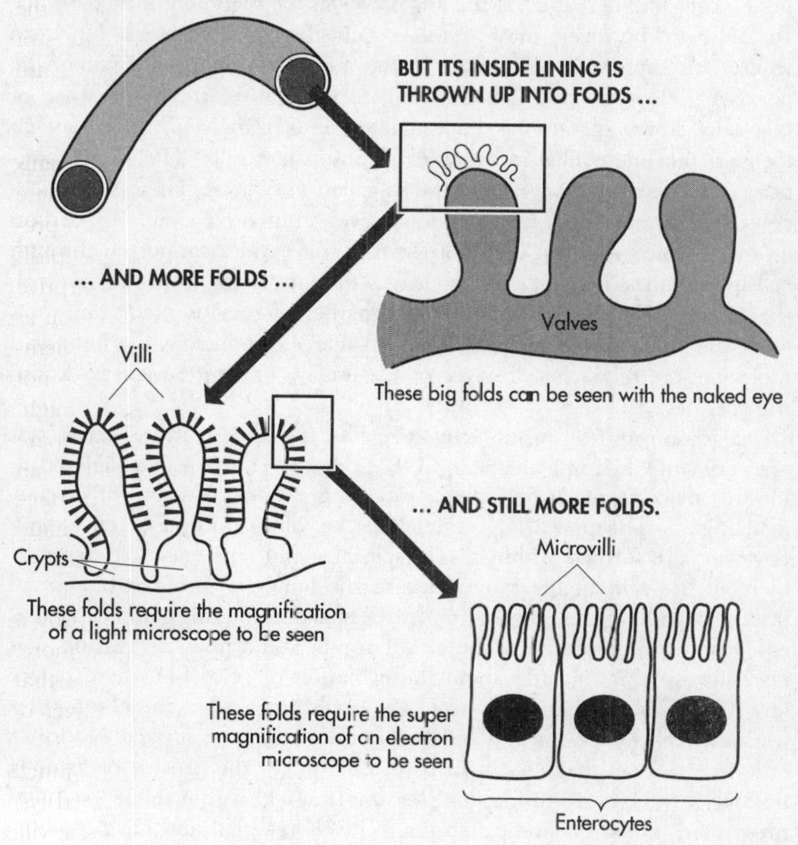

over the twenty-two feet of small intestine and the average person probably has enough intestinal lining membrane to wallpaper an apartment house, although that particular material is unlikely to be much in demand as wallpaper.

Taking in the Fat

The absorption of fats involves a mechanism that is far more byzantine than that which is responsible for the absorption of the digested products of proteins and sugars. The pancreatic enzyme, *lipase,* which we last spoke of as a deadly agent that dissolves cell membranes when it is released during a bout of pancreatitis, is the key player in the digestion of dietary fats. In this case, however, the enzyme is quite benign, because it functions where it is supposed to function, in the intestinal lumen, and not in the peritoneal cavity, where it is dumped in pancreatitis. In the presence of bile salts, lipase digests fats into smaller, simpler molecules that can go through the microvillus membranes of absorptive cells. The most abundant of the fats that we eat are the so-called *triglycerides*. These substances consist of three *fatty acids* (with long, even-numbered chains of carbon atoms) coupled to a backbone of *glycerol*. Trigycerides cannot go through cell membranes, but after lipase cleaves the fatty acids from the glycerol, the liberated fatty acids diffuse through the microvillus membrane and enter the villus absorptive cells. Once in the cells, however, the fatty acid molecules are reattached to glycerol, essentially converting them back into triglycerides.

Digestion and absorption involving the resynthesis of molecules that are very similar, if not identical, to those that were eaten and digested in the first place seems at first glance to be a great deal of work for no particular gain. The new triglycerides that the villus absorptive cell makes, however, are formed within a very special membrane-enclosed compartment of the cell. In the same sense as the lumen of the gastrointestinal tract is outside of the body, so too is the space inside this compartment is equivalent to the outside of the cell. To understand how this can be, it is necessary to think a little about the evolution of cells. Eons ago, before any multicellular organisms stalked the earth, the surface membrane of a primitive cell folded inward and broke away from the surface to form a little internal sac that floated in its cytoplasm. In the process of forming this sac a bubble of outside fluid became trapped within it. As evolution progressed, the sac of the primordial cell was retained and has become a permanent entity that is present in almost all modern cells. The derivative of the primitive sac is now no longer a simple little bag. In most cells, it is a highly complicated series of membrane-enclosed sacs, cisternea, and tubules, which together are called the *vacuolar space* of the cell. Intermittent connections form between the various elements of the vacuolar space, but none of them are ever open to the cytoplasm proper.

The components of the vacuolar space function like soap bubbles. They can fuse with one another and with the surface membrane. The inside of the vacuolar space, reflecting its origin, remains topologically equivalent to the outside world. When a vesicle that is part of the vacuolar space fuses with the surface membrane, whatever is loose in its lumen is delivered to the external fluid (secreted).

Although the newly resynthesized triglycerides appear to be produced inside of villus absorptive cells, they are really within the vacuolar space because a wall of membrane perpetually separates them from the cell's true inner self. The triglycerides are confined to the vacuolar space, and, while still inside it, they are also provided with a coating of protein that helps to keep them suspended in water. Keeping fat suspended in water is not a trivial matter to the body. It is important because all tissue fluids are watery. Fat hates water and, like grease in a dishpan, will not dissolve in it. As the proverb goes, oil and water don't mix. Fats that come out of solution may be deposited in arteries and cause *atherosclerosis* ("hardening of the arteries"), with dire consequences including heart attacks and strokes.

The newly synthesized triglycerides, with their protein coats, are transported through the villus absorptive cells, but they stay within the vacuolar space. Eventually, the triglycerides are packaged into vesicles that fuse with the surface membranes at the bottoms of the absorptive cells, and the protein-coated triglycerides are secreted. The resulting particles, called *chylomicrons,* fill the connective tissue of the cores of the villi. Since chylomicrons are too big to enter blood vessels, they must be removed from the connective tissue of the villi by *lymph* vessels, which have holes in their walls that chylomicrons can get through. The lymph vessels carry the chylomicrons away to the liver, which is able to deal with them. Fat is thus digested, absorbed, resynthesized, secreted, and finally removed in lymph, which turns milky white after a meal because it is so full of chylomicrons. Given the complexity of the digestion and absorption of fat, it is no surprise that when the gut becomes sick and unable to perform, it is usually fat absorption that is the first to fail. Fat in the stool is thus one of the earliest and most constant signs of digestive or absorptive trouble.

Fat also affects the enteric nervous system. The products of fat digestion, fatty acids, are very potent stimulators of sensory receptors that monitor what is doing in the lumen of the intestine. These receptors activate sensory nerves that keep both the brain and enteric nervous systems informed. Fatty acids anywhere in the intestinal lumen turn on an interesting reflex called the *ileal brake.* Essentially, when the brake is on, propulsion of the contents of the intestine stops at its final region, the ileum. The presence of fatty acid in the *chyme* (intestinal contents) way up in the duodenum may be enough to cause the ileum to come to a full stop. It is as if the nervous system (both central and enteric; they cooperate) finds fatty stools as repulsive as people do. More likely, the ileal brake evolved to prevent material from leaving the small intestine until digestion and absorption are finished. The brain and the enteric nervous system just want to make sure that the job they are charged with overseeing is done.

When the last of the fatty acid has been absorbed, the ileal brake is released and the chyme flows on. The colon gets what is coming to it.

Conceivably, a failure of the ileal brake explains the diarrhea sometimes associated with the ingestion of food containing Olestra (sucrose polyester). Since Olestra cannot be digested, it provides no fatty acid to step on the ileal brake. This means that Olestra, as it navigates the passage through the stomach and small intestine, resists the worst the pancreas and gall bladder can do to it and emerges full steam into the unsuspecting colon.

7

"IT AIN'T OVER 'TIL IT'S OVER"

DIGESTION AND NUTRIENT absorption are, for all intents and purposes, completed in the small intestine. The large intestine, or colon, has its own tasks to accomplish. Some we are reminded of daily; others we may never think about at all.

The Large Intestine

The large intestine, or colon, appears to have evolved to help land animals conserve water. Drying out is a threat to an animal that has to survive in air. Evaporation is a constant danger. The skin, of course, is a major adaptation to this peril because it provides a watertight barrier. The lining of the bowel, however, has to absorb water and nutrients and thus must allow both to move across it. Not surprisingly, therefore, water can indeed be lost through the intestinal lining. Since the lumen of the intestine is outside the body proper, water that crosses the intestinal lining to enter the lumen of the gut is in danger of being lost. Evolution, or the Creator, thus found it necessary to devise a means of taking this water back. The job of retrieving the water that would otherwise be lost in feces is handled by the colon. Minimizing water loss is a necessary condition for life in dry air.

When the intestinal contents are transferred from the small to the large intestine, they are essentially liquid. In fact, the amount of water that the small intestine delivers to the large bowel is quite impressive. Some of this water has been swallowed with food, but a great deal more water is poured into the lumen of the gut during the digestive process by the various glands and intestinal lining cells that secrete into the bowel. Pancreatic enzymes, acid, base, salts, mucus, and bile are all delivered in watery solutions. In humans, approximately nine liters (more than two gallons) of water enters the colon every day. Despite this massive water load, which arrives at the upper end of the large intestine, only about 100 milliliters (6–7 tablespoons) normally leaves the anal end of the colon in stool. The difference between the immense amount of water that comes into the large intestine and the small amount that leaves it (8.9 liters; also more than two gallons) is the amount of water that the colon reabsorbs. The large intestine may contribute little or nothing to digestion, but it is still a busy organ.

The mechanism the colon uses to retrieve water is to pump salts out of the lumen and back into the body. This process requires quite a bit of energy because the salts are transported from a region where their concentration is relatively low to one where their concentration is relatively high. Water passively follows the movement of salt. The remaining material in the lumen of the large intestine thus becomes more concentrated as water is removed, and as a consequence it becomes increasingly solid as well.

Propelling the congealing mass of stuff through the large intestine is a problem. The effort involved in moving dense blocks of stool toward the anus is far more formidable than that involved in inducing the liquid contents of the small intestine to ooze along. The circular muscle coat of the large intestine is thus much thicker and more powerful than its counterpart in the small intestine. The lining of the colon also lubricates itself by secreting more mucus than the small intestine, so that the resulting coat of slime can grease the passage of hard blocks of fecal matter.

Bacterial Inhabitants

Feces contain more than just the undigested and nonabsorbed residue of meals that have made it through the small intestine. The large intestine also provides a home for vast numbers of bacteria, only some of which, to use the language of the United States Air Force, are friendlies. Others are either outright enemies or insidious villains that stay benign only as long as the immune system of the person who carries them around is strong. Almost as soon as there is a lapse in the colon's defenses, these dastardly bacteria reveal their true nature and give rise to a nasty infection that is often resistant to commonly used antibiotics. The bacteria that normally hang out in the colon are so numerous that they are a major component of the bulk of the stool that is ultimately defecated.

The most common type of bacterium of the colon is *Escherichia coli*, known to virtually everyone as *E. coli* because almost no one can pronounce its first name. *E. coli* is so much a fixture of the normal human colon that one can accurately judge the degree to which any body of water has been contaminated by human sewage simply by collecting a few drops, observing the bacteria that are floating in it, and determining the number that look like *E. coli* (*coliforms*). No sophisticated apparatus or detective work is needed, just a cup to scoop up the water, a glass slide to hold a drop, and a simple microscope to spot the coliforms.

Under ordinary circumstances, *E. coli* is not a dangerous organism. However, circumstances are not always ordinary. As often as not, the organism that finally carries off debilitated victims of cancer or other diseases that devastate the immune system is *E. coli.* Thus, *E. coli* belongs to the insidious villain category of colonic organisms. Given that *E. coli* fills the colon of nearly everyone, we could clearly not have evolved as a species if *E. coli* were routinely harmful.

Evolution of humans took eons. Evolution of bacteria, however, is on a fast track and whips along with disconcerting speed. Our doubling time (the time it takes for any group of us—a family, for example—to double in number) is measured in decades. In contrast, *E. coli* doubles every twenty minutes. There is thus not just one type of *E. coli* but many. Some *E. coli* are very useful. They are employed by molecular biologists, for example, as virtual farms where genetically engineered genes are grown and cultivated. Scientists put genes of interest in tame strains of *E. coli* and use the organism's ability to multiply to crank up lots of genetic material, or to produce a great deal of the protein encoded by the genes implanted in the *E. coli.*

Other *E. coli*, like strain 0157-H7, which really is a refugee from the cow colon, sticks to our cells and puts out a miserable toxin that causes no end of grief. The toxin can harm the brain, the kidney, and blood cells. Strain 0157-H7 contaminates meat and causes food poisoning that is so severe that it can be life-threatening. As little as a single 0157-H7 bacterium can be enough to establish a serious illness. The *E. coli* of the 0157-H7 persuasion have recently been found at the heart of highly publicized epidemics. They strike at the elderly and the young most commonly, but it will not take very many more epidemics before 0157-H7 ends the American love affair with the rare hamburger.

E. coli is, for most of our colons, the tip of a festering iceberg. A number of bacterial colleagues that are too numerous to mention share the colon with *E. coli.* Some of them do nice things for us, like make vitamin K, but most of them are as unspeakable as they are unpresentable. The mixture inside the colon of dung, mucus, and fermenting germs is thus as dangerous as it is unattractive. We are taught, for good reason, very early in life to stay away from it. The content of the large intestine thus has to be handled by the body with circumspection. It has to be propelled, but it also has to be confined.

When all goes well, the anus is the colon's only exit. Except for what leaves through that portal, whatever is in the large intestine stays in the large intestine. Propulsion is always anally directed, and there is no reverse movement. Nothing is ever driven backwards into the small intestine. Salts and water are transported through the lining of the colon, but nothing awful leaks between the lining cells to get into the body. The infectious stuff in the lumen of the colon stays there. The junction between the small and large intestines, the *ileo-cecal valve,* is thus an important landmark. Above this junction, the lumen of the bowel is normally sterile, while below it, the lumen of the gut is a bacterial heaven.

One reason that the bacteria in the lumen of the colon do not break out and infect the body is that they are at war with one another. No one kind of germ gains ascendancy and takes uncontested possession of the colonic turf. The constant competition between otherwise nasty germs helps to keep the bacterial population under control. It really is a case of "the enemy of my enemy is my friend." Our bacterial companions in the colon may be repulsive, but we still have to take good care of them. Taking an antibiotic, therefore, is not without risk. Killing germs in the colon can be a hazardous venture. Antibiotics that annihilate our colonic friends can get us into big trouble very fast.

Many kinds of bacteria are killed by antibiotics, but not all. Those that are not killed are those that are most resistant. As Charles Darwin pointed out, natural selection is a potent force. When applied to organisms that double in minutes, natural selection is not just potent but fast. Selection by antibiotics is thus not a good move for modern medicine. The therapeutic value of drug after drug has been lost as bacteria adapt to them and circumvent their effects.

Antibiotic resistance is a fact of bacterial life, and it is a difficulty that the indiscriminate use of antibiotics has made far more serious than it used to be. Since antibiotics are routinely added to chicken feed and other agricultural products, the proportion of resistant organisms increases every year. This problem is compounded by the large number of doctors who prescribe an antibiotic without first determining whether the disease they wish to treat is due to a susceptible organism. By killing some bacteria and not others, the administration of an antibiotic may eliminate the competition between germs in the colon, so that one strain, which happens to be resistant, achieves dominance. In essence, the drug selects a bug. The resistant strain is thus dangerous for two reasons. One is that it has been liberated from the restraint imposed on it by the other bacteria that normally compete with antibiotic-resistant organisms. The other is that the resistant bacteria are hard to eliminate because it is difficult to find a nontoxic drug that can kill them. The resistant organisms, therefore, are likely to cause a rip-roaring, florid *colitis* (infection of the colon). They can also escape from the colon and invade the body. Some of the antibiotic-resistant strains of bacteria, such as *Clostridium difficile,* make toxins that peel the lining of the colon right off the organ and lead to an explosive, debilitating, and frequently lethal form of diarrhea.

The Immune System and the Gut

The existence of life in the lumen of the large intestine requires that the bowel maintain an effective line of defense. The fact that bacteria compete with one another helps to keep them distracted, but we still need a range of well-armed protectors to stand guard and patrol the border. The first and most obvious form of protection is provided by a formidable array of defensive cells that reside in the loose connective tissue under the lining of the colon. These cells keep an inventory of the so-called normal *flora,* the bacteria we habitually harbor. This inventory consists of large numbers of immune-competent cells that collectively have learned the identity of all the potential villains and remember who they are. A lethal welcome is thus prepared and ready to greet any of the usual suspects that wander beyond the confines of the lumen of the large intestine (the only space that has been conceded to them). The immune-competent cells also learn to recognize the toxins that colonic bacteria produce, and a subset of these cells stores that knowledge in memory. These toxins, too, are neutralized before they can cause any harm.

Because the immune-competent cells learn and remember, they are superb at dealing with the status quo. As long as the colon's inventory of flora remains unchanged, the modus vivendi we have established with our germs persists, and we live out our lives without distress. We go about our daily affairs, happily unaware of the massive armies of warring germs in our large intestines that march through life with us. In this case, it is surely true that ignorance is bliss. Life at its best is lived when the colon remains, as the politicians say, in deep background. We want to hear from the large bowel only on those occasions when it is ready to void itself of its contents; moreover, although many of us demand that these occasions take place regularly, there are sharp limits on how much frequency and urgency are tolerable. Even a little news from the colon upsets our equilibrium. Clearly, this is an organ from which progress reports are not only unnecessary but unwelcome.

Our balance tends to be upset when almost anything intervenes to change the composition of our normal flora. Antibiotics are obvious perturbants, but something as benign as a vacation in Mexico (or any other country where the local organisms are different from those close to home) can also be unsettling. Just a simple change in the strains of seemingly harmless bacteria that hang out in the large intestine can cause the colon to misbehave. The immune system needs time to cope with new bacteria. Although it is prepared to deal immediately with the old culprits and their toxins, it has to learn the identity of new perpetrators and gear up to mount an effective response. The immune system will remember the insult and be prepared next time, but while it gets its act together, the dreaded "Montezuma's revenge," "tourist two-step," "turista," or whatever one calls the diarrhea of tourists, gets a chance to happen. Depending on the toxins produced by the offending bacteria, the problem can involve much more than diarrhea. One can become quite sick.

The diarrhea associated with tourism that ruins otherwise wonderful vacations in spots that are often warm, expensive, and beautiful seems like a bodily activity that one would want to stop at all costs. Actually, attempts to stop this kind of diarrhea are not always wise. New organisms are fighting to gain a purchase in the bowel and they are secreting toxins that wreak havoc. Diarrhea, especially watery diarrhea, is thus a logical and even helpful response to their presence. In essence, the colon has been dirtied and literally needs to be washed out. Fortunately, the enteric nervous system works with immune cells to get this done. In response to the infection in the colon, intrinsic nervous reflexes are triggered that lead to secretion. Secretomotor nerve cells in submucosal ganglia spring to life and cause cells in the lining of the colon to transport chloride ions from the tissue fluid of the bowel wall to the lumen of the colon. Sodium follows the chloride and water follows the salt. The normal function of the large intestine to conserve water is abandoned as the colon is turned into a secretory machine. The resulting deluge has cleansing properties. Any germ that is not battened down by a sufficiently adhesive attachment to the intestinal surface is washed away. The festering pool of toxins is cleaned out as well.

Stopping the motility of the colon by taking an opiate, such as paregoric (camphorated tincture of opium), loperamide or Lomotil, is effective in putting an end to diarrhea, but cleaning is brought to a halt as well. What might have been a disturbing but relatively short-lived episode can be prolonged by these drugs and made worse. It is better to let nature take its course and get rid of what needs to be expelled. One medicine that can help is bismuth subsalicylate (Pepto-Bismol), which, if you take enough (you need plenty—read the label) absorbs toxins but does not affect either the enteric nervous system or intestinal secretion. Pepto-Bismol can thus antagonize the effects of bacterial toxins on the colon without getting in the way of natural curative mechanisms. Eventually, almost no matter what you do, the immune system will come on-line and bring tourist diarrhea under control. As long as the distress is simply due to a change in the flora of the colon, the immune cells catch on to the new germs and handle them as they handled the old ones.

Between them, the enteric nervous system and the immune cells of the bowel are highly effective in keeping the microbiological world of the colon at bay. Evolution has provided us with the means to overcome the worst attacks of the vast majority of the bacteria and viruses that we are likely to encounter. Bacteria, however, have evolved as well. We have adapted to their presence by acquiring the ability to suppress them, but some bacteria have learned how to deal with our counterattacks, or even to use our defensive mechanisms against us. Microbes as a group are superb cell biologists. They have learned what makes us tick. When our evolution deals these germs a lemon, they make lemonade.

Bon Appetit: The Perils of Food and Water

Consider, for example, cholera, an illness that has remained with humanity from ancient times and has worked, over the course of many centuries, against the overpopulation of the earth by humans. *Vibrio cholera* is an organism that has clearly won the respect of our species, if not our love. Whenever cholera appears, people pay attention.

Why, you may think, should you be asked to consider cholera? This may be a deadly organism, but surely it is not, by and large, a present-day issue. Unfortunately, it is. It is one of those storied diseases, like the plague, that has recurred throughout human history and is still recurring, leaving in its wake huge numbers of deaths and vast amounts of suffering. Clearly, this is an organism that is not to be taken lightly.

Cholera and other food- and waterborne diseases of its kind are very much a present-day concern, and as each day passes, the concern grows stronger. Cholera is the *grand seigneur* of the group only because its outbreaks are described in megadeaths, while the others produce epidemics of more modest proportions. There is a disturbing similarity in the way the entire group of diseases spreads around, and the susceptibility of modern societies to them is increasing dramatically. A by-product of globalization is that no one's disease is theirs alone. Microbes travel with the world's commerce, and they are not stopped by customs inspectors.

The second brain and the immune system are powerful defenders of the body's enteric frontier. All of us harbor in our colons something like five hundred species of potentially deadly organisms. We are able to do this because of the efficiency of the nervous and immunological troops that hold the line at the colonic border. A small number of microbial species, however, have got our number, and when we let them in with the food or water we drink, we are in big trouble.

The problem is that, for a variety of reasons, modern life has caused new diseases to emerge from unanticipated reservoirs, and old diseases, which have been provided with new ways to spread, to reemerge. Anyone who travels to destinations in the developing world, particularly those in tropical regions, is accustomed to hearing the advice from physicians and travel agents alike: "Boil it, peel it, or don't eat it." In other words, be careful not to let the evil germs that are thought to lurk in these locales get into your gastrointestinal tract. Although we Americans follow this advice, sometimes religiously, when we travel to developing countries, we now import food in huge quantities from just those places. Mexico and Central and South America, for example, have become major exporters of food to the United States. Some of the people who grow food in these countries are not accustomed to using toilet paper.

Cyclosporiasis is a disease that nicely illustrates the force of globalization in the dissemination of agents that, like cholera, are able to infect the human gut. Before 1996, infection by *Cyclospora* (the causative agent) was rare and restricted mainly to persons who had been traveling in developing countries. In the spring of 1996, however, a major outbreak occurred in the United States and Canada. This epidemic gave rise to a fair amount of panic until some elegant medical research established that *Cyclospora* is a foodborne agent of diarrheal disease, and that it was being spread by Guatemalan raspberries. Cutting off the importation of raspberries from Guatemala stopped the outbreak, as it did again in the spring of 1997 when a second outbreak occurred. Raspberries are not now being imported to the United States from Guatemala, but even so, new outbreaks of *Cyclospora* have occurred. One was spread by mesclun lettuce and another by fresh basil in pesto sauce. The nature of the foods spreading *Cyclospora* illustrates why foodborne illnesses are everyone's concern. First, they cross borders, and second, they find their way even into the diets of the most affluent.

The fact that much of the food that we obtain from the developing world is produce magnifies the risk. As a result of his observations, Michael Osterholm, the chief of the Acute Disease Epidemiology Section of the Minnesota Department of Health and the world's leading authority on foodborne disease, has pointed out that the heart-healthy diet that we increasingly consume has been great for our hearts but hell on our guts. There has been a huge increase in the American consumption of veggies. These items, which are readily contaminated, are extraordinarily effective vehicles for transmitting the small but intrepid band of microbes that has learned how to penetrate defenses of the bowel.

As our diet has changed, so too has our tendency to eat out. Eating in restaurants or buying "take-out" to eat at home is increasingly taking the place of home cooking. As a result, there are over nine million people employed as food handlers in the United States. These people tend to receive low wages, poor benefits, and few (or no) opportunities for advancement. By and large, the education of food handlers is minimal, and many of them do not speak English (which complicates educational campaigns to change their sanitary practices). Next time you contemplate the elegance of the salad bar, you might want to give a thought to who prepared it.

According to Osterholm, stomach cramps, abdominal pain, and diarrhea have become the *number-one* cause of visits paid to hospital emergency rooms in the United States, moving ahead of chest pain, which holds the number-two position. Food- and waterborne illnesses have become so widespread that we do not even know their full extent. Public health authorities have trouble defining them and tracking them down. Surveillance is questionable, in part because medical laboratories do not routinely test for some of the microbes that are known to cause food- or waterborne disease, and in part because we do not yet know all of the responsible organisms. Estimates of cases per year in the United States range from six million to eighty-one million. Osterholm is sure these figures underestimate the problem. In his state, Minnesota, which is not atypical, there are over six million cases of diarrhea every year, producing nearly half a million visits to doctors. A new monitoring program called FoodNet has been set up as a joint effort of the Centers for Disease Control, the Food and Drug Administration, and the United States Department of Agriculture (USDA) to determine the incidence and prevalence of various foodborne illnesses. Results have been discouraging. Despite the best of intentions and improvements in inspections and food processing, the safety of food in America has clearly not improved since the 1950s.

Gastrointestinal disease in America is on the increase. Not all of this new burden of disease, of course, is due to foodborne illness; however, our changes in diet as well as the sources and handlers of our food have contributed their share. The difficulties of the United States, moreover, cannot all be blamed on the developing world. The industrial handling of farm animals in America has also contributed its fair share. Poultry and swine, for example, are packed in huge numbers into confined spaces, which makes it easy for epidemics to spread. As a result, more than 80 percent of American chickens now arrive in supermarkets contaminated with *Campylobacter*, an agent that can cause diarrhea, vomiting and, in people with a compromised immune system, a lethal infection of the blood. Eggs often contain *Salmonella enteriditis* (which causes a disease that is similar to that caused by *Campylobacter*), and this organism, which is very hard to kill, has even been found alive in 1 percent of pasteurized egg products surveyed by the USDA. Over half of the herd of pigs in the United States now carries *Toxoplasma gondii*, and surveys show that more than 40 percent of American adults have antibodies to this microbe, which means that they have encountered it. As our fear of trichinosis has receded, our consumption of raw or undercooked pork has increased. *Toxoplasma gondii* is the cause of toxoplasmosis, an illness that is usually mild and similar to mononucleosis. Toxoplasmosis, however, can lethally affect the brain in AIDS victims and cause severe birth defects when it infects pregnant women.

The spectacular foodborne agent, *E. Coli* strain 0157:H7, which gives rise to a sometimes lethal combination of kidney failure and blood disease, is only one of the forms of disease-causing *E. coli* that naturally inhabit the colons of American beef cattle. These bovine *E. coli* are a far cry from the relatively tame organisms that we carry. When *E. coli* 0157:H7 contaminates meat at the packing plant, it is a serious issue. *E. coli* 0157:H7 from rare hamburger sold by Jack in the Box restaurants in the Seattle area affected 732 people and caused 195 hospitalizations and four deaths. *E. coli* 0157:H7 is not found only in hamburger. It has also turned up in salad bars and unpasteurized cider prepared from apples that fell from trees into a cow pasture. In fact, the organism has been spread from cows to wild birds, who disseminate it widely in nature.

As frightening as these illustrations are, they represent only the tip of a very large iceberg. We are clearly vulnerable to foodborne diseases and becoming more so. The food supply is also an invitation to a bioterrorist. Certainly, it would help if we were to subject the national food supply to the sterilizing effects of gamma irradiation, as we do the food supplies of astronauts. Our fear of radiation, however, is so strong that it exceeds the limits of reason, and food irradiation is not likely to be introduced anytime soon.

None of the common microbes that get by what the enteric nervous and immune systems put up to stop them is a match for cholera in pure scariness. Given our current vulnerability, cholera is an American disaster waiting to happen. In 1991, just a moment ago in the long human relationship with cholera, the disease reappeared in epidemic proportions in Peru. About four hundred thousand people are known to have contracted cholera during the first year of that epidemic, and by 1994, after cholera had worked its way north and south from Peru, it had laid waste to about one million souls. Eleven cases of cholera linked to that South American outbreak were seen in New York and New Jersey, where the problem arrived in crabmeat that was brought from Ecuador and consumed in America. In the world of the Boeing 747, no one, anywhere, is totally safe from cholera.

The death rate from cholera in South America was over 30 percent of cases during 1991 because physicians did not recognize the illness for what it was. Doctors are much better at diagnosing diseases that they expect to see. Cholera is one disease that physicians expect only to read about, not actually encounter. This is both a pity and a potential American tragedy, because cholera arrives in the bellies of travelers to the United States with considerable regularity. A major study, moreover, has shown that cholera is frequently misdiagnosed by American health professionals and, once diagnosed by them, poorly managed. This situation is both sad and frightening, because cholera can be treated, and when it is, patients do quite well. For example, after the epidemic that started in Peru was publicized and doctors expected to see cases of cholera, the mortality rate fell to less than 1 percent. Since diagnosis precedes treatment, cholera is always most lethal at the beginning of its outbreaks when it is unexpected. Since cholera is never expected in America and northern Europe, treatment in these places will be at its worst.

The Peruvian epidemic of cholera hit hardest the people who are usually hit hardest by whatever social or political problem happens to be rampant at any given moment in time: the poor. In the old days, cholera was much more egalitarian. When ignorance enveloped most of humankind, cholera was as happy to infect the bowels of the wealthy as those of the impoverished. It is not that the *Vibrio cholera* is now engaged in a war of reverse Marxism. In fact, the organism is an equal-opportunity bacterium. It is delighted to infect whatever gut it reaches. The particular problem poor people have with cholera is that the bacteria work sewage like politicians work crowds. Cholera plays out a deadly cyclic game, passing from the colons of its unfortunate victims, to wherever their unsterilized stool meanders, to the mouths of its next subjects. The wealthy should take no solace from this cycle. The poor now process the food of the rich.

The effective collection and treatment of sewage is a great means of preventing an outbreak of cholera. Sewage, however, is mundane and expensive. No politician will ever dominate an election by advocating a good sewage system. The mere mention of the word "sewage" will offend some voters and put the remainder to sleep. For democratic leaders to expend energy and political capital on the treatment of human effluvia, therefore, is for them to perform a truly altruistic act. Since poor people are overlooked and in any case tend not to vote, they are more at risk from raw sewage than the general population.

War and chaos are also friends to cholera and scourges of the poor. Tens of thousands of people died of cholera in Rwandan refugee camps in 1994, and in 1995 hundreds of cases occurred in Romania and the Black Sea states of the former Soviet Union. When public health systems go awry, as they tend to do when an established order breaks down, cholera can be expected to break out.

Of course, one good cholera epidemic changes everything. There is nothing like an outbreak of death to focus the attention of an electorate, an oligarchy, or an autocrat, on what should have been done years ago. Even if an epidemic of cholera in one country is not enough to frighten that country into action, cholera has the kind of ring to it that attracts the attention of neighboring states. Cholera thus comes and goes, following a fairly regular pattern. First, it attacks, killing off susceptible individuals. There are survivors, however, hardy souls who meet this enemy and live to tell about it. These survivors, after their experience, now have immune systems that have learned who cholera is and what toxin it makes. These individuals are thus prepared when cholera comes to them again and they do not get sick. If the population that cholera hits remains in place, therefore, an outbreak of cholera is ultimately self-limited. The proportion of the group that is available to harbor the organism diminishes. Susceptible people die or become immune. Either way, cholera runs out of people it can infect, and the epidemic ends. Just as farmers cannot farm after their fields are no longer fertile, cholera cannot spread after it has run out of infectable people. Cholera then becomes essentially a pediatric illness. Children who are born into a place where cholera perpetually dwells are the only ones who can get it for the first time.

The 1991 cholera epidemic that struck first in Peru came to an end when world health authorities made common cause with terrified local officials to stop it. That epidemic arrived from the sea, as outbreaks of cholera most often now seem to do. Human sewage is dumped in the oceans, which are often regarded by people as limitless cesspools provided by the Eternal for their convenience. Marine beasts are thus bathed in water that contains material that has come out of a very large number of human large intestines. What looks to the eye to be the clear blue waters of a pristine sea may, at a microscopic level, be a swirling mass of human parasites.

Some marine animals, like clams, oysters, and mussels, are filter feeders. These creatures remove slop from the water and thus concentrate germs within their bodies, essentially turning themselves into packages of consolidated human excrement. They become booby traps, deadly ambushes, ready to poison whoever eats them. Equally bad, other marine animals may eat infected mollusks. As a result, if human sewage is uncontrolled, there is no telling what the fruit of the sea will bring with it. As the cholera epidemic worked its way north from Peru toward the United States and south toward Chile, countries with significant middle-class populations who do not put up with this kind of thing, were threatened. Resources and the will to end the South American epidemic were thus found and, once more, cholera faded from sight, or at least from North American sight.

Cholera and illnesses like it are never quite eradicated. We stop outbreaks of cholera when and especially where it suits us to do so, but the disease smolders on in forgotten corners of the earth. In those unfortunate places, cholera remains out of sight, prevented from causing epidemics by the general immunity of the local population. To live to old age in such places is be resistant to cholera. The disease thus stays below the level of concern of leaders with the means and knowledge to eliminate it. Just enough susceptibles, usually children, are present to maintain the presence of the organisms in the environment, but not enough are around to support an epidemic. We call that kind of smoldering *endemic* (as opposed to epidemic) cholera and treat it as an inevitable fact of life for those poor people who have not had the good sense to be born elsewhere. When it comes to the misfortunes of regions of the world that we do not feel are immediate threats to our well-being, we tend to practice a form of global Calvinism. The troubles of those other peoples are regarded as if they were the result of predestination, and we thus leave them alone.

If we think little about who prepares the salad bars over which we graze, most of us devote even less thought to who staffs the ships that sail the sea, and still less to what may have made its way into the colons of these people. The citizens of New Orleans, therefore, would probably be totally shocked, and not a little insulted, if someone were to tell them that the vibrios that cause cholera (and many other germs that spread just like cholera) are regular visitors to their city. Diarrhea is a symptom that has reverse status, especially if it is due to cholera. Cholera is certainly not an illness that any chamber of commerce will countenance. New Orleans thus would be as unthinkable a site for an outbreak of cholera as Zermatt would be for an epidemic of typhoid fever. Still, chambers of commerce are not able to control the flow of events. Cholera occurs in New Orleans, and typhoid fever has occurred in Zermatt. In New Orleans, the cholera problem is caused by the love affair that the city and its surrounding region has with oysters from the Gulf of Mexico. The threat of cholera is thus a constant presence in New Orleans, because raw oysters from the Gulf are a way of life. Since raw shellfish from the Gulf cannot be sterilized, and the toilet habits of people who sail on the Gulf cannot be adequately controlled, cholera is, and will remain, at least an occasional presence in New Orleans and elsewhere along the Gulf coasts of Louisiana and Texas.

In December of 1996, a sick fisherman defecated into the international waters of the Gulf of Mexico. People on boats and ships often believe that their waste is so trivial in volume in comparison to that of the sea that surely a little dumping, however illegal it might be, is harmless. In this case, however, the defecation occurred over a bed of oysters that were harvested for sale in New Orleans. Fortunately, this time the agent the oysters filtered out and concentrated was only *Calicivirus,* not cholera. The oysters did their thing. Havoc, in the form of fever, vomiting, agonizing abdominal cramps, and diarrhea, spread out from raw oyster bars in New Orleans. The epidemic thus came quickly to the attention of people whose profession it is to worry about communicable diseases and who know what to do about them. The Louisiana Office of Public Health was notified about a cluster of six people who had become ill after eating raw oysters on December 25. From December 30, 1996, to January 3, 1997, news arrived of three more such clusters. By January 9, there were sixty clusters comprising 493 people. The responsible *Calicivirus* was quickly isolated, typed, and traced. The molecular typing of the organism created a trail that led investigators to the contaminated oysters, the markets where they were purchased, the fishing vessel that harvested the oysters, and, finally, to the fisherman whose indiscretion caused the trouble in the first place. All it took to start an epidemic was a single person doing what he should not have done in a strategic place. *Calicivirus* is bad, but, to paraphrase Lloyd Bentsen, it is no cholera. There was an epidemic, but it was not a cholera epidemic . . . this time. A warning shot was fired across our bow; we ignore it at our peril.

Cholera is an example, more striking than any other, of an organism that has learned the secrets of the enteric nervous system. Perversely, cholera takes over our own mechanism of defense and turns it against us. Cholera kills its victims with their own antibacterial weapons. Like an expert in judo, cholera leverages the attack by the enteric nervous system and the bowel and turns it into a suicidal thrust that can leave a person dead before the immune system even has a chance to come to the rescue.

Cholera's Use of Human Beings

The *Vibrio cholera* is not totally killed off by stomach acid, and the germs that get through the stomach are not digested as they move into the small intestine. Gastric acid, however, does provide a defense. When a cholera epidemic hit Europe in 1971, it became clear that the use of antacids and antihistamines to treat heartburn and ulcers facilitated the spread of cholera. When the vibrios reach the small intestine, they pierce the mucus and find congenial receptors on the lining cells, to which they attach. The organisms use these receptors as hooks to which they adhere securely. They just turn the lining cells of the small intestine into domiciles that they inhabit for the duration. In this location, the vibrios stay out of the connective tissue, with its array of defensive cells, that lies beneath the lining of the gut, and they do not have to mess with competing organisms in the colon. In their new home, the vibrios are semiprotected, and they can stay put long enough to make a toxin that does a great deal of harm.

Cholera toxin is able to turn one of the body's natural defensive mechanisms into a bizarre form of self-immolation. The secretory process, triggered by the second brain, that normally washes less adapted organisms and toxins out of the colon, is converted by cholera toxin into a lethal flood. The form and color of stool is lost as the contents of the colon are almost completely lost. The stool becomes a colorless, slightly cloudy liquid, called euphemistically the "rice-water stool" of cholera because it looks like the water used to wash rice paper. Nothing solid is in the stool, and even the odor has been lost, so that the smell is characteristically nonoffensive. It is, however, laden with infectious organisms. The diarrhea of cholera is utterly dehydrating. Water is lost from the body so fast that it may be impossible to replace it quickly enough by drinking.

In places where cholera is frequent and centers exist to treat it, patients are placed on special beds, called "cholera cots," which have been designed for the purpose. These beds have holes in the middle. The patient is positioned so that his or her anus is over the hole. A measuring cylinder is placed below the hole to catch the flow of diarrhea coming from the patient, which is nearly perpetual. This way, an accurate record of how much water the patient is losing can be obtained. Treatment consists of replacing that fluid with an appropriate amount of intravenous solution. If one keeps pace with the water loss, antibiotics are not required (although they can shorten the duration of the illness), and the patient will eventually recover. Natural immunity kicks in, neutralizes cholera toxin, and cures the disease. The trick in treating cholera is simply to prevent dehydration anyway one can. In an epidemic, when everyone who has diarrhea can be presumed to be infected with cholera, doctors are not even needed to treat it. All that is required is a supply of willing nursing assistants who can rapidly be taught what to do, a supply of sterile needles, and enough intravenous solution. Effective therapy for cholera can thus be cheaply provided, which is a good thing because the cheapness of the treatment enables it to be given in poor countries.

To understand how cholera turns our colons against us, it is necessary to consider the process that cholera subverts. This process is secretion, the nerve-stimulated assist to the immune system that is designed to wash dirty rotten invaders out of the colon. It is a complicated mechanism, but if a dumb organism like cholera can comprehend and pervert it, surely a higher organism, like one of us, can apprehend its basic principles.

Secretion in the colon is stimulated by signaling molecules that turn on switches, represented by receptors on the surfaces of specialized cells of the lining of the large intestine. The cells that do the secreting that normally washes out unwanted organisms are found within infoldings of the colonic lining and are called *crypt* cells. Ironically, new lining cells are born in crypts, while cell death occurs elsewhere (at the tips of villi in the small intestine and at the luminal surface of the colon). Crypts are simply the names of the infoldings of the lining of the small or large intestine. The molecules that initiate secretion are neurotransmitters or hormones. Molecules that are known to be released by the nerves that stimulate secretion include acetylcholine and *vasoactive intestinal peptide,* abbreviated to VIP (surely the most wonderful nickname of all enteric neurotransmitters). Whatever the signaling molecule, the activating event is always the same. The signaling molecule binds to its receptor on cell surfaces. That phenomenon causes interesting things to happen because it turns on the arcane, Rube Goldberg–like signal transduction apparatus of the cells. The critical element in signal transduction is the generation of a second messenger (the first messenger is the neurotransmitter or hormone) that informs the cells' inner self that their receptors have just been activated. The cells thus know that the time has come for them to do their thing. The second messenger is responsible for setting in motion a complicated cascade of events that eventually leads to secretion. Pumps are activated, molecules are modified, ions are transported from the wall of the gut to the lumen of the bowel, and water finally follows the ions.

One of the second messengers that is used by the secretory crypt cells of the colon is a molecule with a name, *cyclic 3',5"-adenosine monophosphate,* that is so vicious that even the most hard-bitten of scientists quails and refers to it only by its initials, *cAMP.* To generate cAMP, an activated receptor (the membrane switch in the "on" position) initiates a chain reaction. The first step in this chain is for the stimulated receptor to cause a nearby triplex of proteins to come undone. These proteins are called *G-*

proteins, and their components are distinguished with Greek letters, *a* (alpha), *b* (beta), and *g* (gamma). The G in the nomenclature of G-proteins comes from the ability of a component of the G-proteins to bind two high-energy molecules, *GDP* (guanosine diphosphate) and *GTP* (guanosine triphosphate). GDP has two phosphates while GTP has three. When the G-protein complex is at rest (not being stimulated by a receptor), GDP sticks tightly to *a,* which in turn is held on to by *b* and *g.* An active receptor sidles up to *a* and causes it to lose its affinity for GDP. As soon as that happens, GDP drops off *a* and is replaced immediately by GTP.

The simple act of binding GTP instead of GDP transforms *a.* When GDP is attached to it, *a* is a respectable Dr. Jekyll, a protein that hangs out calmly in the membrane and complexes with its partners, *b* and *g,* which appear to exert a restraining influence over it. The binding of GTP hits a like a dose of crack cocaine, converting it into a rampaging Mr. Hyde. The formerly quiescent *a* becomes an energized marauder. Breaking away from *b* and *g,* the GTP-charged *a* prowls along the membrane to find and turn on the enzyme *adenyl cyclase,* which generates cAMP.

As long as GTP remains attached to *a, a* is insatiable and will not let adenyl cyclase out of its stimulating grasp. Adenyl cyclase becomes an enzyme possessed and cannot stop itself. It is driven by the GTP-maddened *a* to pour out cAMP continuously. The perpetual presence of the second messenger locks cells in their secretory mode. Normally, of course, this does not happen. Secretory cells cannot be allowed to be forever "on." If they were, then the evolution of the organism that owned those particular cells would have been brought to an abrupt halt. Cells thus have "off" switches as well as "on" switches. In this case, there are multiple "off" switches. One of the "off" switches is an intrinsic property of *a,* which doubles as an enzyme that causes GTP (which is not, in any case, very stable) to lose one of its phosphate groups and turn into GDP. Since the GTP that loses its phosphate is bound to *a* at the time the phosphate is removed, the surgery that *a* performs on GTP leaves behind a residual molecule of GDP that is still hanging on to *a.* The reappearance of GDP on *a* calms *a* down, causing it to let go of adenyl cyclase and go back to its Dr. Jekyll–like existence in a nice quiet assembly with *b* and *g.* The production of cAMP stops, and the remaining cAMP is destroyed by an enzyme designed for this purpose (another "off" switch, this one with the unpronounceable name *phosphodiesterase*). The secretory activity of the cell thus comes to an end. Under ordinary circumstances, therefore, the G-protein-driven chain reaction initiates secretion, but it is a self-limited phenomenon that does not outlast the activation of the receptor by the neurotransmitter or hormone that first signaled the cell to go into action.

Cholera understands and exploits the role of G-proteins in cell signaling. Cholera toxin has been cleverly designed by the evolution of the organism to enter cells and interact with one particular G-protein. This G-protein, called *Gs,* happens to be the one that is linked to the activation of adenyl cyclase and the generation of cAMP. In fact, it was the sophistication of cholera toxin that helped lead the Nobelist Alfred Gilman to the discovery that Gs is only one of many G-proteins. Other G-proteins activate other signal-transduction pathways or even work to stop the stimulation of adenyl cyclase by Gs.

Cholera toxin, the poison the *Vibrio cholera* synthesizes to carry out its deadly purpose, consists of two separate proteins, or *subunits,* linked together by a chemical bond so that they can function in tandem. One subunit of cholera toxin acts as a molecular chaperone that enables both it and its partner to go through cell membranes. The other subunit, carried into cells by its accomplice, is an enzyme designed to do the toxin's dirty work. This chain of cholera toxin causes the cell to attach a molecule, *ADP-ribose,* to the *a* component of Gs. The attachment of ADP-ribose to *a* blocks *a*'s ability to convert GTP back to GDP. As a result, as soon as a neurotransmitter activates its receptor, and the *a* component of Gs turns on, it never turns off. In essence, cholera toxin induces cells to make a molecular shim that becomes stuck in the cell's switch, jamming it in the "on" position. The result is a virtually unending secretory flood that persists until the patient dies or the immune system learns about cholera and its toxin and mounts a life-saving counterattack.

While a victim waits for the immune counterattack to happen, he/she can be turned into a fatally dried-out lump of clay. The cholera-induced flood of diarrhea also serves the organism in another way. It ensures that cholera has little bacterial competition. Germs that are in the lumen of the colon tend to be flushed out. They are flushed out so well, in fact, that the flushing explains why the usual stench of the products of bacterial fermentation is missing from the rice-water stool of patients with cholera. Cholera, however, does not live in the lumen of the colon with the banal organisms that make up the large intestine's normal flora. Since cholera is securely attached to small intestinal lining cells, it is spared the effects of the flood it has caused. The organism can thus remain serenely in place while germs that might compete with it are washed away. Cholera thus forces the body, contrary to its own best interests, to provide an arc to keep the invading vibrios safe from the onrushing—or rather, in this case, outrushing—water.

Cholera toxin goes even further. It is as if it has been designed by an evil genius to be fail-safe and leave nothing to chance. The way the toxin works, it cannot do its dastardly deeds without an activated *a* component of Gs. By attaching a molecule of ADP-ribose to *a*, the toxin prevents an activated *a* from ever turning off; the GTP that is bound to *a* cannot be cut down and converted back to GDP. But what would happen if GTP failed to bind to *a* in the first place? In fact, nothing bad would actually happen. Clearly, if a cell never turned "on," cholera toxin would not be able to jam its signaling mechanism in the "on" position. In a stimulated cell, an ADP-ribose-poisoned *a*-GTP is endlessly driven to rape adenyl cyclase and churn out cAMP because *a* cannot get the GTP off its back. In contrast, in an unstimulated cell, an ADP-ribose-poisoned *a*-GDP will remain placidly in the grasp of *b* and *g*, because there is no GTP to get off the back of *a*. Cholera toxin thus requires that a hormone or a neurotransmitter stimulate a receptor on a cell it is attacking in order to poison that cell. Stimulation of a receptor causes the Gs complex to dissociate and induce GTP to replace the GDP that is bound to the resting *a* component. Once that happens in the presence of cholera toxin, the signal transduction cascade is turned on, and the rest is history.

If nothing were to stimulate a poisoned cell, it might conceivably escape the clutches of cholera toxin. Cholera, however, provides for that. Its toxin poisons more than just secretory cells. Cholera also co-opts the gut's nerves to be sure that they provide an appropriate neurotransmitter to stimulate secretion. By causing secretomotor nerves to activate receptors on secretory cells, the evolution of the *Vibrio cholera* makes certain that its toxin will not poison these cells in vain.

Serotonin is a major factor in initiating secretory reflexes. This role is very easy to demonstrate in a normal piece of gut. If you stroke the mucosa that lines the luminal surface of an isolated segment of colon, for example, the level of cAMP will rise inside of crypt cells, the cells will secrete chloride ions, and water will follow the chloride into the lumen of the gut. The effect of stroking is due to nerve stimulation; therefore, if nerves in the wall of the gut are anesthetized experimentally with the puffer fish poison tetrodotoxin, the mucosa can be stroked into oblivion without provoking the crypt cells to secrete. Blocking the action of serotonin accomplishes the same end. Stroke away, and secretion does not occur in the presence of an appropriate serotonin antagonist. Stroking the mucosa causes certain of the lining cells, known as *enterochromaffin,* or *EC,* cells to secrete serotonin. These are the cells that together manufacture and contain over 95 percent of the body's serotonin. An increase in pressure inside the lumen of the gut, which also provokes serotonin secretion, is the natural stimulus that is mimicked experimentally by stroking the surface of the isolated colon.

EC cells, it seems, are actually sensory receptors. This is a relatively recent and exciting discovery. Mechanical stimulation, such as stroking or distortion by pressure, makes EC cells secrete serotonin. In fact, I now spend long hours in the laboratory trying to discover the molecular mechanisms that are responsible for coupling the mechanical stimulation of EC cells to serotonin secretion. At the moment, cholera toxin knows how EC cells work, but I do not. In any case, EC cells are endocrine cells, like the gastrin-secreting G-cells of the stomach, which the EC cells resemble. They thus do not secrete serotonin into the lumen of the gut but into its wall. Serotonin, therefore, is delivered to the connective tissue space under the lining of the bowel. There are many nerve fibers as well as blood vessels within this space.

Some of the serotonin in the connective tissue, like gastrin, gets into the bloodstream. This serotonin is quickly inactivated, either by cells of the blood or by the liver (to which blood flows from the gut). Serotonin in the connective tissue of the bowel, however, also comes into contact with the nerve fibers that course through it. Some of these nerves, of course, are secretomotor, on their way to the crypt cells, which they stimulate. Other nerves, however, are not motor, but I have found that they are sensory. One kind of sensory nerve is intrinsic, wired to nervous pathways that lie entirely within the enteric nervous system. Other sensory nerves are extrinsic, wired to the central nervous system. Both sets of sensory nerves have receptors on their surface that respond to serotonin. When the serotonin that has been released from EC cells acts on these receptors, the sensory nerves become stimulated. The intrinsic sensory nerves initiate secretory and peristaltic reflexes. The extrinsic sensory nerves initiate nausea and, if the stimulation is strong enough, vomiting and cramps as well.

Cholera toxin gets into EC cells as well as into crypt cells. Promotion of chloride secretion by crypt cells is thus not the only thing that cholera toxin does in the gut. It also causes serotonin to be secreted by EC cells. The secretion of serotonin provides cholera with the fail-safe mechanism that it needs to ensure that receptors on crypt cells will be active. It is not that the serotonin cholera toxin released from EC cells acts directly on the crypt cells. It does not have to. The released serotonin turns on nervous pathways that end in secretomotor nerves. The secretomotor nerves squirt their neurotransmitters at receptors on crypt cells. As noted earlier, the neurotransmitters used by secretomotor nerves are either acetylcholine, which stimulates muscarinic receptors on crypt cells, or VIP, which has its own receptors. Administration of drugs that block serotonin receptors, therefore, decreases the amount of secretion that follows the poisoning of the bowel by cholera toxin. Taking out the action of serotonin deprives the toxin of its nervous backup. Cholera toxin needs to pervert the second brain or else it loses some of its power.

In a sense, cholera toxin uses for its nefarious purposes mechanisms that have evolved over centuries for our body's own defense. Secretion is a process that usually has protective value. The colon is a filthy place and requires a cleansing mechanism. The major function of the colon is to absorb water, but a little secretion of water at a critical time is a good thing. The operative word here, of course, is "little." Clearly, the secretion evoked by cholera stands the evolution of the colon on its head. Cholera, however, is relatively uncommon. Unfortunately, even the "normal," commonplace activation of secretory mechanisms in the colon can sometimes be as devastating as cholera. When the colons of otherwise healthy tourists secrete and purge themselves of the toxins produced by an unanticipated new set of organisms, the owners of those colons are temporarily inconvenienced but protected from problems that are more serious and long-lasting. Secretion is beneficial and, ultimately, harmless. When the colons of little babies secrete under similar circumstances, however, secretion is anything but harmless. Even a modest degree of stimulation of secretion in the infantile colon may produce a disease that resembles cholera far more than the tourist two-step.

Babies, small children, and old people have very little reserve volume. They are not prepared to lose a great deal of water. Just a small loss may cause them to dry out and go into shock. Diarrhea in infants is thus a far more dangerous phenomenon than diarrhea in adults. Add vomiting to the condition and danger becomes imminent peril. In an advanced society, children suffering from the severe loss of fluid and volume due to diarrhea and vomiting can easily be treated by oral or intravenous rehydration with appropriate solutions. In a less advanced society, which I define as one that does not provide adequate health care to defenseless children, infantile diarrhea with, or even without, vomiting can easily be a cause of death.

Less advanced societies are abundant on our planet. In fact, they are so abundant that simple diarrhea is the world's second leading killer of infants. Almost three million children are slaughtered every year by diarrhea; only pneumonia kills more. Societies that cause children to die in this way are found in the developing world, where ignorance and lack of resources impede health-care delivery. They are also found in rural areas and inner cities of the United States, where a different kind of ignorance and the same lack of resources deprives children of health care. In both cases—developing world or backward parts of America—the needless death of small children due to secretory diarrhea is so preventable that its continuation can only be explained by the meanness of people who have the resources to stop it and do not.

The End Game

All good things come to an end. So too for the transit of intestinal contents. Whatever is swallowed that is indigestible and/or unabsorbable passes through the intestines and is delivered to the rectum for disposal by defecation. The bulk of this material is composed of the cellulose walls of the plant cells that we eat in fruits, grains, and vegetables. We call this stuff fiber, and nutritionists often extol its value. Unlike mice, rats, and cows, whose stomachs have special compartments that provide a home for microorganisms that digest cellulose for them, we have nothing in our gut that can digest cellulose for us. Cellulose, therefore, goes right through us, and except for some grinding and maceration, the cellulose that leaves our bowel at the anus is virtually identical to that which entered at the mouth. Salad may not look as appetizing on its way out as it did on its way in, but all of its fiber is still there.

Fiber gives us the sensation of eating a lot, but it provides no calories. In a society such as ours, in which obesity is the leading nutritional illness, food that lacks calories is something to be treasured. Fiber also gives the colon something to squeeze down on, which is nice because it keeps the organ fit and trim. A colon that lacks any exercise can get to be flaccid and weak. After many years of "good old-fashioned meals of meat and potatoes," a low-fiber diet, the colon may become as flabby as its owner. A weak-walled colon tends to develop little outpouchings called *diverticuli*. Essentially, these diverticuli represent vulnerable spots, places where, as a result of weakness, the colonic wall begins to "blow out," the way a rubber tire does when there is a defect in its sidewall. The disease associated with the presence of diverticuli is known as *diverticulosis*. Diverticuli are blind sacs where the intestinal contents can pool and become stagnant. The normal interaction of the enteric nervous system and the immune system is compromised within diverticuli because the ability of the colon to wash them out is limited. The muscles of the colon are oriented to propel luminal contents in an oral to anal direction, not to pump the contents out of side pockets. If anything, when they contract, the colonic muscles may act as valves on the mouths of diverticuli, holding stuff in, not clearing it out.

Because diverticuli are so difficult to clean out, they often become infected, causing pain, fever, and intestinal obstruction. When that happens, the name of the condition changes from diverticulosis to *diverticulitis*. Diverticuli can even perforate, and in fact they frequently do so. When diverticuli perforate, the lumen of the gut is no longer totally outside of the body proper. Perforated diverticuli provide an opening through which a festering mass of colonic bacteria pour into the body proper. If one is lucky, an abscess may form in the body wall where the infection remains local. Alternatively, the bacteria that escape from the confines of the colon may give rise to a catastrophic infection of the peritoneal cavity and, by entering the bloodstream, infect the body as a whole. Treatment of these conditions, which are respectively known as *peritonitis* (infection of the peritoneal cavity) and *septicemia* (infection in the bloodstream), must be prompt and aggressive. Antibiotics are critical and have to be chosen with great care, to be sure that they are accurately aimed at the bacteria causing the trouble. Time is of the essence. In the face of peritonitis resulting from a perforated diverticulum, a doctor does not have many chances to get the drug right. The selection of an antibiotic to which the organisms are resistant may thus condemn the patient to a premature death.

Diverticuli are common in the elderly (occurring in up to half of the population by age eighty), but they are not benign. Diverticuli have to be considered threats to life itself. Prevention, beginning before diverticuli form, is certainly worthwhile. Including fiber as a regular component of the diet, therefore, is a little like jogging. What running does for our bodies, dietary fiber does for our colons.

Regularity

Fiber can absorb water and expand in volume, thereby increasing the pressure inside the lumen of the gut. Since increasing the pressure inside the lumen of the bowel is the stimulus that triggers the peristaltic reflex, a fiber-filled colon is likely to be one that is propelling its contents toward the anus at warp speed. The ability of cellulose to stimulate the propulsion of material in the colon is a property that people who have lusted after bowel movements have known and treasured for centuries. Prunes are legendary in this regard, and Metamucil is a best-seller. If regularity is viewed as nirvana—and the profitable trade in laxatives suggests that it is—then fiber is nature's way to provide grace.

Metamucil is a concoction of concentrated fiber, which gives the colon a good workout. There is not much of a downside to using Metamucil in order to stay regular, except for people with wafer-thin diverticuli. A Metamucil-full colon running at full throttle, crunching down on an expanded mass of fiber, is an invitation to a perforation.

Other products that are used as laxatives may be less safe than fiber. Mineral oil is not too bad. The guiding principle behind mineral oil is to coat stool with so much slime that even a sluggish colon does not hold it back.

The hope is that if you can turn the fecal pellets in a constipated bowel into the equivalents of greased pigs, they will wiggle their way to daylight no matter what the colon does, or rather does not do, in the way of propulsion. Problems presented by mineral oil include the intolerable social nature of the surprise presented by slippery stool passing uncontrollably out of the anus. Mineral oil can also dissolve and carry away fat-soluble vitamins like vitamin K. Women who take mineral oil to relieve the mechanical constipation that occurs in pregnancy are at risk of giving birth to babies who have a condition known as hemorrhagic disease of the newborn. These vitamin-K-deficient babies bleed because they are unable to make adequate amounts of clotting proteins. In American hospitals, therefore, it is just as routine to give every newborn baby an injection of vitamin K as a slap on the bottom.

Another approach is to provide the bowel with a load of salt that it is unable to absorb. The laxatives that work this way are called *saline cathartics*. Milk of magnesia (magnesium sulfate) is a well-known example of such a laxative. The salt, which stays in the lumen of the gut, attracts water by osmosis. The result is to increase the pressure inside the intestinal lumen, which stimulates the peristaltic reflex, just as it did in dog intestines for Bayliss and Starling way back in 1899. The intestinal contents are driven by the activated enteric nervous system to race down the bowel.

Saline cathartics are reliable and not too stressful, but they cannot be used repeatedly without risk. The gut seems to become tolerant of the laxatives, requiring, as a result, more and more of them to do the job. A laxative abuser can also become addicted to laxatives and may be almost unable to move his/her bowels without them. In extreme cases, the gut may manifest all of the symptoms of obstruction, even though the path through the intestinal lumen is perfectly clear. This phenomenon is called *pseudo-obstruction,* although there is nothing "pseudo" about the distress it causes. Surgery, even removal of the colon, may be required to treat it. At autopsy, the enteric nervous system of a chronic laxative abuser can look very much like the fields of Agincourt after Henry V finished dealing with the French army, a turf littered with the rotting remains of dead soldiers. The soldiers in the bowel, of course, are fallen enteric nerve cells, sacrificed for the cause of "regularity."

Why enteric nerve cells disappear in people who abuse laxatives is not totally clear, but nerve cells can become overly excited and die because of it. The neurotransmitter glutamate is famous for exciting nerve cells in the brain to death. This action is called *excitotoxicity*. Annette Kirchgessner and I recently discovered that glutamate is a neurotransmitter in the enteric nervous system, as well as in the brain and spinal cord. Annette, moreover, has gone on to demonstrate that glutamate can cause enteric nerve cells to drop dead from excitotoxicity. Whether the glutamate-driven excitotoxic death of its nerve cells is what happens to a bowel that has chronically been overly stimulated by laxatives remains to be demonstrated. Whatever the cause of laxative-associated nerve cell death, however, a saline cathartic, like a good Burgundy, is best taken in moderation.

Some situations call for more bowel emptying than fiber, mineral oil, or saline cathartics can provide. For example, the fiber-optic examination of the lumen of the gut (*endoscopy*) requires that nothing be in it that might block the view. The colon that is about to be subjected to a colonoscopy had better be empty. One approach to "preparation" of the bowel is to give a patient a dose of a drug that directly stimulates those circuits of the enteric nervous system that cause propulsion. Castor oil is an example of a natural "medicine" of this type, and Dulcolax is one that acts similarly but is person-made. Stimulating propulsive circuits brings out giant migrating waves of contractile activity in the intestines. These waves plow relentlessly onward, driving before them everything that is in the intestinal lumen. The giant drug-induced contractions of the gut are dangerous if diverticuli are present because they can cause the colon to perforate. Even if there are no diverticuli, giant contractions, and the resulting high pressure in the gut, are very painful. The pain can be so intense that it even attracted the Gestapo, which used the administration of castor oil as an instrument of torture during the Nazi period. Drugs of this type have lacked enthusiastic patient acceptance.

The more modern method of intestinal "preparation" is to have patients drink vast quantities of a slithery, nonabsorbable liquid. The bulk of the ingested fluid stimulates the peristaltic reflex, while its character acts as a lubricant. This material, distantly related to the antifreeze put in automobile radiators, is found in a preparation with the brilliant name Golytely that recalls for prescribing physicians the rigors of the intestinal contractions produced by castor oil or Dulcolax. All a patient has to do is drink a gallon of Golytely, and everything flows from that. Drinking the stuff is the major challenge. Once you get it down, the stool is delivered without agony.

Some of the drugs that are used as laxatives and that work by tweaking enteric nerve cells have been recognized as threats to the enteric nervous system. These compounds bring excitotoxicity to mind, but the mechanism by which they damage the bowel has not been worked out. One of them, *phenolphthalein,* which used to be the active ingredient in Ex-Lax, has recently been removed from the market with the consent of its manufacturer. Senna, a natural product, is present in Senokot, which is hyped in advertisements as nature's way to promote regularity. Senna also stimulates enteric nerve cells. As I noted earlier, "natural" is not necessarily synonymous with good, or even safe. Plants are out there making all kinds of perfectly natural things that are toxic to people and animals. These toxins may be an adaptation that the plants evolved to keep them from being eaten. I am sure that the victims of castor oil administered by the Gestapo drew little or no solace from the fact that the instrument of their torture was all natural. Senna, however, has a following and carries a relatively low medical profile. There have not been many studies of its effects, but one that I remember well, presented at a gastroenterology meeting in England, showed slide after slide of distorted and dying enteric nerve cells removed in biopsies from the colons of patients who developed pseudo-obstruction after taking senna. Certainly, there is not yet any conclusive evidence that proves that senna causes harm. Still, when it comes to "regularity," I think that a bowl of crudités—or, if worst comes to worst, prunes—has a lot to recommend it.

The Finish

By the time the intestinal contents wind up in the rectum, the conservation (reabsorption) of water that has taken place in the colon should have left them in a satisfyingly solid condition. Besides the dietary fiber and other, more exotic materials that we eat but cannot absorb, the stool contains the bacteria that have lost their hold on the colon, mucus that the gut has secreted, and bile pigments, which provide the stool with its proper color. In a healthy individual, the feces should contain a minimal amount of fat, no blood, and no eggs of any parasites. The appearance of any of these things in feces is a sure sign of disease somewhere in the gut. A normally functioning bowel digests pretty much all of the fat that one can eat, does not bleed, and is free of parasites. If the enteric nervous system has not operated optimally, and propulsion in the colon has been too slow, the stool content may become overly dry, assuming the consistency of rabbit pellets. This, however, is not normal and means that an individual is constipated. Constipation thus occurs when the enteric nervous system is malfunctioning, but the malfunction of the enteric nervous system may not necessarily be its fault alone. Constipation can also result from an abnormal input to the bowel from the central nervous system or from an abnormal interaction between the enteric and central nervous systems. In any case, whether normal or pelletlike, the fecal material that enters the rectum has considerable heft.

Once food leaves the mouth and begins its descent in the gut, we lose track of it. Under ordinary circumstances, the magic worked by the enteric nervous system, in coordination with nerve cells in the brain and spinal cord, stays well out of our consciousness. Oenophiles savor their wines, carnivores their steaks, and vegetarians their eggplants, but only at the front end of the digestive tube. Foods may leave an aftertaste that lingers on the tongue, but there is normally no reprise. If a recapitulation of what goes down does occur, it is never pleasant, and it is never normal. A trip through the gut is normally both a one-way passage and a silent one. Unless the pressure within the bowel is abnormally high, sufficiently so that it becomes noxious and painful, we have no way of knowing what is going on inside our gut. The intestinal contents, however, do normally return to consciousness when they leave the bowel at its anal end.

Perception returns with a sense of urgency when feces enter the rectum. As the rectum fills, it delivers a "call to stool." This sensation is usually not painful, although it may become so as it rises to a crescendo; nevertheless, the "call to stool" is not a sensation that is easy to ignore. When felt, it tends to lead to action sooner rather than later. Contributing to the urgency of the situation is the downward curve the rectum takes as it sweeps toward the anus. Gravity thus works together with the propulsive movements of the muscles of the rectum itself to promote expulsion of stool.

Although it helps in expelling stool, gravity is not a threat to our social well-being. The rectum is well endowed anatomically to handle the heft and bulk of the stool it contains. Storage of stool within the rectum is thus possible for limited periods of time. This storage, of course, is utterly necessary to preserve the veneer of civilization. The lining of the rectum is folded into three horizontal shelves so that the full weight of the fecal contents do not have to be borne by the sphincters that close the nondefecating anus. These sphincters, moreover, are not slaves that open as soon as the "call to stool" is issued. They can hold stool back and resist opening until the moment is right. In fact, a guiding principle upon which the smooth working of our society is based is that the anal sphincters of every individual who is not an infant will indeed hold stool back until the right moment.

To help ensure good sphincter control, evolution has provided us with not just one but two anal sphincters. The first of these, the *internal anal sphincter,* is composed of smooth muscle and is regulated only by the enteric and autonomic nervous systems. The internal anal sphincter is thus on autopilot and out of our volitional control. It only does what reflex-driven nerves tell it to do. The internal anal sphincter is really just a thickening of the circular layer of smooth muscle that is found in most of the bowel.

Backing up the internal anal sphincter is the *external anal sphincter,* which is made of skeletal muscle and *is* under volitional control. To be sure, even the external anal sphincter responds to reflexes, but our brain learns to override these reflexes and control it. Victor Marshall, a professor who taught me urology in medical school, beautifully expressed the awe in which surgeons hold the external anal sphincter when he said: "Poems are made by fools like me, but only God can make a sphincter."

The elegance of the sophistication of the control that can be exercised over the external anal sphincter strains credulity. One is tempted to think of a sphincter as a simple purse string that can be loosened or tightened. But no purse string, when loosened, can pass gas under pressure while simultaneously holding back masses of liquids and solids straining to break through. Only under the most extreme duress does this trustworthy, eminently reliable, and utterly essential tool fail. A raging infection of the colon, upstream from the sphincter, may cause the propulsive movements of the bowel to go into overdrive. Accelerated transit is a by-product of the defensive interaction of the enteric nervous system and the immunologically competent cells of the bowel. Nerve-induced secretion produces its cleansing flood, while chemical signals released by the immune cells irritate nerves and increase the frequency of peristaltic rushes. Like the herd of bulls that stampede each year through the streets of Pamplona, the intestinal contents can be turned into an irresistible force. There comes a time when no sphincter can hold any longer.

Excessive pushing by the enteric nervous system, as in instances of infection, is not the only circumstance that can lead to sphincter failure. Severe fright is another. Fright may distract the brain and interrupt its concentration on the external anal sphincter. Artillery shells landing next to occupied foxholes have been known to cause this to happen. In fact, soldiers frequently exchange greetings on the battlefield with the admonition: "Keep a tight sphincter." A weakened external anal sphincter itself, resulting from surgery, accidents, the stretching and ripping of multiple pregnancies, and even old age can lead to the escape of fecal contents. The central nature of the control of the anal sphincter means that control over it becomes lost when the brain becomes damaged by degenerative diseases. Alzheimer's disease and other forms of senile dementia, therefore, often bring back the wearing of diapers.

We are not born knowing how to control our external anal sphincters. The handling of this muscle has to be learned, and as is universally understood, the expertise must be carefully taught. In fact, the successful transmission to a child of the knowledge of how to exert authority over his/her external anal sphincter is one of the great thrills of parenting. Defecation is, first of all, a reflex. The reflex occurs without learning; it is "hard-wired" into the nervous system. The defecatory reflex involves initially the propulsion of stool into the rectum, a process that involves mainly the enteric nervous system and the musculature of the colon. The arrival of feces in the rectum, however, is sensed by extrinsic sensory nerves that relay the information to the sacral (lower) spinal cord, where the defecation reflex is coordinated.

In part, the motor portion of the reflex involves sacral parasympathetic nerves that accelerate downward propulsion of stool within the rectum and relax (and thereby open) the internal anal sphincter. In part, and in coordination with the autonomic activity, the reflex involves the action of skeletal motor nerves to relax and open the external anal sphincter. The muscles of the floor of the pelvis also contract, correctly positioning the rectum for expulsion of what is in it. All of this *Sturm und Drang* occurs without thought and is managed by the sacral spinal cord. The internal drive to push the intestinal contents downward can be supplemented by contractions of the abdominal muscles and diaphragm to increase pressure within the abdomen. This is a familiar activity that we know politely as "straining at stool."

The brain has the ability to exert inhibitory influences over the sacral parasympathetic reflexes. When a child acquires the ability to use the descending fibers that the brain sends to the spinal cord, he/she can learn to suppress defecation or, in the case of the bladder, urination. In the absence of this inhibition from higher brain centers, the rectum and bladder are totally owned spinal subsidiaries and operate only as reflexes dictate. Stool in the rectum triggers a reflex and defecation occurs. The inevitable happens and diapers fill. Before the brain can take over, however, and even have a possibility of reigning in the sacral spinal cord, the nervous pathways that descend from the brain to the spinal cord must become mature. The descending nervous tracts of the lower spinal cord do not finish acquiring the fatty coats (myelin sheaths) that envelope them until sometime during the second year of life. Until that happens, conduction of nervous signals from the brain to the spinal cord is inadequate to support toilet training. The child who is trained before age one has really trained his/her parent. The parent has learned the frequency at which stool is delivered to the child's rectum and thus is able to anticipate the child's defecatory reflexes. A well-trained adult can thus position a child appropriately over a toilet at the key moments of a day, but it is the adult and not the child who is exhibiting "toilet training."

In adults who suffer, for whatever reason, an injury that severs the spinal cord, the descending influence that the brain exerts over defecation and urination is lost. The infantile situation is re-created, and the spinal reflexes take over unopposed. Diapers must be used (or other devices are employed in their place). In these patients, the sacral spinal center that normally follows commands from the brain now operates as a unit with the colon. The colon senses that it contains stool and relays that information to the sacral spinal cord, which then launches the defecatory reflex, whether the patient wants it launched or not. In contrast, if there is an injury to the sacral spinal cord itself or to the nerves that connect the sacral spinal cord to the gut and muscles of the pelvic floor, the defecatory reflex is completely lost. In the absence of the defecatory reflex, only the propulsion of the enteric nervous system is left to expel stool. When this happens, stool either constantly flows out of a flaccid anus that is no longer guarded by a now-paralyzed external anal sphincter or *fecal impactions* form. Impactions are congealed masses of dried-out stool that become embedded in the rectum and/or anal canal. Impactions function as corks, holding back the upstream intestinal contents. A fecal impaction may be so firm and stuck in place that it cannot be dislodged by the amount of force that can be generated by the intestinal musculature without assistance. Pressure in the gut above an unyielding fecal impaction may thus mount to levels that cause excruciating pain that persists until the impaction is mechanically cleared away by a merciful attendant.

I will never forget a patient I saw in the summer of 1957 when I was working as an orderly before going to medical school. The patient had suffered from Parkinson's disease for many years and was totally incapacitated. The beneficial effects of L-DOPA (Levodopa), the miracle drug that dramatically brought functional movement back to many victims of Parkinson's disease, had not yet been introduced. Therapies that were then in use were tried but were ineffective. Moreover, the patient's brain cells had been degenerating for many years, and he had relatively few left. He was rigid as a board and highly demented. He never left his bed, and, in fact, he hardly moved at all. He would let me know from the amplitude of his moans when he needed to have me clear an impaction. To do this, I would put on a rubber glove, lubricate a finger, and manually dislodge the hardened feces. On good days, a flow of stool followed the loosened impaction in a bedpan.

In retrospect, after I learned more about Parkinson's disease, I was puzzled by Mr. X's constant fecal impactions. Parkinson's disease affects nerve cells in the brain that participate in the coordination of the movements of skeletal muscles. The loss of these brain cells leads to the signs and symptoms of Parkinson's disease: tremor (involuntary shaking movements), a shuffling gait, a masklike facial expression, and rigidity. After a time the continued degeneration of brain cells causes dementia. I could imagine the disorder interfering with the descending nervous pathways from the brain that control spinal reflexes; however, I did not see why the patient's spinal reflexes themselves would fail. Parkinson's disease does not affect the sacral spinal cord or peripheral nerves. I thought it might be possible that the external anal sphincter had become rigid, like the patient's limbs, but Mr. X appeared to have lost even the spinal reflexes that regulate defecation.

Very recently, it has been discovered that the enteric nervous system is affected in at least some patients with Parkinson's disease. The same type of *lesion* (tissue damage) that is found in the brains of patients with Parkinson's disease is also found in their enteric nervous systems. It thus stands to reason that Parkinson's-induced degeneration of nerve cells in the bowel will affect the propulsion of intestinal contents in the same way that it affects higher integrative function in the brain. Parkinson's disease may thus eventually cause an enteric as well as a cerebral dementia.

I should not have been surprised to learn that Parkinson's disease is a disorder of the enteric nervous system as well as the brain. Like many surprises that, in retrospect, seem unsurprising after the shock of discovery is past, this one too might have been anticipated. The enteric nervous system is, after all, a very close cousin of the brain. As I explained to the Society for Neuroscientists at the workshop in Cincinnati, the enteric nervous system has more in common, both chemically and structurally, with the brain than with the remainder of the peripheral nervous system. The enteric nervous system looks as if it were the brain gone south. As a result, it is to be anticipated that illnesses of various types that occur in the brain will also involve the enteric nervous system.

Alzheimer's disease, as much as Parkinson's disease, illustrates the validity of this idea. Alzheimer's affects the enteric nervous system in the same way as it affects the brain. Moreover, just as occurs in Parkinson's disease, the characteristic lesions that neuropathologists use to diagnose Alzheimer's disease are found in the enteric nervous system as well as the brain. Since these lesions occur in the bowel, it seems likely that the function of the enteric nervous system is compromised in at least some patients with Alzheimer's disease; nevertheless, I do not know of a single systematic study of the bowel problems of Alzheimer's patients. I suppose that in the face of overwhelming dementia, constipation, and even fecal impactions, pale as medical problems. When a patient's mind is failing, it may seem perverse to focus on his or her bowel habits. Someday, however, I keep telling myself, I should do just that. Clues to the acquisition of Alzheimer's disease, a means of getting a definitive diagnosis, and following the efficacy of treatment might all be facilitated by investigations of the enteric nervous system. Biopsies of the enteric nervous system, for example, are much easier to obtain than biopsies of the brain. Who knows? Perhaps one day the ease of obtaining rectal biopsies will cause the rectum to be called "the window on the brain."

8

A BAD BOWEL

THE TOP AND BOTTOM ends of the digestive system are regions where the brain in the head plays the critical role in determining what the bowel does. Input and output are centrally determined phenomena. These are also attention-grabbing activities that dominate a good part of a normal person's existence. Some people live to eat, while others, who rarely admit it, get their tingle from defecation. Between ingestion and egestion, however, a great many events occur in the bowel, which are ignored by the brain and left to be managed primarily by the enteric nervous system. Because the second brain thus works in obscurity, it is often overlooked as a source of disease.

The Gut Is Not Immune to Mental Disease

The behaviors that evolution has made the prime responsibility of the brain can clearly and obviously be influenced by psychopathology. We should expect, therefore, to find that the enteric functions that are most affected by the central nervous system are also those that are conscripted by the brain to participate in particular forms of mental illness. That there might be enteric symptoms of psychiatric illnesses, therefore, ought not to be a surprise to anyone. Eating disorders such as *anorexia nervosa* and bulimia are examples of conditions that are recognized as neurotic, or even psychotic, in origin. Similarly, certain personality disorders have for many years been associated with the effort devoted to learning to bring the external anal sphincter under socially acceptable control. The anal-retentive or obsessive personality has passed long ago from the annals of medicine into popular literature and common usage. People are thus accustomed to the belief that thought, conscious or otherwise, can affect behaviors that they associate with the gut. The notion that thinking can alter what goes on in the bowel leads easily and comfortably to the belief that neurotic or psychotic thoughts in the head can raise enteric havoc. Psychosomatic disease of the gut, therefore, is a concept that has never encountered much difficulty in gaining acceptance.

Personal experience is another factor that has enhanced the idea that the central nervous system can be the cause of intestinal malfunction. Butterflies in the belly and diarrhea, for example, are frequent accompaniments to anxiety. Some people are so used to the attendant music from their gut that if it were missing, they would not even know that they were anxious. The enteric consequences of strong emotions are thus not esoteric or theoretical concepts to most people but real problems of everyday life. The brain thus seems like an obvious perpetrator of intestinal grief. The psychosomatic origin of enteric illness is, as one of my former teachers of mathematics used to say after filling a blackboard with obscure equations, "intuitively obvious, even to the most casual observer." Unfortunately, being intuitively obvious does not necessarily mean being correct. When behaviors of the bowel that are essentially under enteric control are disturbed, it is by no means clear that the output of the enteric nervous system is aberrant *because the brain has made it so.* Since the brain in the head affects the second brain, it is, of course, conceivable that a disturbed mind may transmit its problems to the enteric nervous system, thereby upsetting even the functions delegated to the second brain. There is nothing, however, to prevent the enteric nervous system itself from giving rise to enteric misbehavior, independently of any influence the second brain receives from the first.

The theory that the brain must be responsible for almost any malfunction of the bowel that cannot be explained by an anatomical lesion in the gut has attracted enthusiastic support. The alternative possibility—that an abnormality of the gut's own nervous system might be the cause of intestinal grief—has, until recently, not been seriously entertained. The reluctance of the medical community to accept the enteric nervous system itself as a cause of intestinal disease is, in part, due to the fact that the enteric nervous system normally functions as an éminence grise. A smoothly functioning enteric nervous system never reaches the threshold of perception. As long as it does its thing right, life is good; everything comes out well, and we neither know, nor generally care, why. Many physicians have thus found it unreasonable to blame enteric troubles that they cannot diagnose on the out-of-sight-out-of-mind enteric nervous system, even though its location is, with respect to intestinal difficulties, right at ground zero. In contrast to the enteric nervous system, the brain is anything but occult. It may not be located near the site where enteric misadventures are taking place, but the brain is an *éminence,* not an éminence grise. The brain, therefore, has seemed to some to be a more logical culprit than the enteric nervous system to blame for illnesses of the bowel that lack evident explanation. Where primitive peoples use a variety of gods to explain the inexplicable, modern humans use psychiatric illness. When all else fails, invoke a psychoneurosis.

Since hypotheses involving the psychosomatic cause of disease are hard to test, it is easy to attribute aberrant enteric function to unconscious thought processes. Unvented anger, for example, has often been invoked as a cause of intestinal malfunction. According to this view, intolerable feelings that cannot be expressed, or even consciously thought, become internalized and find expression in enteric havoc. Physicians are only now beginning, some reluctantly, to consider seriously the possibility that the brain has received over the years more blame than it has earned. When an activity that the brain has delegated to the enteric nervous system misfires, the fault may lie in the enteric nervous system, and not in the brain.

Pain and/or discomfort that a patient reports is coming from his/her abdomen may actually be the result of real pain and real discomfort triggered by undetected physical or chemical abnormalities in the patient's gut. This is a concept that seems less than revolutionary to me, but I have discovered that it is very revolutionary to others. Recently, I was discussing the second brain with a physician I met while on vacation. He was an infectious diseases specialist from Canada who was explaining to me why he had decided not to become a gastroenterologist. It turned out that while on a rotation in gastroenterology in medical school, he was called upon to see a great many patients who were suffering from functional bowel disease. He said that he "hated" these patients because they were so fixated on their intestines. Every little pain in the gut, he believed, was blown out of proportion by these neurotic, chronic complainers. Surely, he posited, gastroenterology is a field that attracts the most flagrant of psychoneurotic people. When I asked him how he knew the pain reported by his patients was "little," he had no answer. Instead, a look of surprise came over his face. To assume that a patient's pain is minor and psychogenic in origin in the absence of evidence to the contrary is both unjustified and unhelpful.

In dealing with other organs such as the heart, doctors believe patients when they report that they feel pain. It is not clear to me why the gut should be treated differently. If a lesion in the gut is obvious, doctors become sympathetic, grant a patient the right to be in pain, and seek to alleviate the problem. If no lesion can be detected, enteric pain tends to be belittled and blamed on neurotic thought processes. Many physicians still have yet to learn the lesson Hippocrates drew from epilepsy in his ancient book *On the Sacred Disease.* Hippocrates had no idea why patients with epilepsy had fits, but that made no difference to him. The cause, he was certain, would eventually become known. In the meantime, however, it was useless to attribute the condition to a supernatural origin. So too with pain from the gut. The fact that doctors do not yet always know why it occurs is no reason either to dismiss its severity or to attribute the pain to what substitutes in the modern world for a supernatural cause, psychoneurosis. It seems very likely that the enteric nervous system will not always operate to perfection. Almost no system of the body does so. I learned early in my career in medicine that when it comes to body parts, if they exist, there is sure to be a disease of them. In some person, at some time, under some circumstances, the part will not work right.

When the enteric nervous system is missing from a section of gut, either because a poor soul is born without it or because a disease destroys it, the function of the bowel comes to a virtual halt. Motility is critically compromised. The operative rule of the bowel is: no nerves, no transit. A segment of gut that lacks the ganglia of the enteric nervous system turns itself into an obstructive barrier. The absence of ganglia is as effective in blocking the propulsion of intestinal contents as a shoelace or a ligature knotted around the gut. When ganglia are congenitally missing or enteric neurons drop dead, there is no doubt that there is a lesion in the enteric nervous system. The wisdom of Solomon is not required to make a diagnosis; however, diagnostic difficulties do arise when the gut malfunctions in a setting of normal-looking enteric ganglia, muscle, and intestinal lining. Clearly, when this situation occurs, the éminence grise is not at work, or, if it is working, it has either itself been corrupted or it has lost control over its minions, muscles and glands.

Functional Bowel Disease

At any given time, about 20 percent of the American population is partially disabled, or at least made miserable, by a condition known as functional bowel disease. It is a disturbing and intractable medical problem that physicians still cannot properly define, let alone cure. The most common variants of this condition are *nonulcer dyspepsia* and, especially, the *irritable bowel syndrome*, sometimes known only by its initials, IBS, or by its epithet-like popular name, the *irritable* (or *spastic*) *colon.*

Most of the early studies of IBS suggested that it was predominantly a disease of the white middle class. It is likely, however, that this impression was gained because the subjects of the first studies of the prevalence of IBS were not drawn from a random sample of the entire population but mainly from middle-class whites, who happened to be the people most likely to be seeking treatment. More recent studies of Hispanic and black Americans, as well as of populations in Japan and China, also report prevalences of about 20 percent, which are very similar to those found in middle-class white America. The prevalence of IBS thus seems to be independent of race, although there is some evidence that it may be affected by socioeconomic class and gender. An Iranian clinic, which treats nomads and industrial laborers, has found that the prevalence of IBS in these people (3.5 percent) is much lower than that reported in middle-class groups. In Western societies, women are more likely to suffer from IBS than men, with a ratio of women to men of up to 2:1. Interestingly, this ratio is reversed in India, where men go to doctors more than women, and also report more IBS. The prevalence of IBS declines with age and, in a study carried out in Omstead County, Minnesota, the prevalence of IBS in people over age sixty-five was only about half that of people in the age thirty to sixty-four group. It is very rare for anyone to be diagnosed for the first time with IBS after the age of sixty. Not that there are not old people with IBS; there are. They just get the disease while young, keep it for years, and carry it with them into old age.

Surprisingly, only about 10 percent of people with IBS present themselves to doctors for medical care. The rest just suffer in silence, often because they are too embarrassed to discuss their illness or because they are not greeted warmly by the medical profession. Others have had the disease so long that they are unaware that life can be lived without it, or even that their symptoms are not present in everyone. Even though 90 percent of patients with IBS thus do not bother doctors about it, IBS alone accounts for about three and a half million visits to physicians in the United States every year. That number makes it the seventh leading diagnosis overall, and the single most common diagnosis reached by gastroenterologists. In fact, IBS accounts for about 25 percent of all patients that gastroenterologists get to see. If you assume an average cost of about $100 per visit, that number of visits works out to about $350 million. This amount of money, however large, is just for openers. The cost to society of functional bowel disease is actually much larger. First, the $350 million includes only visits for IBS, not other types of functional bowel disease. Second, since there is no diagnostic test that is specific for IBS or any other functional bowel disease, all sorts of other conditions have to be excluded before this diagnosis can be made. Other conditions, often life-threatening ones, share symptoms with IBS. That means that many expensive laboratory tests have to be conducted. Third, once diagnosed, a case of IBS that is serious enough to have been brought to a doctor will have to be treated, and since the treatment is only marginally effective and noncurative, it will have to be continued for years. Add to that the fact that patients with IBS have many times more absenteeism from work than the general population, and pretty soon you can apply to functional bowel disease what Everett McKinley Dirksen once said about the federal budget: "A billion here, a billion there, and pretty soon it's real money."

Functional bowel disease is what Winston Churchill called the Soviet Union in 1939: a riddle wrapped in a mystery inside an enigma. The layers of medical enigma that surround this condition are so thick that international conferences have to be held to establish criteria by which it can be recognized. These criteria are named either for the leader who proposed them or for the meeting at which they were promulgated. We thus have the "Manning criteria" and the "Rome criteria." Almost as soon as the various criteria are published, they are inevitably found wanting by at least some "experts" in the field, and the criteria instead of the illness become objects of research and contention. What the lack of definitional clarity illustrates very well is that even today, as these words are written, no one actually knows what functional bowel disease *is*. In fact, whether functional bowel disease is one disease or many has yet to be determined.

Functional bowel disease is diagnosed by the presence in a patient of a set of symptoms that may themselves be self-contradictory and do not all have to be noted simultaneously. Since, as I have said, the symptoms of functional bowel disease are the same as those that accompany many other conditions, those other conditions also have to be actively excluded before functional bowel disease can be diagnosed. These other conditions include carcinoma of the colon, Crohn's disease, ulcerative colitis (the latter two conditions can be grouped as inflammatory bowel disease, or IBD, as opposed to IBS), lactose intolerance, diverticulitis, and a menagerie of parasitic infestations. Adrian Manning's defined criteria for functional bowel disease included stools that are more frequent and looser at the start of episodes of abdominal pain, relief of pain after defecating, a sense of incomplete rectal evacuation, passage of mucus with the stool, and a sensation of abdominal bloating. The Rome criteria added the demand that abdominal pain and alteration of bowel habits be present all of the time and that the patients experience the other symptoms on Manning's list at least 25 percent of the time.

Clinicians who operate in the real world, where they encounter three-dimensional people instead of idealized representations, see many patients who do not meet the Rome criteria but who they nevertheless know have functional bowel disease. For example, patients who complain of chronic painless diarrhea in the absence of an identifiable cause do not meet the Rome criteria but can be considered, nevertheless, to be suffering from functional bowel disease. We thus await Rome II to fine-tune the diagnostic criteria published as a result of Rome I.

While there is still much confusion over definitions, criteria are nonetheless valuable. Their availability facilitates multicenter testing of the efficacy of new drugs as well as communication between investigators. Whatever functional bowel disease ultimately turns out to be will be known sooner rather than later if everyone speaks the same language.

One of the major difficulties that has precluded any effective investigation of the cause or causes of the complex of symptoms that occurs in functional bowel disease is the inadequacy of commonly used techniques for examining the human enteric nervous system. Diseases are recognized as distinct entities when symptoms are compared with the anatomical, and more recently, chemical findings turned up when diseased tissue specimens are examined. The idea is to identify the lesion that gives rise to a given set of functional deficits and thus to a symptom complex. Pathologists are the people who do this sort of thing for a living, and neuropathologists are the subspecialists who concentrate on the nervous system. Neuropathologists are very competent when they work with the brain, spinal cord, or ordinary peripheral nerves. Most neuropathologists, however, do not have a clue as to how to examine the enteric nervous system. The gross examination of the gut reveals nothing; therefore, to get at potential defects, it is necessary to examine the enteric nervous system microscopically. Microscopic examination, as noted previously, involves cutting tissue into very thin slices and staining the sections to add contrast. The critical defect in the investigation of the gut by neuropathologists is that they usually section the bowel in a manner that minimally reveals the layout of the enteric nervous system and they apply stains that do not yield any information about the shapes or identities of enteric nerve cells. In short, the immense strides that have been made in understanding the composition of the enteric nervous system by basic scientists have yet to make any impact on the neuropathological investigation of the bowel.

Worse, even if they knew what they were doing, pathologists would only rarely be able to obtain the tissue they need for the investigation of the enteric nervous system in functional bowel disease. Living patients are reluctant to part with enteric nervous system samples. While it is true that the mucosa of the bowel is easily biopsied by way of a flexible tube illuminated by fiber optics, mucosal biopsies do not include the full thickness of the wall of the gut. A biopsy that is deep enough to include ganglia of the myenteric plexus is one that risks perforating the bowel. Sometimes, rectal biopsies will be taken that look to see whether myenteric ganglia are present or to determine if they are abnormal, but the justification for obtaining this tissue has to be very persuasive. Functional bowel disease is not persuasive enough.

However troubling functional bowel disease may be, and the condition is very disturbing, it is not often fatal. Autopsies of patients who die while in the throes of functional bowel disease are, to say the least, rare. By the time a patient who has suffered from functional bowel disease dies of something else, that person's gut may be old and distorted by a variety of unrelated insults. For good reason, therefore, the hypothesis that the enteric nervous system is defective in patients with functional bowel disease has never been adequately tested by a proper search for associated defects in enteric nerve cells.

Years ago, Michael Schuffler, a young neuropathologist at the University of Washington in Seattle, recognized the inadequacy of existing enteric neuropathology and set out to do something about it. While the majority of pathologists followed a herd instinct and cut sections perpendicular to the length of the gut (cross sections), Schuffler sought either to cut sections through the plane of the myenteric plexus or to dissect the bowel into layers and examine the ganglia as whole mounts in the resulting thin wafers of tissue. The trouble with cross sections of the gut is that they do not reveal the pattern of enteric ganglia, an extensive chicken wire–like network of nerves and nerve cells lying between the muscle layers of the bowel. Cross-sections cut through the branches of the network so that only a few nerve cells and their fibers are present in any given section. The nerve cells, moreover, appear only in profile. Cross sections, therefore, which do not even contain entire nerve cells, reveal almost nothing of their actual shape. Imagine a fishing net stretched flat on the beach and a child standing over it with a sharp pane of glass. If the child were to pass the glass straight through the net into the underlying sand, the child would cut the net in a cross section. If the child were then to peer through the glass at the cut surface of the net, all that he/she could see would be a series of dots at the points where the pane of glass passed through the strands of the net. The resulting image would give the child no sense of what the net really looked like, or any idea of how the strands were knotted together at the sites where individual strands joined one another. The problem of the net and the pane of glass is exactly analogous to the difficulties faced by neuropathologists attempting to examine the enteric plexuses in cross sections of tissue. The task is virtually impossible. Schuffler's approach, therefore, avoided the limitations inherent in what almost everyone else was doing.

In addition to preparing the bowel in a manner that was different from that of most of his colleagues, Schuffler also stained tissue, not just with the usual dyes that added contrast to structures viewed through the microscope but with chemical solutions that deposited metallic silver on constituents of nerve cells. Silver staining, as used by Schuffler, was an antique method, refined very little from the early days of the twentieth century, when it was employed by the great neuroanatomist Ramon y Cajal to establish that the nervous system is composed of separate and distinct nerve cells and not, as suggested by the rival doctrine of Camillio Golgi, a fused reticulum of cells in actual continuity. Silver stains, we now realize, cause metallic silver to become deposited onto the filamentous skeleton of nerve cells. This toughness of this skeleton is derived from fibers that are known as *neurofilaments,* which form an internal array of cables that support the cells. By thus silver plating the neurofilamentous skeleton of nerve cells, silver stains enable the shapes of enteric nerve cells to be visualized and make it possible to trace their meandering processes for long distances though the maze of the myenteric plexus.

Armed with an approach that for the nervous system as a whole seemed primitive, but which for the enteric nervous system was so novel as to be revolutionary, Schuffler began to bring order out of the chaos of enteric nervous disease. Unfortunately, Schuffler was able only to begin this job. He was never able to complete it or really to advance the task of classifying the diseases of enteric nerve cells to a point where others could easily duplicate his results. Schuffler's research funding dried up, and the resulting shortage of money stopped him in his tracks. It is not totally clear why this happened to Schuffler, but a contributory factor appeared to be the lack of support he received from an unappreciative scientific community. Many scientists, at the time, looked down at what Schuffler was doing because they thought that his methods were not sufficiently "molecular." There is a tendency among some biologists to consider that the value of any scientific discovery is directly proportional to its closeness to the expression of genes. These people thus consider that work that identifies the genes responsible for a given disease is far more worthwhile than work that simply describes what is wrong in anatomical, chemical, or even functional terms.

Schuffler had still another problem. His work was not only insufficiently molecular, it was also descriptive. Schuffler did not frame hypotheses and test them with functional experiments, the approach that stirs the blood of grant reviewers. He simply took diseased tissue and described it, hoping that patterns would emerge to define specific illnesses. This kind of investigation, however necessary it might have been, lacked either the electricity of molecular biology or the suspense of a test of a high-flying hypothesis. In fact, in many scientific circles, particularly those that comprise the peer review panels that control the funds disbursed by the National Institutes of Health, the word "descriptive" is, when applied to the critique of a grant, a kiss of death. Scientific projects can be killed by stating that they are no good, poorly controlled, overly ambitious, or descriptive. Even today, the effect of judging work to be descriptive has the same effect on continued research funding that the thumbs-down sign given by the crowd in the Colosseum of ancient Rome had on the continued life and good health of defeated gladiators.

The rumor in the trade is that Schuffler gave up after failing to get his research grant renewed. I really do not know what ultimately became of

Michael Schuffler. He has vanished into the mists of clinical practice. I would like to think that he retired from research and went out to make a fortune as a doctor providing health care to a set of wealthy patients. In the absence of silver stains, there should at least be a silver lining. No matter, though, whether Schuffler's life turned out to be wonderful or full of pain, it remains clear that his departure from the field of enteric neuropathology left it bereft and devoid of absolutely necessary action. The vacuum created by the premature stifling of Schuffler's promise has created a void that still exists, and the nature of his fate has discouraged anyone else from stepping into this particular breach. There thus is no clear categorization of enteric nervous system disease, and, to my knowledge, no one is now trying to provide one.

The More You Know, the More You Know You Know Nothing

The enteric nervous system has now attracted a devoted and effective coterie of basic scientists who study it in very modern ways. The anatomy, chemistry, and function of the enteric nervous system are being probed superbly well, and a gratifyingly "molecular" understanding of its components has emerged. Clinicians, both gastroenterologists and gastrointestinal surgeons, are making real progress in coping with recognized disease entities. A new field, "neurogastroenterology," has been recognized, but it is a field that will continue to have a crippling gap in its knowledge base until someone again takes up the neuropathology standard that Michael Schuffler was forced to throw down.

The might-have-beens, in terms of opportunities missed, are accumulating and becoming more glaring. When Michael Schuffler was trying to assemble a logical classification of the diseases of the enteric nervous system, the basic information that was available to him was scanty and incomplete. In fact, it could be argued that Schuffler was ahead of his time and lacked the tools he needed to do what he proposed. Back in 1981, when Jackie Wood, Marcello Costa, Alan North, and I argued about serotonin as a neurotransmitter at the workshop in Cincinnati, the four of us may have accounted for a quarter of the field's scientists. Serotonin, which became accepted at that meeting, moreover was only the first of what has turned out to be a vast trove of novel transmitters present in the gut. Today, I could not hope to list all of the scientists who are contributing to neurogastroenterology because my memory is not large enough. As I now look out at my colleagues in the field, moreover, I see myself as an elder scientist, surrounded by the bright and energetic faces of youth. The preconceptions and absences of critical bits of information that held back my generation, and the generation that came before me, are no longer there. The effort I have to put out merely to keep up with the outpouring of new knowledge resulting from the work of the new young stars of neurogastroenterology leaves me with my tongue on the floor. If Michael Schuffler were now to resume his work where he left off, the molecular tools that his critics faulted him for not using would finally be there for him.

Progress

To be able to recognize a disease of the enteric nervous system, should one be present, it is necessary to be able to compare the inventory of nerve cells that one finds in the putatively diseased bowel to that of the normal gut. The appearance of various enteric nerve cells in the potentially abnormal bowel must also be compared with the appearance of their normal counterparts. If a given type of nerve cell should happen to be present in numbers that are too small or too large, or if their shapes are distorted and bizarre, then evidence of disease is present. This is the essence of neuropathology. We (by that I mean the entire field of neurogastroenterology, not just my laboratory) have now produced what appears to be a complete inventory of the type of nerve cells that normally comprise the enteric nervous system of the guinea pig gut. These cells are categorized by the unique combination of molecules, such as neurotransmitters or enzymes, that they contain. The shapes of the cells are also known from work done not just with silver stains but from elegant experiments in which individual nerve cells have been impaled and filled with opaque or fluorescent markers. These markers enable the entire injected cell, and all of its processes, to be visualized. Work with species other than the guinea pig has revealed that although there are some differences between various animals, the basic organization of the enteric nervous systems of all mammals is pretty similar.

Clearly, the information derived from the guinea pig gut cannot be applied to the human bowel without at least some modification; nevertheless, the general themes of enteric nerve cell organization that have emerged from animal studies should make it a great deal easier and faster to get a complete inventory of the types of nerve cells present in the human enteric nervous system than it was to get the equivalent data for the first time in guinea pigs. Once it is known which nerve cells belong in the normal human enteric nervous system, and what they look like, a proper examination of the diseased bowel will provide the neuropathological classification of diseases of the human enteric nervous system that Schuffler was trying to obtain. The identification of specific diseases is the first step that will inevitably lead to effective therapies. The idea of producing a complete inventory of nerve cells in the human bowel, as well as a list of the diseases to which they are prone, looks quite feasible today in comparison to the advances made in the field since its coming-out party at the meeting of the Society for Neuroscience in 1981.

Today, functional bowel disease is a complex of symptoms lacking a link to pathology. Tomorrow, I am sure that list will vanish and be replaced by an assortment of evident disease entities. Crohn's disease was once a form of functional bowel disease, but it was removed from this fraternity when its pathology was discovered. In just this way, individual diseases will be whittled out of what I think is clearly a large group of problems that simply give rise to similar symptoms.

Part III

THE ORIGIN OF THE SECOND BRAIN AND ITS DISORDERS

9

THE ENTERIC NERVOUS SYSTEM NOW

SINCE 1981, THE DYNAMIC Australian duo of John Furness and Marcello Costa has separated. I was sad when I heard of their separation, as I am whenever I learn that a married couple that I like has become divorced. I would no more have imagined this happening than a duel between Batman and Robin. Still, since their breakup, both John and Marcello each have found a younger colleague with whom to work. As a result, there are now two dynamic Australian duos, where once there was only one. A Melbourne team headed by John, and an Adelaide team led by Marcello, compete with one another in a more or less friendly way. These two Australian laboratories, together with their bands of fellow travelers, deserve the most credit for assembling the catalog of the enteric nerve cells of the guinea pig gut.

John Furness works with Joel Bornstein, a sharp-thinking, physiologically sophisticated young scientist. Between the two of them, John and Joel have assembled an aggressive group of scientists who pour out new discoveries at a rate that easily transcends that of the majority of other groups in the field. Marcello Costa has established a stable collaboration with Simon Brookes, a dashing English expatriate whose ordinary conversation comes across like a Mozart divertimento—elegant, to the point, and just a little bit archaic.

Current Frontiers

At the present time, John Furness, Joel Bornstein, and their group are concentrating on defining the properties of the subset of intrinsic sensory nerve cells that live in the myenteric plexus, and on working out the nature of the connections that identified nerve cells make with one another. Marcello Costa and Simon Brookes have focused their attention more on the output units, the excitatory and inhibitory motor nerve cells that talk directly to the smooth muscle. It is, of course, the muscle that is the effector that actually mixes the stuff in the gut and moves it toward the anus. To identify which of the many nerve cells of the mélange found in ganglia are motor, Simon has successfully adapted a technique called retrograde labeling (the method that takes advantage of the natural phenomenon of retrograde axonal transport, or the shipment of material from the end of an axon back up to a nerve cell body) that Gary Mawe, Annette Kirchgessner, and I first introduced to the study of the enteric nervous system.

My own experimental life has touched upon anatomy, as my collaborations with Gary Mawe and Annette Kirchgessner illustrate. Through the years I have investigated other things, such as the development of the enteric nervous system and, as a family project with my wife, the cellular biology of the virus that causes chicken pox and shingles; nevertheless, my original interest in serotonin has never flagged. Actually, our understanding of the role played by serotonin in gastrointestinal reflexes has increased greatly since agreement was reached in Cincinnati that serotonin is an enteric neurotransmitter. Even so, we still do not know everything serotonin does for (or to) the bowel.

The more we know about serotonin's role in the enteric nervous system, the closer we come to clinical applications for this knowledge. Manipulating the action of serotonin with drugs appears to be a very promising means of providing relief from the depredations of functional bowel disease. Several pharmaceutical companies are now exploring drugs that affect serotonin for this purpose and are even conducting clinical trials of some of them. As a result, publications describing the actions of serotonin and drugs that affect it have become a growth industry.

Back in 1982, Marcello Costa and John Furness published their papers confirming my suggestion that serotonin is an enteric neurotransmitter. Terri Branchek joined my laboratory at about the same time. Since then, we and others have determined that serotonin is not only a neurotransmitter but also a signaling molecule that is secreted by specialized cells (not nerve cells) of the gut's lining and works within the mucosa to stimulate the intrinsic sensory nerves that initiate peristaltic and secretory reflexes. Serotonin, moreover, even acts as a growth factor during fetal life.

The gut has now been found to contain at least seven different receptors that respond to serotonin. This molecular cornucopia could not have been anticipated, and since each of these receptors is a unique molecule with an action all its own, the multiplicity of enteric receptors for serotonin enables serotonin to evoke a bewildering variety of responses. I like to say that no pharmacologist ever went broke throwing serotonin on the gut. Whenever you apply serotonin to a piece of bowel, something happens. The intestinal music of serotonin has turned out to be Wagnerian in nature, filled with leitmotivs, counterpoint, and unexpected depth.

Together with Terri Branchek, I set out to identify the receptors that serotonin uses to act on enteric nerve cells. At about the same time, I also began a project with a graduate student, Steven Erde, to visualize the nerve cells that respond to applied serotonin in order to determine whether these cells are actually innervated by serotonin-containing nerve fibers. In effect, Steve and I wanted to put an exclamation point on my recent victory on the serotonin front in Cincinnati by getting a direct look at serotonin's neurotransmission unit. The two projects, receptors and neurotransmission, were, of course, related to one another.

Synapses Revisited

Steve Erde was an agreeable and confident young man. Recording from enteric nerve cells with sharp microelectrodes was, at that time, a relatively new technique that had been mastered by only a small number of investigators. For a student like Steve to say "OK. No sweat. Let's get it done . . ." struck me as extraordinary. The problem of impaling enteric nerve cells in my laboratory, however, could only partially be solved by Steve's courage, willingness, and self-confidence. I also needed to find some way to get Steve the requisite training.

Jack Wood came through for me. Since I had not myself recorded from enteric nerve cells, I thought that I should send Steve out to a laboratory to see how enteric recordings were obtained by a real professional of the field. Jack Wood was the nearest such person, although Reno, where Jack was then the chairman of the Department of Physiology at the medical school of the University of Nevada, did not exactly qualify as the laboratory next door, and Steve was the kind of person who rarely ventured west of the Hudson River. As it turned out, however, Steve adapted to the West with surprising ease. Even better, he learned the technique. By the time Steve returned home, Jack was asking me to send more such people, and Steve was impaling enteric nerve cells with ease.

Once home, Steve set up the apparatus he needed and introduced me to a new set of methods that I have relied upon ever since. His project also went along very well. He readily found nerve cells that behaved as if they were being driven by nervous inputs that used serotonin as a neurotransmitter. After characterizing the responses of the cells he had impaled, Steve injected them with horseradish peroxidase, an enzyme diverted from the fixings of roast beef, that scientists use to produce an opaque reaction product that is easily recognized when viewed by electron microscopy.

The contacts, or synapses, that nerve cells make with one another are very close. The synaptic gap between these cells is so small that it cannot even be seen as a space in a light microscope. The thin processes of nerve and supporting cells, moreover, are themselves beneath the level of resolution of the light microscope. At the light-microscopic level, therefore, it is easy to be fooled; many close contacts that appear to be synapses are not, and many scientists have come to grief by drawing conclusions based on what they thought they saw in a light microscope. Electron microscopy thus is required to really be sure that what seems to be a synapse actually is one.

Unfortunately, electron microscopy adds a quantum leap of difficulty to any experiment that employs it. Although electron microscopy provides an extremely close look at whatever it is used to look at, the field of view is very limited. A single ganglion, for example, assumes the breadth of a football field. The sections one needs to cut are also very thin, because an electron beam is not nearly as penetrant as light. Thousands of such sections are required to serially cut through a single nerve cell; therefore, section thickness is measured in billionths of meters. Once Steve injected a nerve cell, moreover, he had to cut sections through all of it so that he could examine each of the thousands of synapses on its surface.

At the same time that Steve looked at nerve endings on the cells that he injected with horseradish peroxidase, he also needed to be able to determine whether any of the many synapses he encountered contained serotonin. We decided to use radioautography to identify the synapses that used serotonin as their neurotransmitter. Since serotonin is turned off after it acts by the very specific reuptake of serotonin into the same nerve fiber that released it in the first place, it is possible to use radioactive serotonin to label serotonin-containing nerves. The radioactive serotonin is added in tiny amounts that mix with the natural transmitter. After the nerve cells take up the radioactive serotonin, they become radioactive and take their own picture (the radioautograph) when a photographic emulsion is layered over the tissue. This method is similar to the one that I used originally to find the serotonin-containing nerve cells of the gut (see chapter 1), except then I "fed" the nerves a radioactive precursor of serotonin (instead of serotonin), and let them make the radioactive serotonin themselves.

Steve's project thus was to record from nerve cells, identify those he thought were being driven by serotonin-containing nerves, and then inject the cells with horseradish peroxidase. The tissue was subsequently incubated with radioactive serotonin, fixed, and prepared for electron microscopy. An electron-dense reaction product was obtained from the horseradish peroxidase, the appropriately large numbers of ultrathin sections were cut, and each section was coated with photographic emulsion. The coated sections were then exposed for weeks to months, developed, and finally examined in an electron microscope.

All of this work took years to accomplish and would not have been done had not Steve been helped by Diane Sherman, a technician whom God had sent me many years previously as a gift for some good deed that I never knew I had done. With Diane's help, Steve found that every cell that behaved as though it received the input of serotonin indeed was actually covered by serotonin-containing synapses.

Steve set out to write his thesis but had an unexpected crisis along the way. A general rule of my laboratory is that data books are never to leave it. I am a bit paranoid when it comes to data, which I regard as more valuable than gems, gold, or money. Steve has many virtues, but respect for rules is not one of them. He was writing his thesis at home and wanted the data books handy for reference. He therefore put them into the trunk of his car and headed home for what turned out to be a disaster. The data books were stolen, along with his spare tire, jack, and other tools.

The theft of his data was almost a catastrophe of epic proportions. Fortunately, it was only an ordinary disaster. Steve's electrophysiological records were all on tape, and each of his anatomical preparations had been photographed. Since he had lost his index, however, Steve had to replay and relive all of his experiments. Fortunately, Steve did it and the observations were published.

Enteric Serotonin Receptors

While Steve Erde was demonstrating that serotonin is present in the synapses where we presumed it is working, Terri Branchek was succeeding in identifying the receptors upon which serotonin acts. Terri developed an assay that, for the first time, made it possible for us to characterize directly the kinds of serotonin receptor that are present in the gut. Before this was done, receptors in the bowel for neurotransmitters were studied like Sir Arthur Conan Doyle used Sherlock Holmes in his books to investigate crimes. Inference ("my dear Watson") was everything. Drugs were employed to alter responses of the gut to a neurotransmitter, and conclusions were drawn about the receptors that were responsible for the neurotransmitter's effects. These conclusions were based on the known (or presumed) actions of the drugs. Progress could be made by inference, but all such knowledge had to be governed by an uncertainty principle, like that of quantum mechanics. The presumed actions of drugs might not be correct, and as Sir Henry Dale pointed out, no drug has only one action. The effect of a drug that one knows about may not be the only one it exerts. It is, of course, the actions of drugs that one does *not* know about that are likely to be lethal to one's interpretations of data.

Serotonin receptors provide a good illustration of the limitations of inference. In 1957, J. H. Gaddum and Z. P. Picarelli reported that there are two receptors for serotonin in the gut. They called these receptors "M" and "D." The letters stood for the drugs *m*orphine and *d*ibenzyline, upon which their inferences were based. Gaddum and Picarelli defined "M" receptors as those that were responsible for mediating responses of the bowel to serotonin that were blocked by morphine, while "D" receptors were defined as those that mediated responses to serotonin that were abolished by dibenzyline. As it turned out, neither morphine nor dibenzyline does anything specific to serotonin receptors. In the context of their own paper on the subject, therefore, Gaddum and Picarelli were totally wrong in their view of serotonin receptors.

The data of Gaddum and Picarelli were fine. Their observations were accurate and well described. The system, however, was much more complicated than they imagined, so that the inferences they drew were based on erroneous presumptions. Gaddum and Picarelli were measuring the contraction of the gut. When they applied serotonin, which stimulates intestinal nerves, the nerves caused the bowel to contract by squirting acetylcholine at smooth muscle cells. Morphine interferes with the secretion of acetylcholine by those enteric nerves; therefore, after the application of morphine, the contraction of the bowel in response to serotonin is diminished in amplitude. After the nerves that serotonin stimulates have been exposed to morphine, they simply do not work well. Morphine thus does not affect the contraction of the gut evoked by serotonin because morphine affects serotonin receptors. Instead, morphine prevents the manifestation of receptor stimulation that Gaddum and Picarelli were investigating. The enteric nerves that serotonin continued to stimulate, despite the presence of morphine, were prevented by the morphine from making their activation evident. Morphine thus did not distinguish a specific subtype of serotonin receptor, as Gaddum and Picarelli thought.

Rather, morphine distinguished nerve-mediated effects of serotonin from those exerted directly on the smooth muscle cells themselves.

Many years after Gaddum and Picarelli published their work, I introduced the use of tetrodotoxin to produce deliberately exactly the same result that they accomplished without meaning to. Like morphine, tetrodotoxin inhibits the contraction of the guinea pig intestine in response to serotonin. Tetrodotoxin stops nerve conduction and thus shares with morphine (but via a different mechanism) the ability to decrease the release of acetylcholine by stimulated enteric nerves. In the sense that morphine is an antagonist of "M" receptors, therefore, so too is tetrodotoxin. I was, however, not using tetrodotoxin as a serotonin antagonist (which it is not) but as an agent that could be relied upon to distinguish the indirect nerve-mediated actions of drugs in innervated smooth muscle preparations from those that are direct effects of the drugs on the muscle itself. Morphine was thus no more specific an antagonist of the receptor-mediated actions of serotonin than tetrodotoxin. Similarly, dibenzyline, which is a highly reactive compound related to the mustard gas that was used in the First World War, inactivated almost all of the receptors on smooth muscle. After muscle cells are exposed to dibenzyline, serotonin is just one of many compounds to which the muscle cells no longer respond. Almost all of the muscle cells' receptors have been done in by dibenzyline.

Gaddum and Picarelli thus misinterpreted their data; nevertheless, their paper has become a classic in the field, because even though Gaddum and Picarelli were actually mistaken, they really were the first authors to show that the bowel contains more than a single kind of serotonin receptor. This bit of history thus illustrates not only the perils of inference but the importance of luck. Making the right statement, even for the wrong reasons, can be the key to long-lasting fame.

In place of inference, Terri Branchek used very highly radioactive serotonin to assay serotonin receptors. Neurotransmitters literally bind to their receptors in order to turn them on. Compounds that bind to receptors are called *ligands,* and those ligands that activate their receptors are, as we saw earlier, called agonists. Similarly, antagonists are also ligands that bind to receptors; however, antagonists bind to a receptor without causing the change in the receptor's chemical configuration that is necessary to turn it on. The binding of an antagonist to a receptor thus is silent, but by sticking to the receptor, the antagonist interferes with the binding of an agonist.

Antagonists come in two flavors, *competitive* and *noncompetitive. Competitive antagonists* bind reversibly to receptors and thus can be displaced from them by an agonist if the concentration of the agonist is high enough. If the antagonist is present in a concentration that is greater than that of the agonist, the antagonist wins; reverse the situation and the agonist wins. Competitive antagonism is surmountable. A *noncompetitive antagonist* cannot be displaced. No matter how much agonist is added, its effects are irreversible. After having been hit by an irreversible antagonist, the receptor is finished, and if all of the receptors on a cell are occupied by such a drug, the cell will not be able to respond to an agonist until it makes new receptors.

The quantities of ligands (agonists or antagonists) that bind to receptors are very small and usually cannot be detected by purely chemical means. When a highly radioactive ligand sticks to a receptor, however, the sensitivity by which radioactivity can be detected makes the binding both observable and measurable. By using a very "hot" ligand, therefore, it is possible to study not only how much ligand binds to a receptor but also how long that ligand takes to stick and how fast it comes off. It is also possible to detect directly the effects of agonists and antagonists on the binding of the ligand and, most importantly, to investigate the chemical properties of ligands that determine whether or not they will bind to their receptors. Binding assays with radioactive ligands thus permit receptors to be characterized without having to make inferences that depend on what one thinks one knows about the actions of drugs. In combination with radioautography, the binding of a radioactive ligand to its receptor can even be used to visualize receptors in the sites where they live.

Terri used her binding assay with radioactive serotonin to confirm that membranes isolated from enteric nerve cells do indeed contain serotonin receptors. That is, the membranes contained a limited number of sites that were able to bind serotonin. Once all of these sites were saturated with bound radioactive serotonin, no more could bind. The concentration at which serotonin saturated its binding sites, and the rate at which serotonin stuck to these sites and then came off again, indicated that the receptor to which serotonin was bound loved serotonin and had a very high affinity for it. The receptor, however, appeared to be unique in that serotonin could not be induced to relinquish its hold on the receptor by any of the drugs that were known to be agonists or antagonists at other serotonin receptors. In other words "classical" serotonin antagonists all failed to compete with serotonin for binding to this receptor.

The failure of known serotonin antagonists to block the binding of radioactive serotonin was, at first, discouraging. At the time Terri obtained these results, we did not even suspect that there are vast numbers of different serotonin receptors. If a drug was an antagonist, therefore, I expected that it should be able to compete with radioactive serotonin for binding to bona fide receptors. The fact that they did not do so made me wonder whether the binding of radioactive serotonin Terri had observed was an artifact. Fortunately, my discouragement was premature and stemmed from the same kind of mistake that Gaddum and Picarelli had made earlier. I drew erroneous inferences from what I thought I knew about drugs. I later learned that the "classical" serotonin antagonists that are effective in blocking the effects of serotonin elsewhere in the body (remember LSD and the rat uterus?) are impotent in blocking the physiological actions of serotonin on enteric nerve cells. As far as the enteric receptors for serotonin are concerned, therefore, the "classical" serotonin antagonists are not antagonists at all. Nature was playing a game with us. It was, as it so often is, being coy. The observations, both from receptor binding studies and physiology, coincided, and each was telling us that serotonin acted on a novel receptor that was present in the bowel, but not in any of the other sites where serotonin receptors had previously been found and where the "classical" serotonin antagonists had been defined. It just is not true that if you've seen one serotonin receptor, you've seen them all.

What we needed at that point was a new antagonist that would be our magic bullet, a compound that would see our enteric serotonin receptor but ignore others. The prospect of finding one was daunting. Pharmaceutical companies employ legions of organic chemists to produce new compounds that they tailor to meet the specifications of the pharmacologists they work with. If you have such a legion of chemical compatriots at your beck and call, you can act like a designer (on the basis of the structure-activity relationship that you first investigate), get a bunch of new drugs synthesized, and have a chance to find a new antagonist. If you are like me and have no chemists at your beck or call, making a new antagonist is a serious problem. I solved it the old-fashioned way. I got lucky.

Terri's binding studies had indicated to us some of the structural properties a chemical would need in order to bind to the novel enteric serotonin receptor that we had just discovered. While I fantasized a variety of approaches to drug companies that I dreamed might get them interested in making a set of new compounds for me to try, my colleague, Hadassah Tamir, called me for some advice about studies she was doing with a peculiar drug she had recently been asked to evaluate for a friend in Israel who made it. Naturally, I was both busy and distracted when she called, but when Hadassah calls, I forget whatever else may be on my mind because I know that I am going to do what she wants anyway. Hadassah fought in the Israeli war for independence, and as a person who once skitted around English soldiers with hand grenades concealed in her skirt, she is fazed by nothing. As it happened, the peculiar drug she was evaluating as a potential painkiller for reasons I no longer remember appeared to have all of the chemical properties I was looking for.

Jerusalem Juice

Because of its origins, I first called Hadassah's compound "Jerusalem juice," but after it began to work I decided to abandon the profanity and give it a proper name. Of course the drug was already known, at least to its makers, by a chemical term, but this term was too full of letters to be a viable option as a name that anyone could actually use. Except in a word game, no one can take seriously anything called *N-acetyl–5-hydroxytryptophyl–5-hydroxytryptophan amide.* We chose to go with initials, which were based on the chemical structure of the compound. The active drug molecule is essentially formed by sticking together two molecules, nose to nose, of serotonin's precursor, in a configuration called a *dipeptide.* The initials of serotonin's precursor (5-hydroxytryptophan) are 5-HTP, so we called the drug, 5-HTP-DP, which most people can remember. When they forget, they call it simply "the dipeptide," a usage that makes me think, whenever I hear it, of journalists who refer to Donald Trump as "the Donald."

The first clue that 5-HTP-DP was going to be interesting was Terri's observation that it competed nicely with radioactive serotonin for binding to the novel enteric serotonin receptor. The next clue was turned up by Miyako Takagi, a visiting whirlwind from Japan who was spending a year in my laboratory and had taken over the microelectrodes from Steve Erde. Miyako discovered that 5-HTP-DP blocked the action of serotonin on enteric nerve cells (this was the serotonin antagonist Jack Wood needed back in 1981). Moreover, best of all, 5-HTP-DP also blocked the response of enteric nerve cells to nerve stimulation that we thought was mediated by serotonin. We had thus discovered that 5-HTP-DP was a serotonin antagonist, but more than that, it was a unique antagonist that had the ability to bind to the novel 5-HT_{1P} subtype of serotonin receptor that we had found in the gut. Further studies, in my laboratory, and those of other investigators, revealed that 5-HTP-DP was quite specific, in that it confined its effects to the 5-HT_{1P} receptor and did not bother receptors for other neurotransmitters or even other serotonin receptors in sites outside the enteric nervous system.

Although it was specific, 5-HTP-DP was only effective when it was applied at high concentrations. It was also hard to make and available only in tiny quantities; therefore, 5-HTP-DP was not going to cure any plagues, and no drug company became interested in it. A few companies have synthesized it for research use, and Hadassah tries to give it out to scientists who need it, so that our work can be duplicated. Nevertheless, despite its limitations, 5-HTP-DP is still the only specific 5-HT_{1P} antagonist available, and its use has enabled that receptor to be operationally defined.

Their Loss Was Our Gain

We later obtained another set of drugs that we found were active on the 5-HT_{1P} receptor. This one came from a failed trial in Europe of a drug intended to be an antidepressant. This compound, called *indalpine,* is a member of a class of drugs, which includes Prozac, that inhibits the reuptake of serotonin by the nerve fibers that release it. These compounds clearly work in the brain to relieve depression, and except for people who make it their business to worry about the "side effects" of drugs, no one had thought very much about the actions of antidepressants outside of the brain.

In all its preclinical trials, indalpine had seemed to be a very good and specific type of antidepressant. It had passed its tests in animals with flying colors, and no important toxicity was revealed. The rats who had been given indalpine, moreover, had done all of the things expected of rats on a potent antidepressant. One cannot ask a rat if it is happy, but there are behaviors that correlate well with antidepressant efficacy, and the indalpine-treated rats had exhibited all of them. The first patients to receive indalpine, however, developed severe diarrhea that cut short the study of indalpine's efficacy in the treatment of depression.

That clinical failure would probably have ended matters, except that I found the structure of indalpine interesting. Indalpine itself lacked what I thought a compound would need to bind to the 5-HT_{1P} receptor; however, with the addition of a single chemical group, which would make it 5-hydroxyindalpine instead of indalpine, I thought it might stick nicely to the receptor. I decided to write to the company, Pharmuka (since acquired by Rhone-Poulenc), that made indalpine and ask if they might like to make some 5-hydroxyindalpine for the good of science. It turned out that they had already done so for metabolic testing and that they had 6-hydroxyindalpine as well. In fact, the scientists at Pharmuka were pretty sure that the problem of diarrhea in the clinical trials had come from the conversion of indalpine in the bodies of the patients who took it to 5-hydroxyindalpine. The villain of the story, as far as Pharmuka was concerned, was the very compound that I wanted to investigate. The folks at Pharmuka were happy to send us their stocks, which might well have now seemed worthless to them, and we acquired two bright new drugs to test. We applied the initials 5-OHIP to 5-hydroxyindalpine and 6-OHIP to 6-hydroxyindalpine and went to work.

Luck was still with us. Terri Branchek found that both 5- and 6-OHIP competed with radioactive serotonin for binding to the 5-HT_{1P} receptor. Pharmuka then got the European Atomic Energy Commission to make radioactive 5-OHIP for us. Sure enough, the radioactive 5-OHIP did indeed bind to the 5-HT_{1P} receptor, and it could be displaced from the receptor by serotonin and by 5-HTP-DP. These observations confirmed that all three of these compounds competed with one another to bind to the same receptor. 5-OHIP, however, turned out to be very different from 5-HTP-DP.

Gary Mawe had by then joined my laboratory, taking up the microelectrodes from Miyako Takagi. Gary soon discovered that, in contrast to 5-HTP-DP, which was a silent antagonist that caused nothing to happen to enteric nerve cells when it was applied to them all by itself, 5- and 6-OHIP mimicked the action of serotonin on the nerve cells to which they were applied. These compounds thus bound to the 5-HT_{1P} receptor because they were agonists, not antagonists. Gary's data also explained why the clinical trial of indalpine had failed. The 5-OHIP, made from indalpine by the patients who took the drug, had hit their bowels like a runaway freight train. By inadvertently making 5-OHIP, they had let loose a pure 5-HT_{1P} agonist, which, until they finally excreted it, had excited enteric nerve cells without respite in every part of their gut.

In thinking about the syndrome associated with the ingestion of indalpine and its metabolism to 5-OHIP, I was reminded of a nasty gastrointestinal cancer that secretes serotonin. This tumor is called a *malignant carcinoid*. It comes from the enterochromaffin, or EC, cells of the intestinal lining. These are the same cells that release serotonin in response to the prodding of cholera toxin and that are normally believed to be sensory receptors that detect pressure or mucosal distortion. EC cells are found in the stomach as well as the small and large intestines, and each one packs in a huge amount of serotonin. Collectively, the serotonin present in EC cells makes every other source of serotonin in the body pale. These are the cells that hold over 95 percent of the body's serotonin.

When they become cancerous, EC cells secrete serotonin without control and often for no apparent reason. If this aberrant secretion occurs within the confines of the bowel, nothing awful usually happens. The serotonin that gets into the blood flows to the liver, which wipes it out. The wall of the gut prevents mucosal serotonin from directly reaching the nerve cells in the myenteric plexus. On the other hand, when the tumor spreads (metastasizes) to the liver and sets up shop in that organ, the carcinoid tumor now secretes serotonin into the blood from a perch where the liver can no longer remove it. Through the blood, the metastatic tumor thus exposes the whole body, including the nerve cells of the bowel, to an excess of serotonin. Results are many, and all are bad. The right side of the heart (for reasons that are still unclear) becomes scarred, and patients flush and wheeze. Worse, the intestines are driven to do a version of the Indy 500. Motility becomes so rapid that the peristaltic waves are visible through the skin of the patent's abdomen. Patients become emaciated from malabsorption. The gut simply does not have time to digest and absorb what the patient eats. If the tumor or the serotonin is not stopped, starvation is the inevitable result. The serotonin, set loose on the 5-HT_{1P} receptors of the victims of metastatic carcinoid tumors, had its counterpart in the 5-OHIP set loose on the unwitting subjects of an ill-fated drug trial. Sometimes, in a clinical trial, it really is good to get the placebo.

Terri used the radioactive 5-OHIP that the Europeans had given her to locate the 5-HT_{1P} receptors by radioautography. She found them on nerve cells, especially those in the myenteric plexus of the gut, thereby confirming our physiological data as well as the results of her own binding studies. However, 5-HT_{1P} receptors were also located on what appeared to be mucosal nerve fibers just under the lining of the gut. The location of the 5-HT_{1P} receptors in the mucosa made me think that this receptor might be involved in enteric sensory reception as well as neurotransmission. I was to return to this theme a few years later.

There Is More to It

While investigating 5- and 6-OHIP, Gary Mawe observed that the action of serotonin was more complicated than previously believed. At least two components could be recognized within an excitatory response to serotonin. The first one was manifested immediately but was transient to the point of evanescence. This component of the response to serotonin was over and done with before the second component even got started. Jack Wood had made a similar observation, and (as you'll remember) he called the two components "fast" and "slow" responses, respectively, to keep them straight.

The electrical properties of fast and slow responses to serotonin, as well as their timing, were quite different from one another. During the fast response, the electrical conductance of the membrane of the responding nerve cell increased, suggesting that ion channels were opening and that ions, carrying current, were moving through the channels. In contrast, during the slow response, the conductance of the nerve cell membrane actually decreased, suggesting that ion channels were closing and that a current had been interrupted. At the time, we did not know what ions were moving or what channels were involved in each response; nevertheless, it was obvious that a single receptor could not evoke two such disparate responses. Both the time courses of the two components of the response to serotonin and their wildly different nature implied that they had to be the result of two different serotonin receptors.

Gary soon found that he could distinguish the fast and slow responses with the drugs that we had found were active at 5-HT_{1P} receptors. The slow response to serotonin was the most interesting to us because it was mimicked precisely by 6-OHIP and abolished by 5-HTP-DP. This meant that 5-HT_{1P} receptors are responsible for the slow response to serotonin. In contrast, the fast response was totally resistant to inhibition by 5-HTP-DP and was not elicited by 6-OHIP. The fast response, therefore, has nothing to do with 5-HT_{1P} receptors.

There are thus two completely different receptors on enteric nerve cells, and each responds to serotonin. One, the 5-HT_{1P} receptor, also binds 6-OHIP, which turns it on, and 5-HTP-DP, which keeps it turned off. The fast response is mediated by an entirely different receptor that also binds serotonin but has no affinity for either 6-OHIP or 5-HTP-DP. (5-OHIP was a little different from 6-OHIP in that it was more like serotonin itself and elicited both fast and slow responses.) Since we already had a 5-HT_{1P}, we decided to call the receptor for the fast response "5-HT_{2P}," following what were then the rules for naming serotonin receptors. We used the "P" in the nomenclature to indicate that the receptor was in the peripheral, rather than the central, nervous system. Unfortunately, the name 5-HT_{2P} was wrong. We did not realize at the time we discovered it that the 5-HT_{2P} receptor was going to be discovered again, independently, at a later time by a different group who would give it a new, and probably better, name that it would keep.

Two years after Gary, Terri, and I had published our work, a new set of compounds was introduced that blocked those serotonin-induced contractions of the gut that were nerve-mediated. We soon discovered that one of these new drugs, now called *tropisetron*, also antagonized the receptor we had called 5-HT_{2P}. Since the number of known serotonin receptors was then escalating without precedent in front of the eyes of an amazed scientific audience, a group of pharmacologists, including those who helped to introduce tropisetron, self-assembled to publish an orderly serotonin receptor classification scheme. They decided to call the receptor that was blocked by tropisetron and related antagonists "5-HT_3." That name is now official, and 5-HT_{2P} is forgotten ancient history.

At first glance, it may seem odd that serotonin should need two different receptor molecules, which act in completely different ways to excite enteric nerve cells. Besides, since the 5-HT_3 receptor is primarily responsible for the nerve-induced contraction that is elicited by adding serotonin to an isolated segment of gut, what is left for the 5-HT_{1P} receptor to do? In thinking about these issues, you should recall that nerves do not operate at all like endocrine glands. They do not spew out a neurotransmitter and have it work on every receptor that is within reach. Instead, nerves very precisely deliver their neurotransmitter to particular synapses at specific spots on the surface of nerve cells, where accumulations of appropriate receptors are found.

Adding serotonin to an organ bath that contains a segment of gut cannot be compared to stimulating serotonin-containing nerves. The effects are not the same at all. Serotonin in an organ bath is an artificial situation in which the neurotransmitter gets to act as if it were an endocrine hormone. The serotonin is allowed to slosh about and bind to whatever receptor it finds. Under these conditions, the serotonin is very liable to stimulate receptors that would never, during life in the animal, receive serotonin from a self-respecting serotonin-containing nerve. For example, the final motor nerve cells that make the intestinal muscles contract happen to have 5-HT_3 receptors on them; thus, the addition of serotonin to an organ bath excites these cells, whether or not they are actually innervated by serotonin-containing nerves. The resulting muscle contraction is entirely 5-HT_3-mediated and is completely blocked by 5-HT_3 antagonists. Since the nerves that talk to the muscle have been directly stimulated in this example, there is no way to know what else might have happened to other nerve cells located upstream from the motor nerve cells. Stimulating the output cells short-circuits the enteric nervous system and masks what is happening within enteric ganglia.

During life, in the gut of a human or an animal, serotonin does not slosh around. To know what serotonin actually does for the bowel, it is necessary to learn what happens when serotonin is placed specifically on the sites that normally receive their serotonin from serotonin-containing nerves or other cells that contain serotonin. Recordings from single nerve cells impaled with a microelectrode come much closer to this ideal than gross recording of muscle contractions. Our work, and that of our Australian colleagues, has shown that serotonin-containing nerve cells are interneurons, which means that their function is to both listen and talk only to other nerve cells. Putting serotonin in an organ bath is to override and thus ignore the interneurons.

The Most Important Enteric Serotonin Receptor

When serotonin-containing nerve cells are stimulated, the excitatory response they provoke can be seen (by recording with microelectrodes) to be identical to the slow response to applied serotonin. Most of the time, the serotonin released by serotonin-containing nerves does not, like the exogenous serotonin applied by an experimenter, evoke a fast response. You'll recall that the fast response is mediated by 5-HT_3 receptors, while 5-HT_{1P} receptors are responsible for the slow response. In addition, the effects of the messages from serotonin-containing nerves are blocked by 5-HTP-DP (the 5-HT_{1P} antagonist). This provides more evidence that the primary receptor responsible for serotonin's ability to act as an enteric neurotransmitter is not 5-HT_3 but 5-HT_{1P}. The 5-HT_3 receptor is important to the bowel, but its importance is subtle (and will be discussed later). Neurotransmission mediated by 5-HT_3 receptors, if it occurs, is a very hard event to find. At best, it is rare. The 5-HT_{1P} receptor is therefore the most important subtype of serotonin receptor in the enteric nervous system.

Unfortunately, the 5-HT_{1P} receptor is also one of the few serotonin receptors for which the sequence of the genetic code remains unknown. I have tried to determine it, but my skills in molecular biology have not yet proven to be up to that task. Five years ago, my colleague at Columbia, Richard Axel, who is to molecular biology what Beethoven is to music, told me he could clone the DNA that encodes the 5-HT_{1P} receptor in about a month. He thus predicted that it would take me about four months. I proposed to try to clone the receptor in a grant application that I submitted to the National Institutes of Health. The study section liked the rest of my application, which was funded, but they told me to forget about the part in which I proposed to determine the genetic sequence that encodes the 5-HT_{1P} receptor. Even though I had taken and successfully completed a course at the biological laboratories at Cold Spring Harbor in "Cloning Neural Genes" (which I took to avoid passing like a dinosaur into scientific oblivion), the study section did not take seriously my credentials in molecular biology. I cannot complain about that, but it is a shame that the genetic code for this important receptor still has not been found.

The Antibody to the Antibody That Finds Serotonin Receptors

More recently, Hadassah Tamir produced an unusual sort of antibody that recognizes serotonin receptors. To obtain this antibody, Hadassah could not simply take the straightforward approach and immunize an animal with purified receptors. Since there have now been over fifteen different serotonin receptors that have been cloned and sequenced at the molecular level, it would have been an overwhelming job for Hadassah to purify each one and then try to raise over fifteen different antibodies. Instead, she decided to make a single antibody that would bind to virtually every kind of serotonin receptor.

To make her super probe, Hadassah first immunized a rabbit against serotonin itself. (The process she used, which involved coupling serotonin to a protein carrier, already has been described in an earlier chapter.) As expected, the rabbit obliged by making very good antibodies that could be used to seek out serotonin in any of its hiding places in tissues. These antibodies, however, were for Hadassah only an intermediate step in obtaining the probe she really wanted. She took the antiserotonin antibodies that her rabbit had just produced and used them as an antigen to immunize a second set of rabbits. The goal of the reprise was to have the second rabbits raise antibodies to the first antibody. Specifically, she wanted an anti-antibody that was able to recognize the particular spot on the antibody to serotonin that did the serotonin binding.

In a very real sense, the site where an antibody combines with a molecule like serotonin functions like the ligand binding site of a serotonin receptor, because serotonin sticks specifically to each. Both molecules—the serotonin receptor and the antibody to serotonin—thus have serotonin-binding domains. There are only a limited number of molecular configurations that make serotonin binding possible. An antibody to a serotonin-binding domain is thus able to recognize the serotonin-binding domains of many different molecules, including those of serotonin receptors. If you recognize one such domain with an appropriate antibody, you can spot many. When raised successfully, these receptor-recognizing probes are called *anti-idiotypic antibodies.*

The production of anti-idiotypic antibodies is a little like finding the people who really know the news in official Washington. First, you invent a story and publish it on the Internet. That is analogous to raising the first antibody. The story on the Internet is then taken up and reported as a leak or a rumor by the so-called "mainstream" or "responsible" press (if that is not an oxymoron). This is analogous to raising anti-idiotypic antibodies. The dissemination of the leak/rumor in the media brings out all kinds of governmental folks who are in the know and who divulge the real news. They can be thought of as the receptors who stick to the mainstream press like glue. Hadassah had raised an anti-idiotypic antibody using antibodies to serotonin as a starting material, and she gave us the opportunity to test her novel re-agent to get a firsthand look at the serotonin receptors in the bowel.

When Hadassah gave me her anti-idiotypic antibody to examine, Gary Mawe had left my laboratory to pursue the gall bladder on his own in Vermont. Terri Branchek had also left and was leading a group at a biotechnical company that was seeking (with considerable success) to decipher the genetic codes of all of the many varieties of serotonin receptor. Upon leaving, Gary passed the laboratory's microelectrodes on to Paul Wade, who took them up to pursue Hadassah's anti-idiotypic antibodies. Paul was a frail young man who quickly established himself as the person everyone turned to whenever a piece of apparatus failed to function as expected. He fixed everything and also set up everyone's computer. Soon after Paul arrived, I found myself relying on him to keep the laboratory running smoothly. Paul worked slowly and deliberately. He was also distracted a great deal by the crowds who sought his help. By the time his data finally emerged, however, it was always superb. When Paul spoke, people listened.

Paul found that Hadassah's anti-idiotypic antibodies did surprising things. When he squirted some onto an enteric nerve cell, he was shocked to find that the initial effect of the anti-idiotypic antibodies was to produce a response that was just like that elicited by serotonin. Paul, who was by nature a skeptic, really expected the antibodies to do nothing, but if they did anything, he thought that they might antagonize responses to serotonin. Instead, he found that the anti-idiotypic antibodies evoked an excitatory response that had fast and slow components, just as did the response to serotonin. Other antibodies, raised against irrelevant molecules, did nothing. The fast component of the response to the anti-idiotypic antibodies, like the fast response to serotonin, was inhibited by tropisetron and was thus due to activation of 5-HT_3 receptors. The slow component was inhibited by 5-HTP-DP and thus was due to stimulation of 5-HT_{1P} receptors.

In retrospect, we should not have been surprised by what the anti-idiotypic antibodies did. Since the anti-idiotypic antibodies stick, just like serotonin, to the serotonin-binding domains of serotonin receptors, there is no reason why the initial binding of the antibodies should strike the receptors as different from that of serotonin. Tropisetron and 5-HTP-DP were able to protect, respectively, the 5-HT_3 and 5-HT_{1P} receptors from the anti-idiotypic antibodies, because these antagonists occupied the serotonin-binding domains of the receptors and prevented the antibodies from gaining access to them.

The anti-idiotypic antibodies did not act as agonists for long. The initial excitement they caused wore off very quickly, and once it did, the receptors were dead to the world for hours. The anti-idiotypic antibodies had treated the receptors like roach motels: Once they checked in, they never checked out. The anti-idiotypic antibodies attached themselves to the serotonin receptors and stayed put. As the receptors desensitized, the effect of the bound anti-idiotypic antibodies changed from that of an agonist to that of an antagonist. This antagonist, moreover, was essentially irreversible. Once the receptors had bound the anti-idiotypic antibodies and desensitized, enteric nerve cells not only failed to respond to the serotonin that Paul squirted at them, they also failed to respond to stimulation of serotonin-containing nerves. The anti-idiotypic antibodies had thus marshaled the exquisite specificity of the immune system to confirm that serotonin is indeed an enteric neurotransmitter.

Having demonstrated that the anti-idiotypic antibodies bound to serotonin receptors in tissue, Paul set out to use them to locate the receptors in preparations that he could examine microscopically. Finding the bound antibodies was not a problem. The techniques of immunocytochemistry were designed for just this sort of thing and were easily adapted to looking for the sites to which the anti-idiotypic antibodies were attached. What was a problem was the fact that the antibodies bound to many different types of serotonin receptors. Paul thus had too much of a good thing. If his probe was going to light up every serotonin receptor under the sun, how was he to know which serotonin receptor was which?

The solution was suggested by Paul's physiological experiments. The antagonists had protected specific subtypes of serotonin receptor from the anti-idiotypic antibodies. As a result, Paul could use tropisetron and 5-HTP-DP to find 5-HT_3 and 5-HT_{1P} receptors. Sites that bound anti-idiotypic antibodies in the absence, but not the presence, of tropisetron were 5-HT_3 receptors, while those that bound anti-idiotypic antibodies in the absence, but not the presence, of 5-HTP-DP were 5-HT_{1P} receptors. Both types of receptor were observed only on nerves and nerve cells, and a great deal of the localization of the two receptors appeared to be in the same place. The location of the 5-HT_{1P} receptors was also identical to the location that Terri Branchek had previously found by radioautography for sites that bound radioactive serotonin with high affinity. Again, there was a localization of binding sites on mucosal nerves that was intriguing.

Receptor Overload

Serotonin receptors in the gut have been a little like ants at a picnic. First you have one, and then before you know it, you are swimming in them. After we had learned to accept that there are two receptors for serotonin on enteric nerve cells, and one more on muscle, we opened our eyes and there were seven. The 5-HT_{1A} receptor was the next to be found. Like the 5-HT_3 and the 5-HT_{1P} receptors, the 5-HT_{1A} is located on nerves, but unlike the other two, the 5-HT_{1A} receptor is inhibitory. Since some enteric nerve cells express all three of these receptors, the electrical record that one obtains from such a nerve cell after it has been challenged with serotonin looks like the old "Cyclone," the roller coaster at Coney Island. First, there is an excitatory hill (5-HT_3), then a deep inhibitory dip (5-HT_{1A}), and finally a long slow ascent and even slower decline to the finish (5-HT_{1P}).

After the 5-HT_{1A}, the 5-HT_4 receptor joined the group. This receptor was at first mysterious (and perverse) to people who impale nerve cells. They knew the 5-HT_4 receptor was present, but no matter how hard they looked, it was impossible to stick a microelectrode into a nerve cell and find one that responded to serotonin in a way that could conceivably have been mediated by a 5-HT_4 receptor. On the other hand, the people who crudely dump serotonin and other drugs onto segments of intestine in an organ bath had no trouble at all in finding 5-HT_4 receptors. When they electrically stimulated enteric nerves in the gut, they found that the addition of serotonin would cause the amplitude of muscle contraction to increase. Pharmacological studies showed that the 5-HT_4 receptor was responsible for this response.

The first drugs used by the 5-HT_4 receptor's discoverer, Joel Bockaert, to define it were relatively nonspecific. For example, the key drug that made the initial identification of the 5-HT_4 receptor possible was the 5-HT_3 antagonist, tropisetron, which was employed at a concentration that was at least tenfold higher than that needed to block 5-HT_3 receptors. Since its initial discovery, the 5-HT_4 receptor has come clean. Both specific agonists and antagonists have now been developed, and the receptor's genetic code has been deciphered. How the receptor works in the bowel is also now known.

The nerves that make intestinal muscle contract do so by releasing acetylcholine. To potentiate smooth muscle contraction, therefore, 5-HT_4 receptors increase the amount of acetylcholine that comes out of stimulated motor nerve fibers. More recently, the 5-HT_4 receptor has also been found to be able to increase the amount of acetylcholine coming out of the acetylcholine-containing nerves that excite cells in enteric ganglia. The 5-HT_4 receptor thus appears to be restricted to nerve endings, which explains why it does not participate in mediating the effects of serotonin on nerve cell bodies.

The other serotonin receptors that have been found in the gut have the names 5-HT_{2A}, 5-HT_{2B}, and 5-HT_7. It used to be thought that the two enteric members of the 5-HT_2 family were located only on muscle, but recently they have been discovered to be expressed also by nerve cells. The 5-HT_7 resembles the 5-HT_{1A} in what it does. At the moment, there are more serotonin receptors in the gut than there are known functions of serotonin. Clearly, we still have no idea of what most of the enteric receptors for serotonin are doing. This is a nice situation to a scientist like me. There is still lots more to learn, which is why we took up the trade in the first place.

Edith Bülbring and the Peristaltic Reflex

The location of serotonin receptors in the mucosa caught my attention each time we found it. Now, after Paul had turned it up again, I decided to try to find out what serotonin receptors were doing in the mucosa. Actually, I had what I thought was a pretty good idea about why the mucosa should have serotonin receptors. Way back, between 1957 and 1959, Edith Bülbring, my old sponsor at Oxford, had published a series of papers suggesting that serotonin is responsible for initiating the peristaltic reflex. You may recall that it was these papers that attracted me to Oxford to work with Edith. Edith's experiments had made a notable impression on many people at the time she published them, but then, as so often happens, the revisionists got to Edith's hypothesis, and her work was largely forgotten. It was not, however, forgotten by me, and it is suddenly au courant once again.

Edith was the first to propose that the EC cells of the intestinal mucosa function as pressure receptors. Her idea was that if you squeeze them, they squirt serotonin. The serotonin that the EC cells secrete was, Edith believed, what stimulates the intrinsic sensory nerve cells of the gut that start the peristaltic reflex. Back in 1965, when I went to Oxford, I thought that the evidence Edith had produced in favor of her hypothesis was overwhelming and totally convincing. Now that I can look back at it from the vantage point of a modern perspective, I still do.

Working with an isolated piece of guinea pig small intestine maintained in an organ bath, Edith showed that if she asphyxiated the mucosa or anesthetized it, the peristaltic reflex could not be evoked. These simple observations showed that the mucosa was critical for the manifestation of a pressure-induced peristaltic response. Edith then demonstrated that serotonin would stimulate the peristaltic reflex if she put it into the lumen of the isolated segment of intestine. Again, she could block the response to luminal serotonin by asphyxiating or anesthetizing the mucosa.

In contrast to serotonin's effect after mucosal application, Edith found that if she added the serotonin to the outside of the bowel, the peristaltic reflex would be inhibited. These experiments established that serotonin, which works on the inside but not the outside, must stimulate something in the mucosa to trigger the peristaltic reflex, and that luminal serotonin does not get to the same sites in the gut as serotonin applied from the outside. Edith went on to show that pressure on the mucosa does, in fact, release mucosal serotonin, and that if she desensitized serotonin receptors in the bowel with an excess of serotonin, the peristaltic reflex was antagonized. Finally, anatomical observations that Edith made with Graeme Schofield, an Australian microscopist, suggested that there are intrinsic sensory nerve cells in the submucosal plexus. Edith proposed that mucosal serotonin stimulated these sensory nerve cells, and that the sensory nerves relayed the message to the myenteric plexus, from where the muscle was controlled.

Almost no good scientific theory ever gets accepted until it weathers the attacks of revisionists who try to disprove it. Edith's ideas were challenged on the basis of experiments carried out with animals that were maintained for relatively long periods of time on a tryptophan-deficient diet. I remember being horrified by these experiments when I read about them because they seemed cruel. Tryptophan is an essential amino acid, and when it is deficient in the diet, an animal is unable to make protein adequately. The effects of the tryptophan-deficient diet on an animal, therefore, are multiple and all bad. I assumed that the affected animals would at the very least be uncomfortable, and there is no justification for that. The rationale behind the experiments was to exploit the fact that dietary tryptophan is the ultimate precursor of serotonin. If there is no tryptophan, an animal cannot make serotonin. The gut's stockpile of serotonin, therefore, was expected to become exhausted in the tryptophan-deficient animals because they could not replenish it. The revisionists thought Edith's hypothesis would be tested by the tryptophan-deficient diet because in their simplistic view, Edith's hypothesis predicted that an animal that had no serotonin in its gut would be incapable of exhibiting a peristaltic reflex.

When the revisionists showed that they could evoke a peristaltic reflex in the bowel of tryptophan-deficient animals, Edith graciously gave up. She abandoned her hypothesis that serotonin triggered the peristaltic reflex and postulated instead that serotonin might be a nonessential "modulator" of the reflex, whatever that meant. In fact, Edith surrendered when she did not have to, and really, she should not have given up at all. There were many reasons why the attack of the revisionists on Edith's work was faulty; unjustified cruelty to animals was only one of them.

The tryptophan-deficient diet had reduced the concentration of serotonin in the bowel, but it did not drive the enteric serotonin level to zero. Serotonin was still there, albeit in markedly reduced quantities. When animals (or people) cannot make new proteins from the amino acids they eat, they break down proteins they already have. In essence, animals cannibalize disposable bits of themselves in order to preserve what they cannot do without. Recycling is a concept that was discovered by evolution and applied to animals long before it became popular among conservationists. Protein breakdown might thus have provided the tryptophan-deficient animals with just enough tryptophan to keep up a minimal level of enteric serotonin. As long as some serotonin is present in the gut, it is possible that the small amount of remaining serotonin can fulfill its function. No one knows how much serotonin is needed to initiate peristaltic reflexes or whether the huge amount that is normally present in the bowel includes a reserve. Tryptophan deficiency is a widespread problem in Central and South America because corn (maize), which is the staple of the diet of the poor in these areas, happens to contain very little tryptophan. Tryptophan-poor diets, therefore, may very well have plagued not only our ancestors but those of other mammals as well. As a result, the gut may have been prepared by evolution to cope with a tryptophan deficiency and thus be endowed by nature to foil just this type of experiment.

In addition, the hypothesis that serotonin is *a* substance that triggers the peristaltic reflex does not necessarily mean that it is *the only* substance that can do so. By eliminating the possibility that serotonin may be one of many substances that can be recruited to get the reflex going, the revisionists had set up a straw man that would be easy to tear down. The destruction of a simplistic version of a sophisticated concept, however, does not invalidate the concept. The studies with tryptophan-deficient animals were not really quantitative. They were interpreted on an all-or-nothing basis; for serotonin to be considered important, the reflex would have to completely disappear from the intestines of the tryptophan-deficient animals. Since the reflex could be evoked, even if only with difficulty, and not in every tryptophan-deficient animal, serotonin was concluded to be unimportant. The possibility was never considered by the revisionists that serotonin might be the normal initiator of the peristaltic reflex, but that if serotonin were rendered unable to perform, another substance might step into the breach and take over the job. In the absence of serotonin, for example, more powerful stimuli might be required to elicit a peristaltic reflex, because the greater intensity is required to release a backup molecule. The body has many such fail-safe mechanisms. Actually, it would seem highly improbable that evolution would have devised only a single means of initiating something as basic as the peristaltic reflex.

EC Cells and the Peristaltic Reflex Revisited (Ecclesiastes Returns)

Many years after I returned from Oxford, I set out to test Edith's hypothesis once more, this time with modern methods. The first trials were carried out in collaboration with Annette Kirchgessner. We used newly developed histological probes to identify the nerve cells in the gut that became active after we had stimulated the bowel in a variety of ways. One of our methods was to assess the level of an enzyme, *cytochrome oxidase,* that is involved in fulfilling the energy needs of activated nerve cells. The other, described previously, was to detect the transcription and translation of the *c-fos* gene, which turns on when nerve cells are activated. We applied a stimulus to the intestinal mucosa, which would be adequate to evoke the peristaltic reflex. Instead of examining the reflex itself, however, we looked quantitatively at the nerve cells that our stimuli had induced to become active.

Sure enough, as we expected, stimulation of the mucosa caused nerve cells to become active in both the submucosal and myenteric plexuses, just as Edith Bülbring's conclusions predicted they would. Moreover, when Annette blocked synaptic transmission between nerve cells within ganglia, she restricted the number of activated nerve cells to a small set in the submucosal plexus very close to the site where we applied the stimulus. Once we prevented spread from cell to cell, only the intrinsic sensory cells themselves detected the stimulus. Annette had thus identified intrinsic sensory nerve cells of the bowel, and they were located right where Edith thought they would be, in the submucosal plexus. This was the first time intrinsic sensory nerve cells of the gut were ever directly visualized.

Annette went on to apply a tool that Edith Bülbring had lacked: 5-HTP-DP. In doing so, Annette was taking advantage of the many advances in knowledge about serotonin receptors that had transpired since Edith did her work. This knowledge provided Annette with an abundance of serotonin antagonists that were not available to Edith. In particular, the 5-HTP-DP was a gift from Hadassah Tamir, and its use was made interpretable by the observations that Terri Branchek, Miyako Takagi, Gary Mawe, and I had made on the 5-HT_{1P} receptor. Annette found that when she applied stimuli to the mucosa, 5-HTP-DP totally blocked the activation of enteric nerve cells. A few years later, Annette repeated her experiments, but this time she used even newer fluorescent probes that permitted her to identify activated neurons in living preparations. She again confirmed that 5-HTP-DP would inhibit the activation of intrinsic sensory nerve cells, but now she also found that more intensive stimulation would activate sensory nerve cells in the myenteric as well as the submucosal plexus. Annette and I concluded that mucosal stimulation did indeed release serotonin and that the serotonin was responsible for turning on intrinsic sensory nerve cells. The Ecclesiastes principle had applied itself once more. Yet again, there was nothing new under the sun.

After Annette and I described her observations, other scientists confirmed and extended them. The existence of intrinsic sensory nerve cells in the enteric nervous system is not now a concept that anyone would challenge. John Furness, Joel Bornstein, and Wolf Kunze, a young colleague who works with them in Australia, have found another set of intrinsic sensory nerve cells in the myenteric plexus, and they are devoting most of their waking hours to studying them. Wolf Kunze has actually been able to catch intrinsic sensory nerve cells *in flagrante* by impaling them with microelectrodes and recording their electrical responses to various stimuli.

Jack Grider, a young investigator in Virginia (not to be confused with Jack Wood), has also studied intrinsic sensory nerve cells, but he has done so indirectly, mainly by measuring the release of their neurotransmitter and by drawing inferences from the effects of drugs. Jack Grider's experiments are immensely clever. He does complicated studies, but his data are always perfect and support his ideas with great precision. No eventuality is left unaccounted for, and he manages to minimize the random spread of experimental values that plagues the rest of us. Jack Grider has confirmed that serotonin is released by mechanical stimulation of the intestinal mucosa, that serotonin does indeed initiate the peristaltic reflex, and that serotonin works by stimulating the mucosal processes of submucosal intrinsic sensory nerve cells. Everyone thus agrees with the basics, and Edith Bülbring is very much back in vogue. It would, however, be impossible to imagine a world populated by investigators who did not find something to disagree about. Unfortunately, this is another rule of science that is followed by those of us who work on the gut.

Jack Grider has taken issue with some of the conclusions reached by John Furness and his colleagues. Jack believes that the intrinsic sensory cells of the gut respond only to mild mucosal stimuli of the kind used by Annette and me, and not to distention of the wall of the gut, a stimulus that has frequently been employed by the Furness group. He has shown that the response of the gut to distention of the bowel disappears if he cuts the extrinsic nerves to the gut. This observation suggests that extrinsic sensory nerves, rather than the intrinsic sensory nerve cells of the enteric nervous system (as presumed by the Australians), may mediate responses to distention. Jack thus postulates that the gut has two ways of responding to sensory stimuli. Mucosal stimulation is perceived differently from distention of the bowel. He may well be right, but the response of the gut to distention strikes me as a bit academic. I would think that such a stimulus would rarely, if ever, be experienced in the life of a healthy animal or person. Only an obstructed or inflamed gut would be likely to become distended in a way that mimics the stimuli applied by the various investigators.

Jack Grider and I have also had our own differences of opinion. Naturally, I think he is wrong and that my data proves it. Jack basically agrees with Annette and me that serotonin stimulates sensory nerve cells to start the peristaltic reflex. However, we part company on our identification of the receptor that is responsible. Our basic observations are the same, but Jack, who does not impale nerve cells, believes that the 5-HT_{1P} and the 5-HT_4 receptors are identical, and that they are just two names applied to same molecule.

Jack Grider formed his impressions of the 5-HT_4 receptor from studies he carried out on receptors he discovered on single isolated smooth muscle cells rather than on receptors which are expressed by nerves. He has found that 5-HTP-DP, as well as specific 5-HT_4 antagonists, can inhibit contractions of the muscle cells stimulated by the 5-HT_4 receptor. Since Jack has not himself studied nerve cells, he has not been impressed by the observations by many investigators that you can virtually pack an enteric nerve cell in crystals of any one of a number of 5-HT_4 antagonists without affecting the slow (5-HT_{1P}-mediated) response to serotonin. Actually, 5-HT_4 antagonists do not affect any of the components of the response of any enteric nerve cells to serotonin. In addition, it is impossible to mimic the slow response to serotonin with a 5-HT_4 agonist, or to affect the high affinity binding of radioactive serotonin with a 5-HT_4 agonist or antagonist. All of these things are characteristics of the 5-HT_{1P} receptor. Personally, I think this information is conclusive, and that it proves that the 5-HT_{1P} and the 5-HT_4 receptors are different, but then, who would respect me if I did not believe my own data?

Mucosal Signaling

All the agreement about how serotonin starts peristaltic reflexes has been very gratifying. I like it much better than the derision that met my earlier proposal that serotonin might be an enteric neurotransmitter. Recently, moreover, serotonin's ability to initiate peristaltic reflexes has been a model that investigators are finding is applicable to other reflexes as well. For example, Helen Cooke, at Ohio State, has found that stroking the intestinal mucosa stimulates mucosal secretion, and that this phenomenon is a nervous reflex that involves stimulation of intrinsic sensory nerve cells in the submucosal plexus. The sensory nerve cells relay the signal to secretomotor nerve cells, and these in turn make cells in the lining of intestinal crypts secrete chloride and water into the intestinal lumen.

In addition, Helen has found that the secretory reflex, just like the peristaltic, is blocked by 5-HTP-DP and thus utilizes 5-HT_{1P} receptors. The transmitter of the secretomotor nerves is either acetylcholine or VIP (remember vasoactive intestinal peptide?), but not serotonin, and 5-HTP-DP thus does not affect the ability of secretomotor nerves to talk to crypt cells. Also, 5-HTP-DP has no effect on synaptic transmission within enteric ganglia. The secretory reflex, therefore, is abolished by 5-HTP-DP because stroking the mucosa releases serotonin, which stimulates 5-HT_{1P} receptors on intrinsic sensory nerves.

Information regarding a very important set of experiments that adds considerably to the significance of the mucosal action of serotonin has recently been published. The principal author of this work, David Grundy, is a taciturn Briton who has produced it in the grimy Midlands of England. David is an unusual person whose quiet demeanor and mournful expression mask a fine sense of humor and love of an uproarious good time. He has worked for years at the University of Sheffield, a town known more for its cutlery than for its science or the advancement of learning. Unfortunately, the way David Grundy has been valued by his university suggests that Sheffield will continue to be best known for its cutlery. David is now moving to Germany, where research and scientific progress are held in higher regard. This is a direction of immigration that I would never have thought possible, but it is a route that has been opened by years of Thatcherite nonsupport of biomedical science in the United Kingdom and one that Tony Blair has not yet reversed. I guess that the emigration of British brain power to Germany shows that World War II is really over.

David's new studies demonstrate that the mucosal release of serotonin is important not only in signaling within the bowel to initiate peristaltic and secretory reflexes but also in sending messages from the gut to the brain. David records from sensory nerve fibers in the vagus nerves that he identifies as carrying signal traffic from the bowel. He can detect the messages from the gut, which are minor electrical disturbances, as they are carried by the nerves past his recording electrodes. Sophisticated computer assistance allows him to distinguish signals emanating from individual axons within a nerve bundle. David's work clearly indicates that the secretion of serotonin within the gut activates sensory nerve fibers in the vagus nerves and that the serotonin does so by stimulating 5-HT_3 receptors. A single stimulus that causes serotonin to be secreted by EC cells, therefore, can send two very different messages to two very different recipients by acting on nerves that carry distinct receptors. Messages destined for internal, enteric consumption are registered by 5-HT_{1P} receptors, while those destined for the brain are relayed by 5-HT_3 receptors.

The Paradox and Therapeutic Benefit of the 5-HT_3 Receptor

The information content of the messages sent by the bowel to the brain is not entirely known. It may, however, include the kind of input that initially makes one queasy, goes on to be nauseating, and finally is downright sickening. This type of signaling probably once had survival value, and in some cases probably still does. A bowel that is ravaged by inflammation is one that is best kept free of food. Animals or humans turned green by their gut will not eat much. On the other hand, modern medicine has developed a number of counterproductive ways of getting intestinal serotonin to stimulate extrinsic sensory nerves enough to make a great many patients very sick. Cancer chemotherapy and radiation are two examples. The nausea and vomiting induced by these modalities of treatment can be so severe that continued therapy may become intolerable. One patient whom I heard about at a meeting of the FDA became so sick from one of these drugs that she began to vomit whenever she arrived at the hospital parking lot on her way to be treated. Obviously she needed an improvement in her cancer therapy.

Fortunately, it is possible to interfere with the ability of the gut to send upsetting messages back to the brain without interfering with the peristaltic or secretory reflexes. This is the beauty of having different receptors for serotonin on different nerves—and, of course, of recognizing and learning about them. Administration of a 5-HT_3 antagonist to a patient undergoing cancer chemotherapy or radiation can block the associated nausea and allow the treatment to continue. Several such drugs, such as *ondansetron* and *granisetron,* have successfully undergone clinical trials and have been approved by the FDA for this use. The 5-HT_3 antagonists are safe because the 5-HT_3 receptors do not do anything that is critical in regulating the normal motility or secretory activity of the gut.

New and more potent 5-HT_3 antagonists, such as *alosetron* (by Glaxo) have recently been developed and are undergoing clinical trials for the treatment of functional bowel disease, especially for its most common form, the irritable bowel syndrome (IBS). In an animal or a person, 5-HT_3 antagonists do not appear to exert major effects on normal intestinal motility. As we have just seen, some cancer patients have taken them for several years to combat the nausea they would otherwise suffer during treatment, and they have not experienced limiting gastrointestinal side effects. The 5-HT_3 antagonists thus are quite safe. On the other hand, 5-HT_3 receptors may play a role in sophisticated or abnormal forms of gastrointestinal motility that are not well understood, and they may facilitate the perception of intestinal pain. If so, then 5-HT_3 antagonists might exert actions that are not easy to document when everything is normal but that might nevertheless be beneficial when the gut is distinctly not normal, as in patients who suffer from IBS. There is some justification from work in animals for the belief that 5-HT_3 receptors do play a role in intestinal motility, although exactly what that role may be is not easy to pin down.

My own experience with 5-HT_3 antagonists illustrates the complexity of these drugs. I had, of course, found that they block the fast response to exogenous serotonin when the serotonin is squirted onto an enteric nerve cell. On the other hand, I have never seen a serotonin-containing enteric nerve evoke a fast response, nor have I seen its effect modified by a 5-HT_3 antagonist. As we saw previously, this is a bit of a paradox that can be explained by assuming that there are no 5-HT_3 receptors under the terminals of the serotonin-containing nerves that we stimulate. That explains our observations, but it raises the interesting question of why the bowel should have gone to the trouble of evolving 5-HT_3 receptors only to put them where serotonin cannot get at them. Evolution (or the Eternal, if you prefer) is not a practical joker. I think we can thus discount the possibility that the 5-HT_3 receptors on enteric nerve cells are there for sport or decoration. My working assumption, therefore, is that there are serotonin-containing nerves that we have not yet stimulated, and serotonin-mediated effects that we have not yet encountered.

Jim Galligan, a scientist at Michigan State University, has found some responses to nerve stimulation that are probably 5-HT_3 effects, but he has not found enough of them to carry out a proper study. David Grundy's work has established that 5-HT_3 receptors are critical for sending signals, perhaps of enteric distress, back to the brain, but those 5-HT_3 receptors are located on extrinsic sensory nerve fibers. The unexplained issue is what the 5-HT_3 receptors on intrinsic nerve cells are doing and whether turning them off is a good thing to do for a person with a functional bowel disease.

To investigate the potential effects of receptors that I could not find by using electrophysiological methods, I first duplicated the kind of research that the drug companies do to characterize their compounds.

That is, I applied serotonin and blocked it with a 5-HT_3 antagonist. Not that I doubted what I read in the literature, but I wanted to see, firsthand, what the mass of the pharmacological world was looking at. This kind of experiment left me with the same impression that drug company scientists have of 5-HT_3 antagonists; they are potent agents that virtually wipe out the enteric effects of exogenous serotonin. In contrast, when I looked not at *exogenous* serotonin poured out of a bottle onto a segment of gut strung up in an organ bath but at *endogenous* serotonin at work, starting reflexes and acting as a neurotransmitter, the 5-HT_3 antagonists seemed to do nothing at all. The paradox was back. I realized that we needed to investigate the peristaltic reflex in more detail.

A "Clockwork Colon"

A few years ago, Paul Wade and I devised a simple little preparation of guinea pig terminal colon that would allow us to evaluate the peristaltic reflex in this region of the bowel. If one is, as we were at the time, interested in an irritable colon, this would seem to be the right part of the gut to study. By the time the guinea pig's stool reaches the terminal colon, the fecal material has been concentrated into very hard little round pellets, not unlike the stool of a person with severe constipation. Paul isolated the guinea pig's terminal colon, maintained it in an organ bath, and allowed it to clear out its fecal pellets at whatever rate it chose. He then inserted an artificial fecal pellet, which he had fashioned out of plastic (and, as a nice touch, colored brown), into the open oral end of the colon. The pellet was transported by the activity of the bowel's musculature to the anal end, where it was expelled. Paul then retrieved the pellet and recycled it by putting it back in the oral end of the piece of gut. As soon as the bowel sensed the reinsertion of the pellet, it transported the pellet again, right back to the anal opening where it was again expelled. What was incredible about the transport of artificial pellets down the isolated gut was its reliability. The pellets moved at an absolutely constant rate that stayed the same through repeated trials for hours and hours. The preparation reminded me of a small machine in a painting by Fernand Léger, a biological equivalent of industrialization, the "clockwork colon."

In characterizing how the isolated colon performed, Paul found that the transport of pellets down the bowel was nerve-mediated, and thus stopped immediately when Paul paralyzed the nerves in the gut with tetrodotoxin. Paul could also stop the pellet from moving by adding 5-HTP-DP, especially when he put that 5-HT_{1P} antagonist into the lumen. On the basis of these and other experiments, Paul concluded that an artificial pellet moved down the bowel because it exerted pressure on the lining of the gut; the pressure applied by the pellet to EC cells released serotonin; the serotonin secreted by EC cells stimulated 5-HT_{1P} receptors on intrinsic sensory nerves, and these nerves triggered the peristaltic reflex that transported the artificial fecal pellets. We thus had developed our own little peristaltic reflex to study. Eschewing overly scientific nomenclature, we called the preparation the IVD, or in vitro dump.

After setting up the IVD preparation, we were joined by Makoto Kadowaki, a scientist from Fujisawa, a Japanese drug Company. Makoto came to learn how to impale and record from enteric nerve cells with microelectrodes, but he could not resist experimenting with the IVD preparation. It is amazing to watch the pellet be whipped along inside the terminal colon. Fujisawa had produced a drug, which we knew only by a numerical code, that was believed to be a combined antagonist at both 5-HT_3 and 5-HT_4 receptors. The company thought that this drug was going to work wonders in the treatment of IBS, and it was reputed to be undergoing clinical trials for this purpose in Japan. To learn more about their drug, Fujisawa contracted with me to study it.

Makoto applied three different specific 5-HT_3 antagonists to the isolated guinea pig colon and was surprised to learn that none affected the transport of the artificial pellet in any way. The pellet neither slowed down nor speeded up; nothing at all seemed to happen. Makoto then did the same with three different 5-HT_4 antagonists and obtained the very same result. Nothing changed in the rate of pellet movement. Finally, when Makoto either added the Fujisawa compound or administered a 5-HT_3 and a 5-HT_4 antagonist together, the pellet stopped dead in its tracks. When the drugs were washed away, the pellet was again transported down the bowel.

Makoto's work suggested that both 5-HT_3 and 5-HT_4 receptors really do participate in intestinal motility, but that one can compensate for the loss of the other. We thought that the receptors in the bowel might be arranged in a manner that is analogous to the way in which electric lightbulbs are wired in houses. The electrical term is that bulbs are connected "in parallel," as opposed to "in series." This means that the current does not have to flow through one bulb to get to the next. If lightbulbs were to be connected in series, the whole circuit would be interrupted and every bulb in the house would go dark as soon as any one bulb was unscrewed. Similarly, if the 5-HT_3 and the 5-HT_4 receptors of the bowel had been inserted in series in the circuit that mediates the peristaltic reflex, then the reflex would have been blocked when either the 5-HT_3 or the 5-HT_4 receptor was antagonized. Since nothing happened until both receptors were simultaneously inhibited, there must be parallel circuits, one with a critical 5-HT_3, and the other with a critical 5-HT_4 receptor. Knock out one circuit and the other suffices. To abolish the peristaltic reflex, you have to take out both circuits at the same time.

Makoto had thus shown that both the 5-HT_3 and the 5-HT_4 receptors do indeed participate in reflex-driven intestinal motility, but he needed to go to some lengths to show it. The potential of this type of effect for therapeutic use is that blocking either the 5-HT_3 or the 5-HT_4 receptor might be able to reduce the irritability of the gut in patients where that is abnormal. Conceivably, the increased motility (or, in some patients, spasticity of the gut) arises because of activity traveling in one of the redundant circuits. Cutting this down might be very beneficial. Since the 5-HT_3 and the 5-HT_4 receptors are involved in parallel circuits, nothing devastating will happen when either is inhibited. In contrast, shutting them both down or administering a blocking dose of a 5-HT_{1P} antagonist would be expected to eliminate the peristaltic reflex entirely. As a therapeutic modality, even in the treatment of IBS where diarrhea predominates, this would be a counterproductive form of overkill.

There is a hidden moral to the story of the interest drug companies are directing at 5-HT_3 and, more recently, 5-HT_4 antagonists. It means that people in the position to do something about functional bowel disease have begun to take seriously the possibility that the problems some individuals have *with* their gut are the result of problems *in* their gut. The facile assumption that if these people could only learn to think straight, then their intestinal difficulties would disappear, is yielding to the desire to find a treatment that really works. Drug companies have a limited tolerance for theories that do not produce safe and efficacious drugs. The potential market, represented by the millions of people with functional bowel disease who need treatment, is a reality they see. They also see the persistent failure of "psychological improvement" and tranquilizers to affect the course of functional bowel disease. They are therefore doing what is reasonable. Since fixing the brain in the head has not worked, they are having a go at fixing the brain in the bowel. Moreover, as new developments occur in understanding the enteric nervous system, they are taking advantage of them, and because of their ability to generate novel and interesting new compounds, the drug companies are actually playing a leading role.

The 5-HT_3 and 5-HT_4 antagonists are likely to be the tip of an iceberg. They are the visible drugs of the present. Underneath is a much larger set of compounds represented by drugs developed to affect mood. Because no one considered the enteric nervous system when these products were developed, the gastrointestinal actions they packed were always encountered as a nasty surprise. But now that the pharmaceutical industry has discovered that there is a second brain, they can look again at their list of gastrointestinal surprises and see if they can find some that they can reclassify from nasty to nice (that is, effective against functional bowel disease). After all, if profit can be found in altering the mood of the brain in the head by drugs, perhaps there is also pay dirt in doing the same for the brain in the gut.

Turning Off

If serotonin is, as everyone's data now seems to indicate, an important signaling molecule in the intestinal mucosa that initiates intrinsic enteric reflexes and sends messages up to the brain, the body must have a way to turn the serotonin off once it has done its job. As we have already seen, serotonin in the enteric and central nervous systems is turned off by the reuptake of serotonin by the serotonin-containing nerves that release it. This mechanism is a conservationist's dream. Recycling, not profligate dumping, is used. The mechanism works very well in the ganglia of the enteric nervous system, which has many serotonin-containing nerve fibers, and in fact is the process that allows serotonin-containing nerves to be identified by radioautography in tissue that has been incubated with radioactive serotonin. There are, however, no serotonin-containing nerve fibers in the mucosa of the gut. The sensory nerves, intrinsic and extrinsic, are affected by serotonin because they contain serotonin receptors, but they do not take up serotonin. I therefore began to worry about our embrace of Edith Bülbring's hypothesis and set out to find a mechanism for inactivating serotonin released in the mucosa.

While I worried about the turn-off of mucosal serotonin, Beth Hoffman at the National Institutes of Health and Randy Blakely, then at Emory University in Atlanta (now at Vanderbilt), independently cloned and determined the genetic sequence of the *serotonin transporter,* the molecule in the plasma membrane of serotonin-containing nerve cells that is actually responsible for serotonin reuptake. The serotonin transporter is an extremely important molecule because it plays the pivotal role in serotonin inactivation. It is also the target of the most effective antidepressant compounds ever devised, which makes it *numero uno* among the many molecules under pharmacological attack in the brain. The most famous of the drugs that inhibit the serotonin transporter is Prozac (fluoxetine), the first of a series of *serotonin-selective reuptake inhibitors,* the so-called *SSRIs,* to be cleared by the FDA for therapeutic use.

Since reuptake is the mechanism by which nerves turn off the action of serotonin, it follows that these nerves do not turn off serotonin very well when serotonin's reuptake is blocked. The initial effect of this failure to prevent the inactivation of serotonin is that the action of the neurotransmitter is potentiated. Responses to stimulation of serotonin-containing nerve cells become larger in amplitude and longer in duration. Subsequent compensatory effects that follow from the initial actions of SSRIs are very complicated. Long-term treatment with an SSRI causes some serotonin receptors to desensitize and fail to respond anymore, while others simply become less sensitive to stimulation by serotonin. In addition, nerve cells in individuals treated with an SSRI may make less serotonin. Because of all these complications, which can be grouped as various kinds of *down-regulation*, no one yet knows exactly why SSRIs, including Prozac, relieve depression. They may act by making brain serotonin more effective, but they could equally well act by causing brain serotonin to become less effective (through down-regulation). How the SSRIs thus accomplish their relief from depression is hard to say and very confusing; nevertheless, their effectiveness in providing this relief is unchallenged.

While the effects of SSRIs on mood can be uplifting and wonderful, their effect on the gut can be a real downer. When a person first begins taking them, SSRIs are very likely to cause nausea and even vomiting. The SSRI-induced intestinal disturbance can then progress on to diarrhea and finally to constipation. It seems as if the SSRIs cause the poor bowel to writhe, then churn, and finally to freeze up. With luck, a persistent patient can wait out the intestinal revolt against the SSRI and then go on to see his/her depression lift. Others find it difficult to get beyond the gastrointestinal "side effects" of the SSRIs, which are a depressing denouement to taking an antidepressant. Considerable attention has thus been paid by chemists, without notable success, to trying to separate the gastrointestinal "side effects" from the antidepressant "direct effects" of SSRIs.

The untoward gastrointestinal "side effects" of SSRIs are not rare and may interrupt, limit, or delay treatment in up to 25 percent of the patients who take them. The frequent occurrence of these "side effects" suggested to me that the serotonin transporter may do *something* very important for the bowel and that the gastrointestinal "side effects" of Prozac and the other SSRIs might be due to their interference with whatever that *something* might be. In other words, "side effects" could be a misnomer. There might be nothing "side" about the gastrointestinal effects of SSRIs. The bad things that go on in the gut of a person taking one of these compounds might well be due to direct effects (the inhibition of gastrointestinal serotonin uptake). If they are, it would be distressing to the manufacturers of

SSRIs, who would prefer to wish away the gastrointestinal effects of SSRIs; nevertheless, the knowledge that it is futile to try to get rid of the gastrointestinal actions of SSRIs would at least save the industry a lot of money, because the effort to separate the antidepressant from the gastrointestinal effects of the SSRIs could be abandoned. Speculation is fun, but I needed to do a little research to find out what the serotonin transporter was doing for the gut. I suspected that in so doing I might learn something interesting about the way serotonin is inactivated in the mucosa.

Randy Blakely and Beth Hoffman each were kind enough to send us molecular probes that we could use to look for the serotonin transporter in the gut. Our first interest was to learn whether the same molecule is expressed in the brain and the bowel. The probes, which allowed us to detect the RNA message encoding the serotonin transporter, revealed that this message is present in the gut and that the message in the bowel is absolutely identical to that in the brain. The next issue was to find out which cells of the gut express the serotonin transporter. We expected that we would find it in the serotonin-containing nerve cells of the myenteric plexus, but if that was all we were going to find, the information would tell us nothing about how the serotonin signaling is turned off in the intestinal mucosa.

We employed a technique called *in situ hybridization* to find the cells that contain the RNA message for synthesizing the serotonin transporter. To use this technique, sections or whole mounts of tissue are first prepared as if they were going to be stained or examined with a labeled antibody. Instead of a stain or an antibody, however, a chemically or radioactively labeled DNA or RNA probe, which is complementary to the RNA message in cells, is applied to the tissue. Each of the four bases that make up the genetic code has a complementary base to which it binds. As a result, complementary strands of DNA and/or RNA stick to one another. This sticking is called *hybridization.* The transcription of DNA, for example, results in the production of a complementary strand of RNA. The complementary RNA will bind to (*hybridize* with) its corresponding DNA.

A labeled DNA or RNA probe can hybridize with either RNA or DNA that has been extracted from cells and presented as a blot on a supporting material. A blot that contains DNA is called a *Southern* blot, while one that contains RNA is called a *Northern* blot. *Western* blots are analogous to their Southern and Northern counterparts, but they contain proteins instead of DNA or RNA, and they are probed with antibodies.

In situ hybridization is a procedure that resembles Northern or Southern blots in principle. The difference is that probes are applied directly to cells or parts of cells instead of to blots on supporting material. We had used a Northern blot to discover that RNA transcripts encoding the serotonin transporter are present in the bowel. We would now use in situ hybridization to see where these transcripts are made.

Paul Wade was working with me on this project, but he needed help. Paul was a fine histologist and great impaler of nerve cells, but molecular biology was all new to him. Paul reminded me of a story attributed to David Tucker, the son of the great tenor Richard Tucker. David was a medical student. When asked to comment on David's prospects as a vocalist, Richard Tucker was said to have advised him to study hard in medical school. If Paul had not been able to get help from a real molecular biologist, I would have advised him to stick to his microelectrodes. As it happened, life was good and help was available.

The help was the result of a friendship I had developed with a colleague at Columbia, Elvin Kabat. Elvin had been one of the truly great figures of immunology, but he was getting ready to retire. Through the operation of chance, I happened to be looking for a person with training in immunology and molecular biology just when Elvin was seeking to find a suitable laboratory for a scientist, Jingxian Chen, whom Elvin had trained and who now needed to move on. Elvin called me, said that Jingxian was good, and recommended that I hire him. I trust Elvin like I trust my rabbi, and did so immediately. Elvin was right, as expected, and Jingxian, who is very talented, has become a close friend and vital collaborator. Paul and Jingxian cooperated, and Jingxian provided the expertise in molecular biology that Paul lacked.

Paul and Jingxian worked with rats because the probes that Randy Blakely and Beth Hoffman had sent were rat probes that did not hybridize well with guinea pig RNA. Paul and Jingxian received a surprise when they employed in situ hybridization to look for the cells in the gut that make the serotonin transporter. RNA encoding the serotonin transporter was indeed present—not only in the myenteric nerve cells, where they expected to find it, but also in cells that lined intestinal crypts.

Randy Blakely then sent us an antibody to the rat serotonin transporter that he had raised. The antibody was important to us because we needed to know whether the lining cells, which do not produce serotonin, actually translate the message they contain and make the serotonin transporter itself. Immunocytochemical studies with the antibodies soon revealed that the cells that line the crypts do indeed translate the message and express the protein. Surprisingly, they also demonstrated that the particular lining cells that express the serotonin transporter are not the EC cells, which make serotonin, but their neighbors.

Experiments with radioactive serotonin showed that the serotonin transporter is every bit as functional in the cells that line intestinal crypts as it is in nerve cells of the brain and gut. As one would anticipate, the expression of the serotonin transporter endows the intestinal lining cells with the ability to take up serotonin. Moreover, when Paul and Jingxian examined the uptake of serotonin by the lining cells in detail, they found that its properties are identical in every way to those exhibited by the nerve cells. The most important feature of this identity is that the uptake of serotonin by the cells that line intestinal crypts is blocked by Prozac and other SSRIs.

The ability of SSRIs to block the uptake of serotonin by intestinal lining cells suggested two very interesting hypotheses to me. One was that the uptake of serotonin by the cells that line intestinal crypts might well be the mechanism by which mucosal serotonin is inactivated. The other was that the unwanted gastrointestinal effects of SSRIs might occur because these drugs interfere with the way serotonin is turned off in the mucosa. Both of these ideas could be tested. A small problem in carrying out the tests, however, was that all of the physiological information about serotonin as a mucosal signaling molecule had been obtained from studying guinea pigs, while the probes for the serotonin transporter only recognized the molecule in rats. Our choice was either to redo a couple of decades' worth of physiology in rats or to obtain a probe that would be useful in guinea pigs, which would require cloning and sequencing the guinea pig serotonin transporter. To Jingxian, that was no choice at all.

It took Jingxian only a few months to clone and sequence the guinea pig serotonin transporter. His work revealed that the molecular difference between the guinea pig serotonin transporter and those of rats, mice, and humans is very slight. Evolution has been quite conservative in shaping this molecule. Most importantly, from our point of view, Jingxian now had a probe that could be used to find the serotonin transporter in the guinea pig. The main question was whether it would again be found in the cells that line the gut. In fact, it was. In this regard, the guinea pig's intestinal lining even surpassed that of the rat. The guinea pig serotonin transporter is not confined to crypt cells as it is in the rat. In the guinea pig, the serotonin transporter is expressed by intestinal lining cells from the bottoms of crypts right up to the tips of villi. Since the serotonin transporter was thus shown to be present in the lining of the guinea pig bowel, we could begin to investigate the role it plays in signaling with serotonin.

Putting the Gut on Prozac

We carried out two types of experiment. In the first, Paul Wade investigated the effects of Prozac on the peristaltic reflex in the terminal colon. He used the isolated (IVD) preparation that he had devised for this purpose. Prozac first caused the transport of the artificial fecal pellet down the isolated colon to speed up. No other drug Paul tried could do this. However, as the concentration of Prozac was raised, the pellet slowed down, and eventually it stopped moving altogether. When the pellet stopped, the gut still responded to acetylcholine and electrical stimulation. Only its ability to respond to serotonin was lost. What had evidently happened was that by blocking the inactivation of serotonin (which is usually accomplished by the uptake of serotonin into intestinal lining cells), Prozac first potentiated the action of serotonin. The gut could not turn off the serotonin it secreted, so the effect of the stimulation of the mucosa by the pellet was greater than before the bowel was exposed to Prozac. The serotonin receptors, however, eventually found themselves awash in more serotonin than they could handle. This caused them to become desensitized. That ended the responsiveness of the gut to serotonin, but only to serotonin. Other stimuli could still get the bowel to move. The gut was not paralyzed, but in the absence of the gut's ability to respond to serotonin, the peristaltic reflex could not be evoked. Prozac had, by desensitizing serotonin receptors, done the same thing as a serotonin antagonist.

The second type of experiment was carried out very recently by Jingxian and Hui Pan, who joined my laboratory a few years ago as a postdoctoral fellow. Hui is a doctor from China who has been forced by circumstance to be a scientist. Her Chinese degree in medicine did her no good in the United States, so Hui made do and learned to impale enteric neurons at Michigan State University with Jim Galligan. Fortunately, her aim is good and the cells have been kind to her.

Hui impaled submucosal nerve cells with a microelectrode and electrically recorded their responses to stimulation of the mucosa. When pressure was applied to the mucosa, the submucosal nerve cells responded. However, their response was blocked by 5-HTP-DP. The site of action of 5-HTP-DP was found to be the mucosa, not the submucosal nerve cells. Therefore, the submucosal response that Hui studied must have depended on the stimulation of nerves in the mucosa by endogenous serotonin. Moreover, when Hui added Prozac, she found that the amplitude of the signals coming from the mucosa to the cells she impaled was increased. This observation showed that the mucosal serotonin transporter actually is important in inactivating serotonin. The potentiated response that Hui had detected occurred because more sensory nerve fibers were activated when the mucosa was stimulated in the presence of Prozac.

Hui went on to confirm the data she had obtained from single nerve cells by using a fluorescent probe that is taken up by nerve cells when they become active. Hui gently stroked the mucosa on one side of an opened preparation of intestine and compared the number of submucosal nerve cells that became active on the side that was stroked to the number that became active on the opposite side. The comparison, of course was no contest; the stroked side won. Almost no nerve cells were activated on the control side, and many were activated on the side where the mucosa was stroked. The really interesting observations were made when Hui repeated this experiment in the presence of Prozac. Once again, almost no cells were activated on the control side, but the number that became activated on the side that was stroked was *eight to ten times greater* than the number that had become active in the absence of Prozac. Prozac thus does not excite nerve cells all by itself (the control side proved this), but it greatly potentiates the signals sent by mucosal serotonin.

The collective force of the data obtained with Paul, Jingxian, and Hui very strongly suggests that the expression of the serotonin transporter by the cells that line the gut is the means by which serotonin is inactivated in the mucosa. Based on this work, and the prior studies dating back to those of Edith Bülbring, our current hypothesis is as follows: The EC cells secrete serotonin when stimulated to do so by pressure or some other perturbation. The serotonin stimulates intrinsic enteric sensory nerve cells to initiate reflexes. After the serotonin has done its job, neighboring cells of the intestinal lining, which express the serotonin transporter, take up serotonin, thereby lowering its concentration around serotonin receptors. This gives the receptors time to breathe, so to speak, and they recover. The receptors thus do not desensitize, and are able to respond to serotonin the next time it is released.

When Prozac or another SSRI enters the picture, pressure or other stimuli still release serotonin, which still stimulates receptors on sensory nerves. Now, however, the serotonin transporter is inhibited and the lining cells are unable to take up serotonin. As a result, serotonin stays in contact with the receptors for longer periods of time and wanders further away from its source. More sensory nerves thus respond, and the effect of stimulation of the mucosa is magnified. The receptors, however, get no respite from serotonin. The serotonin that is in contact with them is not removed quickly enough. The receptors thus desensitize. When this happens the gut loses its ability to respond to serotonin, and with that the bowel also loses its ability to respond to stimuli that are sensed by a serotonin-dependent mechanism.

Our current hypothesis can account for the detrimental gastrointestinal effects of the SSRIs. We would explain the initial nausea evoked by the administration of these drugs as the result of the potentiation of the action of serotonin on the 5-HT_3 receptors that extrinsic sensory nerves express. This early effect of the SSRIs passes away, despite the continued administration of the drugs, because the 5-HT_3 receptors desensitize relatively quickly. The diarrhea can be explained by a potentiation of the 5-HT_{1P}- (and possibly also 5-HT_4-) mediated initiation of peristaltic reflexes. Motility speeds up. Potentiating the action of serotonin as an excitatory neurotransmitter within the enteric nervous system may also contribute to this effect. Finally, when all the serotonin receptors desensitize, constipation results. Ultimately, the problems of SSRIs may be overcome if compensatory mechanisms, as yet unknown, come into play. This explanation still needs to be confirmed, but it is a good working hypothesis.

A New Player Takes the Field

Many years ago, while horses still pulled milk carts around cities to satisfy the needs of a bygone generation's babies and coffee cups, the great Spanish scientist Ramon y Cajal—the same neuroanatomist who, as discussed earlier, used silver staining—took time out from his fathering of the field of neuroscience and turned to the bowel. Naturally, in doing so, he used techniques that were similar to those he had previously used successfully to explore the brain. (If it ain't broke, don't fix it.) His silver stains, now hoary and ancient, but then very new, worked for reasons that Cajal never understood and that still have not been adequately explained. Many of Cajal's stains had actually been devised by his archrival, Camillio Golgi, but Cajal was not shy about utilizing whatever was in the public domain.

When applied to the brain, Cajal's stains had revealed individual nerve cells, complete and glorious. When applied to the gut, however, Cajal's stains revealed a fantastic network of oddly shaped cells that looked like no other cells Cajal or anyone else had seen previously. Cajal's drawings of what he saw look to a modern observer like a coral garden viewed through a snorkel mask. Spines and jagged edges protruded from the sides of every cell. Nothing is seen of cellular interiors, because Cajal's stains precipitated metal on cell surfaces, revealing only shapes in profile. In his drawings, Cajal simply colored the insides of the cells black. Individual cells in the drawings seem to touch one another only at their ends, gently and ever so delicately, as if they dared only to make contact with a tentatively offered cell tip. The oddly shaped cells appear to surround the ganglia of the myenteric plexus and to fill in the gaps that would otherwise exist between nerves and muscle. Because of their position, Cajal speculated that the cells revealed by his stains might be intermediates, relaying instructions from nerves to the muscles of the gut. After Cajal, the position of the cells he discovered in the interstices of the bowel, and Cajal himself, were reflected in their name, the *interstitial cells of Cajal.*

I think that the attribution of these cells to Cajal, which was never left out whenever anyone referred to the interstitial cells, may have been a little tongue-in-cheek. For many years Cajal's successors did not take these cells very seriously. If they had been found by anyone less august than Cajal, they would probably have passed into the trash bin of history, like hula hoops and boiled shirts. There has, however, never been much future in launching attacks on the observations of Cajal. Those who have tried have usually regretted it. Cajal had a way of seeing what should have been impossible for him to see with the techniques that he used, and for correctly interpreting what he observed.

Until 1982, the numbers of people who thought that the interstitial cells of Cajal were real were very small. Most people thought that this time Cajal had finally been fooled. His stains, they believed, were never designed to work in an organ like the gut, which is filled with connective tissue. Besides, no one had ever heard of another cell separating a nerve from its effector. There was thus no need for an interstitial cell. My teachers told me that the interstitial cells were probably artifacts, stained images of connective tissue or muscle cells created by the misapplication of a technique to an organ where it could not properly be used. Still, this was Cajal, so the cells continued to be remembered, but they were given to Cajal in the nomenclature, as if no one else wanted them. The interstitial cells were not trusted, so they were his, like the purchase of Alaska was, in another context, Seward's folly. The simile is apt. The folly in Cajal's identification of interstitial cells was of the same order of magnitude as that in Seward's purchase of Alaska.

In 1982, Lars Thuneberg, a Scandinavian anatomist, published a reexamination of the interstitial cells of Cajal that has become a classic. Thuneberg had the advantage of electron microscopy, which revealed that the cells that Cajal had seen were, in fact, quite real and located right where Cajal had found them. Thuneberg is an astute observer, and it was obvious to him that there are cells in the gut whose appearance is very different from the cells of connective tissue, muscle, or nerve. There was thus no question in Thuneberg's mind (or his manuscript) but that Cajal had indeed recognized a novel type of cell, which is present in the bowel and not elsewhere. Thuneberg noticed, furthermore, that the interstitial cells of Cajal form special junctions with their banal smooth muscle neighbors. Because of these junctions, he ventured the speculation that the interstitial cells of Cajal act as pacemakers that set the slow electrical rhythm that characterizes sheets of intestinal smooth muscle.

After Thuneberg's work became known, the interstitial cells of Cajal acquired respectability in the scientific community. People no longer felt the need to disown them and return them to Cajal. As a result, the long name became shortened to initials, *ICC* (plural ICCs), which bred familiarity, obscured Cajal, and showed that their user was an up-to-date member of the cognoscenti. Thuneberg continued to demonstrate ICCs by different methods, heaping confirmation upon confirmation onto work that was already well accepted. Another investigator, however, had assembled a team that was doing extremely original studies that ultimately proved that the speculations of Thuneberg, and Cajal before him, were correct.

When Kenton Sanders arrives at scientific conferences, stripping off his golfing gloves for action, no one who does not recognize him would suspect what he does for a living. Kent is the chairman of the Department of Physiology of the University of Nevada in Reno. Although Kent is a scientist, educator, and administrator, he looks like he is the Rhett Butler of Reno playing a part without Scarlett O'Hara. His appearance is that of a movie star, with hair that is always incredibly kempt. Unlike my hair, which is mown like a lawn when it gets too long, Kent's hair is like a Teutonic army, with not a single strand daring to break ranks and get out of order.

Appearances are very deceiving. There is nothing histrionic about Kent Sanders when he deals with business, which for him is the ICC. This is a field that has become vicious with the claims and counterclaims of people struggling to establish priority. Given that the ICC was discovered by Cajal (and even carries that discovery in its name), I have never understood the ill feelings that people who study ICCs have for one another. None of them have priority. They are all fighting over crumbs that Cajal has left, but still they fight on. The names they call one another are truly wonderful, ranging from the unspeakable to the "Prince of Darkness." Kent is no actor; he cowers before no one, and his data are very real.

Kent had been studying the electrical slow waves that occur in intestinal muscle, trying to locate the pacemaker cells. He had assembled a potent team for this investigation, including Sean Ward, a cheerful, earthy, Irish scientist who speaks with a brogue so thick that I usually miss every third word he says. Fortunately, Sean also speaks with his hands and his eyes, and he is willing to overlook my dumb looks of noncomprehension during conversations. The approach taken by Sean and Kent was progressively to cut away muscle layers until they were left with cells that paced themselves. The closer they came to the layer of the gut that contains ICCs, the stronger the pacemaking signal became, and when the pacemaking layer was reached, it consisted mainly of ICCs. When the ICCs were cut away, slow waves were lost from the remaining muscle cells. Sean and Kent had thus produced strong but not yet conclusive evidence that ICCs are pacemakers.

Having made these observations, Kent got lucky. He happened to be traveling in Japan, when he heard about the work of a large team of Japanese investigators (H. Maeda, A. Yamagata, S. Nishikawa, K. Yoshinaga, S. Kobayashi, K. Nishi, K. and S.-I. Nishikawa) who had raised antibodies to a receptor that no one had previously associated with the bowel. This receptor, called *Kit* (encoded by a gene called *c-kit*), was known to be critical for development of pigment cells (*melanocytes*), *germ cells* (in the gonads), and the *stem cells* that give rise to various blood cells. Kit had recently been shown to be expressed in the bowel, around ganglia in fact, but what the investigators in Japan had done that was novel was to inject antibodies to Kit into mice. The antibodies proved to be lethal to the animals, and their intestines filled up and became dilated, as if the motility of the bowel had been stopped. When the mice were autopsied, it was apparent that their ICCs had been destroyed by the antibodies.

Kent was fortunate to be in Japan at that time, and the Japanese team was just as fortunate to have him there. He was able to provide critical advice that the Japanese scientists needed, and a lasting transcontinental collaboration was established. Kent, Sean, and the Japanese soon determined that Kit is a marker for ICCs and that it is critical for the development of these cells. Antibodies to Kit can also be used to characterize in beautiful histological detail the network of ICCs. The pictures of preparations immunocytochemically stained with fluorescent antibodies to Kit look like glowing Technicolor versions of the drawings of Ramon y Cajal.

Mice breed rapidly and often enough that their genes mutate with some frequency. These mutations alter genes, thereby providing natural experiments that would be difficult to duplicate artificially. The Jackson Laboratory in Bar Harbor collects mutant mice and maintains a bank of these animals. There is a mouse catalog, which is very extensive. Fabulous experiments have come simply from thumbing the pages of this catalog. The mouse bank is thus a great resource. Mice with mutations in *c-kit* as well as in the gene that encodes Steel factor, the molecule that stimulates the Kit receptor, are both found in the Jackson Laboratory catalog.

When a mutation leaves mice with no Kit at all, its effect is so lethal that study of the bowel is impossible. Mice have to survive long enough to develop intestinal motility before one can tell whether or not motility is affected by the mutated gene. Mice with no Kit thus reveal nothing very interesting about the role Kit or ICCs play in the development of pacemaker cells in the gut. Milder mutations of *c-kit* are also present in the

Jackson mouse catalog, and these have been very useful. A defective form of Kit is produced in mice with these milder mutations, which does not signal normally but allows the animals to survive. The mutant mice are sterile, anemic, and spotted, reflecting defects in germ cells, stem cells, and pigment cells. Ward, Sanders, and their Japanese collaborators were the first group to discover that the intestines of the mice with relatively mild *c-kit* mutations also contain a disrupted network of ICCs and a bowel that lacks intestinal slow waves. The effect of these mutations was later reported by Thuneberg and a collaborator in Canada, Jan Huizinga. Similar effects are seen in mice with mutations in the gene that encodes Steel factor. These observations confirm that ICCs not only express Kit but require that the Kit they express be stimulated by Steel factor in order for them to develop normally. If either Kit or its ligand are knocked out by mutation, therefore, the ICC network becomes deficient. In the absence of this network, smooth muscle cells lose slow waves. The ICCs, therefore, are indeed the pacemakers.

Very recently, data obtained by Sean Ward has suggested that ICCs are more than just pacemakers for muscle. They also seem to receive messages from nerves and transmit them to the muscle, just as Cajal proposed. In fact, when it comes to the inhibitory neurotransmitter that gets muscle to relax, the ICCs appear to act as amplifiers, not only relaying the message from nerves but making it stronger.

One of the most important of the molecules that nerves use to get intestinal muscle to relax is actually a gas, nitric oxide. Using a gas for this purpose is very difficult because the nerve cells are not able to package it and hold it in readiness for when they need it. Nerves are not like balloons that can be filled or tanks that can be packed with gas under pressure. Muscle cells, moreover, do not sit still, like a patient in a dental chair, so that the nerve can strap a respirator onto it to deliver the gas. The nerves thus make the nitric oxide fresh whenever it has to be delivered. When they do so, the nitric oxide diffuses away and signals the ICCs to mimic them and make more. Even other neurotransmitters, like VIP, may stimulate ICCs to make and release nitric oxide. All of this nitric oxide released around the smooth muscle, as well as electrical signals the ICCs pass on to the muscle cells, helps to bring the mass of the muscle under inhibitory control. When nerve cells decide that it is time for muscle to relax, therefore, the nerves and their ICC assistants make quite sure that it happens.

A good model that helps to understand how smooth muscle of the gut is controlled is a flock of sheep. Like the sheep, layers of smooth muscle stay together but can also move about on their own; however, for the good of the whole their movements are better directed by a benevolent controlling force that makes decisions for them. For sheep, this controlling force is provided by a shepherd, who decides what the flock ought to do and directs the movements of the group. If the flock is a large one, a single shepherd needs help to keep all the sheep in order. It is possible to use a number of shepherds for this purpose, but it is more economical to minimize the employment of shepherds and use sheepdogs instead. The shepherd issues order to the dogs, which pass the message on to the sheep and keep them in order. The effect of the message from the shepherd is both transmitted to the sheep and amplified as the dogs spread the word. In the gut, the smooth muscle cells play the role of the sheep, the nerve cells are the shepherds, and the ICCs are the sheepdogs. The nerve cells make the big decisions that determine what the layers of intestinal muscle will ultimately do; the ICCs get the message and amplify it, helping to keep the mass of muscle under control. For evolution, it was probably more economical to devise an intermediary cell (which can multiply and be replaced) to play the role of an enteric sheepdog than to increase the size or number of nerve cells, which are irreplaceable.

The appreciation of the importance of ICCs to the gut dates from 1992, when the effect of antibodies to Kit was published. The studies of the effects of Kit mutations appeared in 1995, and the data suggesting that they are intermediates in transmitting messages from nerve to muscle have just emerged. Little is thus known of what diseases occur because of defects in the ICC network, but that knowledge will be coming soon. The role of the ICC in the pathogenesis of many types of intestinal disease is very much a subject under intensive study in many laboratories. Various forms of pseudo-obstruction, in which the bowel behaves like that of mice injected with antibodies to Kit, are good candidates to be conditions linked to ICC abnormalities. Sometimes the bowel stops propelling even when nerve cells are present and smooth muscle seems normal. This is an obvious situation in which to examine ICCs, but there is no reason to believe that there will not be other, more sophisticated problems to which ICCs also contribute.

Finding diseases that arise because of ICC defects will not make them curable just because their etiology is known. On the other hand, ICCs are not nerve cells, which are lost forever once they are destroyed. It is conceivable that a means can be found, perhaps through stimulation of Kit, to restore ICCs that are damaged or lost due to a disease process. As always in medicine, the march of progress brings with it new approaches to disease entities and new hope of treatment.

To See the Future Is to Envy the Young

The progress that has resulted from the recent concentration of a great many first brains on the second brain has been exciting. I saw the system rise like a phoenix from the dead. I worked on the second brain when to do so was to enter a field that was as lonely as a scene in a painting by Edward Hopper. I was still working on it when all of a sudden there were enough readers to make it possible to publish a *Journal of Neurogastroenterology*. I have seen a new science establish itself because an archaic old system has been rediscovered in the modern age. Now is a great time to be working on the enteric nervous system.

We have only just begun to understand how the enteric nervous system works to control the behavior of the bowel and its accessory organs. Since the simple stuff that got me started was fascinating, I am in awe of the complex material yet to be discovered. It is a pleasure to have the opportunity to train new people to study the second brain. I know that the scientific problems that await their attention and will yield to their probing will enrich their lives at least as much as the enteric nervous system has enriched mine. Great advances are coming. We are going to learn how the bowel knows when to mix and when to propel, how the brain and the enteric nervous system coordinate their respective actions, and what the meaning is of all the messages these two systems send back and forth to one another. We are going to be able to provide doctors with safe and effective drugs that can calm an irritable colon, bring regularity back to those who have lost it, or cool a burning esophagus. We are going to learn how to induce the second brain to cooperate with the massive immune system of the gut to improve resistance to infection and even to control inflammatory bowel disease. Best of all, we are going to put a smiling and carefree face on all the millions of abused people with functional bowel disease. All of these things that I would love to see I know are about to happen. I feel it in my gut.

10

IMMIGRANTS AND THE LOWER EAST COLON

WHEN I WAS VERY little, I used to annoy my mother by constantly asking her where I came from. The issue that troubled me as a child was not the question of what people do to initiate the formation of a new human being but how a human (myself in particular) is produced once that process is set in motion. The primitive words that I was using at the time failed to convey to my mother the true meaning of my question. What I wanted was a simple lecture not on sex, the details of which would have struck me as a bizarre and unbelievable story to put me off, but on embryonic and fetal development. Of course, since that question remains one of the great unsolved issues of biology, I did not realize my question had no answer. Until relatively recently, development could be approached descriptively but not mechanistically. The events that take place from the moment an egg is fertilized to the time of birth have been chronicled with great accuracy. Why these events occur is another matter. What happens has thus been known for a long time. What makes it all happen is only now being investigated effectively.

The problem facing anyone who tries to explain development is easily posed. Every cell in the body has the same set of genes; therefore, all cells work from the same blueprints. No cell has genetic instructions that are not found in every other cell; nevertheless, this single set of plans can be read to produce myriad different cell types. This happens because individual cells do not follow the entire list of instructions (the whole genome) contained in their nuclei. Some genes are turned on (*expressed*), while others are turned off (*repressed*). The design of any given cell is thus shaped by the particular subset of genes that cell happens to express. Development, therefore, is an improbably complex process of determining for every cell in the body which genes are turned on, which genes are turned off, as well as when, where, and in what order the genetic on/off switches are thrown. To understand development, therefore, it is necessary to know how the expression of genes is differentially controlled in the various cells, tissues, and organs of the body. No wonder my question was not explained by my mother. The answer is not child's play.

Many years later, early in my scientific career, I was prompted to ask once again where something came from. The context of the question this time was a plan to gain information from the developing bowel that could not easily be obtained from studies of an adult gut.

Clues to Adult Mysteries May Be Found in Development

My old sponsor, Edith Bülbring, and Graeme Campbell, her old bête noire, had each independently discovered the presence in the bowel of intrinsic inhibitory nerves. These nerves did not appear to be sympathetic, yet they exerted an effect, intestinal relaxation, that was very similar to that elicited by sympathetic nerve stimulation. Graeme's chief, Geoffrey Burnstock, an English investigator who was then working in what he regarded as self-exile (in Australia), had proposed that the intrinsic inhibitory nerves contained neither norepinephrine nor acetylcholine. In fact, in a radical departure from accepted wisdom, Geoff suggested that the inhibitory transmitter these nerves used was ATP, the same molecule that cells also employ internally to get the energy necessary to force chemical reactions to run uphill. Geoff's suggestion was greeted by the scientific establishment with all of the enthusiasm it had used to welcome my earlier proposal that serotonin is an enteric neurotransmitter. For Geoff, as for me, two neurotransmitters (acetylcholine and norepinephrine) was company, but three (apparently any third transmitter) was a crowd.

I think that Geoff may have suffered even more than I did from the slings and arrows of outrageous colleagues. It helped eventually, of course, for Geoff to have been right, both about the intrinsic inhibitory nerves not being sympathetic, and about ATP being a neurotransmitter. Geoff is now the grand old man of the autonomic nervous system, and he was able to return to England in triumph, but his path to vindication was anything but smooth.

Since Goeff began his career as a professional boxer in East London, he does not speak with an Oxford accent. Unfortunately, there are certain circles in Britain where speech and manner (breeding) count for more than actual substance. These circles did not take kindly to Geoff and his radical ideas, and when it comes to snubs, these people are the champions of the world. Goeff was eventually snubbed all the way to Australia, and when he returned, it was to be the professor of anatomy at University College, London, which is one of the most prestigious chairs in the country.[1]

The thought that attracted me to development was the possibility that intrinsic nerves might differentiate in the fetal bowel before extrinsic nerves grow into the gut. I knew that the sympathetic nerve cells are all located some distance from the wall of the bowel. Many things thus had to take place before any sympathetic fibers could enter the gut. For example, the sympathetic nerve cells had to be born, form ganglia, and send out axons. The sympathetic axons then had to find the bowel and invade it. I expected all of these events to take time; therefore, I thought that we might be able to take advantage of a brief period during development when the gut was likely to lack sympathetic nerves. The trick would be to determine whether intrinsic nerve cells could cause the fetal muscle to relax at this time.

I presented my ideas about using development as an experimental tool to my second graduate student, Elizabeth (Tiz) Thompson, who was anxious to try them out. Tiz was incredibly bright but was often depressed. She wanted both to be a scientist and to be a mother. These goals seemed to Tiz to be in irrevocable conflict. It is hard today to imagine an age when women were expected to abandon their careers to take care of children as soon as they had any. Tiz, however, had been brought up to believe that a woman's place was not in a laboratory, a concept that I worked hard to dispel. Tiz thus did not work smoothly or uninterruptedly on the project, but she did work well. Eventually she finished her research, wrote a nice thesis, received her doctorate, and took me to dinner at Lutece to celebrate.

Tiz set out to determine the earliest time in development when the

[1] Goeff's stories about his early career in England bring the American Revolution to mind. It took an extraordinary degree of snobbism for England to lose the American colonies, but as I have said, when misguided British snobs put their mind to it, they are world class. After talking with the British envoy (not even a real minister) during the post-Yorktown negotiations in Paris, John Adams wrote, "The pride and vanity of the nation is a disease; it is a delirium; it has been flattered so long by themselves and others that it perverts everything." The British could never even deign to refer to George Washington as General Washington. He was only a Mister. Washington, of course, had his revenge on Cornwallis at Yorktown. He rubbed it in, too, by having the band play a piece of music called "The World Turned Upside Down" during the surrender ceremony.

smooth muscle of the primordial mouse and rabbit gut acquired the ability to move, and when the muscle could be driven by electrically stimulated enteric nerves. This study involved great delicacy and a spectacularly sensitive instrument to detect the contractions of the primitive muscle. When a fetal smooth muscle pulls for the first time, it does not pull hard. We used a microbalance to record the force of the muscle contractions, which were measured in millionths of grams. Both spontaneous muscle movement and its elicitation by nerves occurred gratifyingly early in fetal life. The primitive nerve fibers, moreover, could either excite the smooth muscle and make it contract or, more importantly from our point of view, inhibit the smooth muscle and get it to relax.

Tiz soon found that the nerves that excited the smooth muscle used acetylcholine as their neurotransmitter, but the nerves that inhibited the muscle clearly could not have used the sympathetic neurotransmitter, norepinephrine. In fact, these inhibitory nerves caused the fetal bowel to relax even before the gut contained, or could make, norepinephrine; moreover, histological studies confirmed that when nerves first made the fetal gut relax, the sympathetic nerves had not yet entered the bowel. Since the fetal gut thus relaxed when it had no sympathetic nerves, the nerves that made it relax could not have been sympathetic. Tiz did not herself identify the inhibitory neurotransmitters (now known in various locations to be ATP, nitric oxide, and VIP), but her work was still very important because it definitively established that the bowel really does contain intrinsic inhibitory nerves that use neither acetylcholine nor norepinephrine as their neurotransmitter. Tiz thus helped bury the old concept that there are only two peripheral neurotransmitters and provided a neat confirmation of a major part of Geoffrey Burnstock's hypothesis. Geoff has liked me ever since Tiz's work was published.

The Critics Were Wrong

I had just finished supervising Tiz Thompson's research when it occurred to me that development might be equally useful in helping me to put my own hypothesis, that serotonin is a neurotransmitter in the gut, to a critical test. You may recall that Marcello Costa and John Furness had cast doubt on the significance of my observation that radioactive serotonin was produced in enteric nerves when I injected a radioactive version of its precursor molecule (5-HTP) into mice. Marcello and John suggested that the enteric nerves in which I had found radioactive serotonin might not normally be serotonin-containing. Instead, they proposed that radioactive serotonin might have been made nonspecifically in sympathetic nerve fibers that opportunistically took up the radioactive 5-HTP that the non-physiological circumstances of my experiment had made available. Their idea, therefore, was that the radioactive serotonin that I found in the enteric nervous system was simply a false neurotransmitter in sympathetic nerve fibers. Clearly, the objection raised by Marcello and John was a serious one. If they were right, then my hypothesis was probably wrong and my experiments would have been rendered meaningless.

I thought of a variety of ways to determine whether the nerves in the gut that became labeled by radioactive serotonin were sympathetic. For example, I could destroy sympathetic nerves with drugs, or I could block their ability to take up neurotransmitters. In the end, I did both of these things without interfering with the accumulation of radioactive serotonin in enteric nerve fibers. Drugs, however, leave room for uncertainty because they may exert unknown effects. A cleaner test, which does not involve drugs, could be made by taking advantage of the late arrival of sympathetic nerve fibers in the fetal gut. Intrinsic nerve cells develop right in the bowel, and at least some of them, I had just discovered with Tiz Thompson, differentiate before the developing gut receives its sympathetic innervation. At that time, I did not yet know how early in development enteric serotonin-containing nerve cells arise, but my observations had already suggested that they are intrinsic. I therefore thought that they might well beat out the sympathetic nerves in a developmental race. If radioactive serotonin were to be made in developing enteric nerves before the most precocious of sympathetic nerves grew into the fetal bowel, then I could dismiss Marcello's and John's criticism of my previous work.

I had Tiz's work in mind when I set out with another graduate student, Taube Rothman, to investigate the development of enteric nerves that make serotonin. Taube had solved the problem of science versus motherhood by marrying early, bringing up her kids, and starting her career. Taube, moreover, was not working just to earn a living. She worked for the pure joy of it. Getting the answers to scientific questions was the primary driving force, and her enthusiasm was contagious. The U.S. Army should have recruiting officers like Taube Rothman.

Taube soon discovered that the developmental race was no contest. She used radioautography to show that enteric nerves became packed with radioactive serotonin that they either synthesized from radioactive 5-HTP or took up from the fluid around them, long before any sympathetic nerve fibers ventured into the bowel. Using the uptake of radioactive norepinephrine to find the ingrowing sympathetic fibers, Taube had no difficulty in distinguishing them from the nerves that labeled with radioactive serotonin. Taube's experiments thus established that the objections of Marcello Costa and John Furness were not valid, and radioactive serotonin was not adventitiously entering sympathetic nerves. The observations were important in accomplishing that much, but they also hooked me on the study of development itself. I began to ask the question: Where does the enteric nervous system come from?

The Source

The first people to study the developmental origins of the enteric nervous system did nothing very useful. The early theories of its origin are more embarrassing than anything else. I think that the problem faced by scientists starting out to study the issue was that the actual source of the enteric nervous system seems such an unlikely one that it took a long time to get to the truth. In fact, the cells that give rise to the enteric nervous system are not even part of the bowel when the gut is formed. Making matters worse, mammals go through embryonic and fetal development locked away in a uterus, where, until recently, a scientist could not easily get at them.

The big breakthrough came in 1954 and 1955, when Yntema and Hammond used chick embryos to provide the first strong evidence that the enteric nervous system, like most of the remainder of the peripheral nervous system, comes from an embryonic structure called the *neural crest.* The neural crest is a transient aggregate of cells that appears next to the structure that will become the brain and spinal cord of all vertebrate embryos, and then disappears as the cells of the neural crest acquire a wanderlust and set out on a variety of journeys through the embryo. In humans, the neural crest arises at the end of the third week of development (in chicks it is during the second and third days). At this time, the embryo does not consist of much. There are no arms, legs, or recognizable organs. The embryo is just a flat three-layered disc (see the accompanying figure). The three layers are the *ectoderm* on top, the *mesoderm* in the middle, and the *endoderm* on the bottom. As difficult as it may be to believe, this simple disc is able, during later development, to fold itself in a complex series of steps to establish the form of the body, build the gut, and give rise to the nervous system.

The ectoderm looks like a simple layer of cells, and they do form the outer wrapping of the embryo, once the embryonic disc folds enough to establish that there is an inside and an outside. Ectodermal cells, however, also have a far more exalted developmental mission than just the forma-

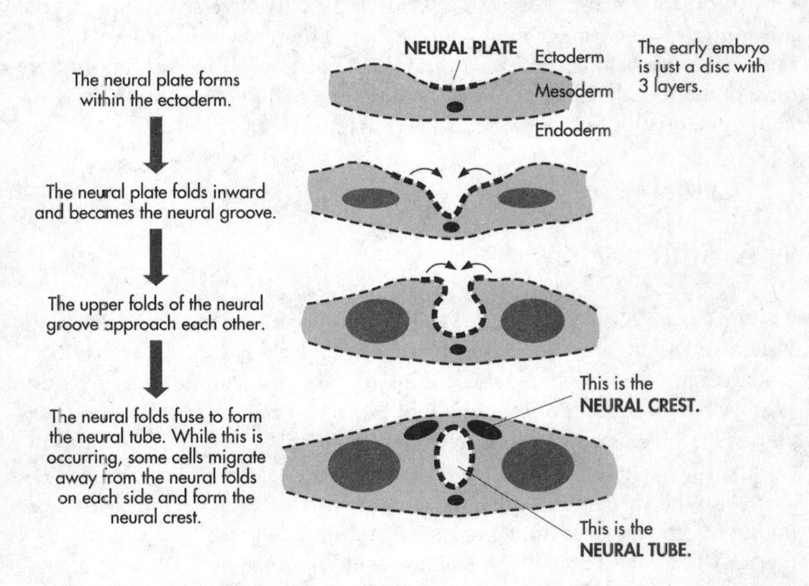

tion of skin. When subjected to *induction* by the right sort of molecules provided by their neighbors, they go through a process called *neurulation,* which give rises to the brain, the spinal cord, and the peripheral nervous system (as well as some glands and many of the bones and muscles of the face). As far as I am concerned, neurulation is the most important event that takes place in anyone's life.

Neurulation begins with a thickening, called the *neural plate,* which appears in the middle of the ectoderm of the embryo while the embryo still is just a disc. The neural plate folds inward to produce the *neural groove.* The tops of the neural groove give rise to *neural folds* that approach one another and ultimately come together and fuse to convert the neural groove into the *neural tube.* The neural tube then separates from the banal ectoderm and goes on to produce the brain and spinal cord. Just before the neural folds fuse to form the neural tube, some cells break off from the tips of the folds and move away to produce wings of cells on either side of the fusing neural tube. These wings of breakaway cells, which acquire an irrepressible urge to travel, are the *neural crest.*

One of the beauties of experimenting with chick embryos is that they can be surgically rearranged. To find out what derivatives are formed by the neural crest, therefore, it is possible to remove the crest early in development and see what is missing in later embryos as a result of the surgery.

That is exactly what Yntema and Hammond did. They removed a large portion of the neural crest, allowed the embryo to survive long enough for the gut to form, and found that it contained few or no enteric ganglia. This simple experiment, which showed that if there is no neural crest there will be no enteric nervous system, provided compelling evidence that the enteric nervous system is derived from the neural crest. Removal of large portions of the neural crest, however, is a pretty shocking thing to do to an embryo. It is thus conceivable that the loss of such a major piece causes embryonic structures to become abnormal, even if the neural crest does not directly participate in their formation. Two kinds of results thus have to be anticipated from large experimental deletions of the neural crest. One is straightforward. Some structures may not develop because they normally arise from the neural crest cells that have been surgically removed. The other is less obvious. Some embryonic structures may be disturbed, not because they themselves are neural crest derivatives, but because their development requires a signal that some other neural crest–derived structure provides. It was thus necessary to confirm the conclusions of Yntema and Hammond by a method that does not provide such a profound insult to an embryo.

The needed confirmation of the neural crest origin of the enteric nervous system was achieved as a result of an extraordinary series of observations that changed the face of modern developmental biology. These observations were made by a French scientist, Nicole Le Douarin, who was working then not only to solve developmental problems but also to overcome the male chauvinism of the French scientific community. I have come to know Nicole very well and have had the opportunity to collaborate with her. She is a most charming woman, and she is elegant in a way that only the French can be. She is gracious to a fault, but she also has a steely determination that it is wise to take seriously. Nicole has great powers of observation, an appreciation of the significance of details that others miss, and a wonderful way of designing just the right experiment to answer important questions. Many studies end with intriguing suggestions; Nicole's tend to finish with definitive solutions. Few major scientific honors have escaped her. Nicole has won the Kyoto medal and the Louisa Gross Hurwitz Prize, and if it were up to me, she would have won the Nobel Prize as well. Nicole inherited the laboratory and the scientific tradition of Marie Curie, but unlike Madame Curie, Nicole has successfully run the gauntlet of males who have traditionally guarded the portals of the French Academy of Sciences and has become a member. She also became a member of ours.

Early in her career, Nicole noticed that she could distinguish cells of quails from those of chicks. The DNA in the nuclei of quail cells is distributed in a funny way. The cells look like targets, dart boards that got lost on the way to a pub. Taken by itself, that discovery sounds mundane, and so it would have been but for the use Nicole made of it. Nicole realized that during early embryological life, before the immune system develops, it is possible to replace the cells of one species with those of another, provided that the donor species is not too distant a relative of the recipient. Chicks and quails are close enough. Since quail cells are so distinctive that they can be recognized at fifty paces in a field of chick cells, they are ideal markers to use to trace the fate of cells that migrate from one place to another in an embryo. Nicole has followed the peregrinations of quail cells through chick embryos to make major contributions to many fields, including the development of both the immune and nervous systems. She is, however, probably best known for her work on the neural crest.

Nicole removed the neural crest from various levels of chick embryos and replaced them with quail equivalents. When she did this, the transplanted quail cells evidently felt quite at home in their new hosts, and they appeared to migrate through the chick embryo, just as if they had been left undisturbed in their normal locations. An animal produced by the combined efforts of two species of cells is called a *chimera,* in this case part chicken and part quail. The beauty of this beast is that all of its quail parts are neural crest derivatives; moreover, these structures are labeled as definitively as if a celestial customs inspector had written on them, "MADE IN THE QUAIL BY THE NEURAL CREST."

Nicole used her technique of making quail-chick chimeras to produce a complete fate map of the neural crest. In doing so, she painstakingly defined what becomes of every portion of the crest. These studies revealed that the enteric nervous system is formed by cells that migrate to the bowel only from three very specific regions of the neural crest. Enteric nerve cells and their cast of supporting cells are mainly derived from a level of the crest that lies just below the developing ears of the embryo. This level, which is located next to the developing hindbrain (being produced by the neural tube), is called the *vagal* region of the crest. Neural crest cells migrate from the vagal area to the bowel along a pathway that is later followed by the vagus nerves. The vagal crest cells colonize the entire oral-to-anal extent of the gut. A second set of neural crest cells from just above the tail of the embryo, the *sacral* region of the crest, also migrates to the bowel, but these cells colonize only the portion of the gut that lies below the umbilicus (belly button). The *hindgut* (last part of the bowel), therefore, receives crest-derived cells from both vagal and sacral sources. The esophagus receives the third infusion of neural crest cells, which come from the level right below the vagal crest. (See the illustration on the following page.)

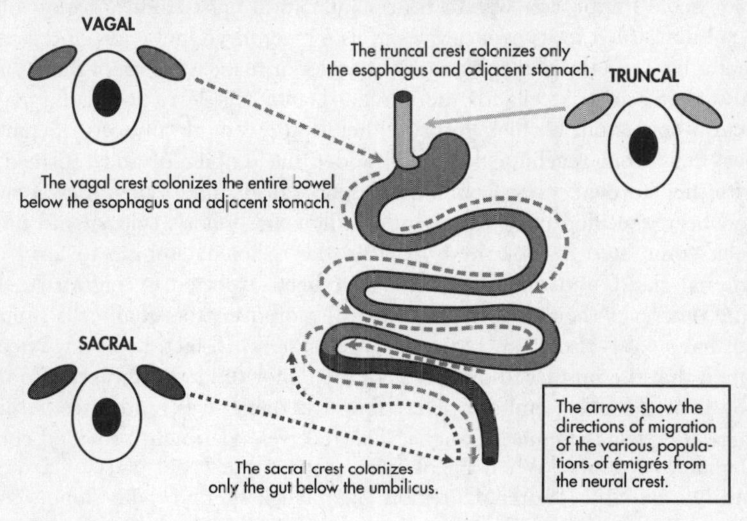

Potential

The nerve cells of the second brain thus have a great deal in common with most Americans, myself included. Like us, enteric nerve cells are the progeny of immigrants. The neural crest is the old country for the nerve cells of the bowel. Once these facts became known, the question arose as to what serves as the Statue of Liberty for the neural crest–derived émigrés. What beckons these intrepid migrating cells to the promised bowel? Do they think "gut" and go off wandering through the wilds of the embryo in search of an enteric ideal, or is it something entirely different? You might think that the limitation of sources of enteric nerve cells to only a small number of sites in the neural crest suggests that the cells in these locations are committed to go to the gut. If that is so, the cells in other regions of the crest presumably lack the necessary "homing" directions for finding the bowel and thus go elsewhere. If "smart" cells endowed with guidance information, "homing" on their targets like the bombs we dropped in the Gulf War, is what you thought, however, you were wrong. Nicole's ability to follow the migrations of quail cells in chick embryos allowed her to test the hypothesis that neural crest–derived cells migrate to the gut because they are committed to do so and know how. The hypothesis failed the test.

Nicole's approach was to transplant neural crest from a region of a quail that, if left to its own devices in its own embryo, never produces cells that migrate to the gut. She chose as her donor tissue a region of the neural crest that provides cells for the adrenal gland. Nicole removed the vagal neural crest from a chick embryo (thereby preventing any of the chick's own cells from reaching the bowel above the umbilicus) and replaced it with the "adrenal" crest from a quail donor. If the fate of neural crest cells had been specified prior to migration, then the donor's cells should have been committed to look for a place in the recipient embryo to form an adrenal gland, and they would not have been expected to migrate to the gut. Wherever they went in the embryo, moreover, the quail cells should not have given rise to an enteric nervous system. In fact, however, Nicole found that the misplaced quail cells of the donor did indeed migrate to the bowel of the chick embryo. They found it themselves, and after getting there they formed enteric ganglia. The reverse experiment provided complementary results. When Nicole transplanted the vagal crest of a quail embryo into the "adrenal" region of a chick embryo, the donor cells migrated, not to the gut of their host (which would have been an appropriate thing for vagal crest cells to do), but to the site where the adrenal develops; furthermore, after reaching this site, they formed a normal adrenal gland.

Taube Rothman and I collaborated with Nicole in further experiments that showed that the ganglia in the gut that arose from neural crest cells from the "adrenal" region of a quail donor were organized in a chimeric bird into what appeared to be a perfectly normal enteric nervous system. When examined by electron microscopy, for example, the ganglia had the typical brainlike structure of the enteric nervous system rather than the structure of peripheral nerves. We also detected serotonin-containing nerve cells in the enteric ganglia of the chimeric birds. All of these serotonin-containing nerve cells, furthermore, were quail cells, confirming that they were derived from the "adrenal" crest. The "adrenal" level of the neural crest does not normally produce any serotonin-containing nerve cells, which are found outside the brain only in the bowel.

These observations make it clear that neural crest cells do not colonize the gut because they have been specified to do so before they migrate. Instead, neural crest cells follow defined pathways in the embryo, which lead to the bowel from specific regions of the neural crest. If you put almost any neural crest cell on this path, it winds up in the gut. The same experiments also suggested that the developmental potential of a population of neural crest–derived cells is very large, or *multipotent*. The fates of the population are not predetermined before its cells leave the neural crest but depend instead on signals these cells receive from the microenvironment they find either along their migratory routes or within the target organs that they ultimately colonize.

The pioneering studies of Nicole and her collaborators established that the population of crest-derived cells that colonizes the bowel is multipotent, but they did not establish why. The population could be multipotent because its individual cells are themselves multipotent. Alternatively, the population may be made up of committed cells, each of which can give rise to only a single kind of mature cell. If this were to be the case, then many different types of committed precursor cells would have to be present, so that the group as a whole would appear to be multipotent. The distinction between these two possibilities is important. If individual crest-derived cells in the gut are multipotent, then microenvironmental signals provided by the bowel wall would be necessary to tell them what to do. The gut would have to provide an instructive influence. On the other hand, if many different types of irrevocably committed crest-derived cells colonize the bowel, then the microenvironment of the gut would have to choose the correct subset of precursors to produce an enteric nervous system. The influence of the bowel would be one of selection rather than instruction. In either case, the microenvironment of the gut, as well as the neural crest, clearly plays an important role in the development of the enteric nervous system.

To ascertain whether individual crest cells are multipotent or restricted to a single fate, it is obviously critical to see what single crest-derived cells can actually do. Individual cells of the neural crest have now been studied by Nicole Le Douarin and many other investigators, including Maya Sieber-Blum, David Anderson, Marianne Bronner-Fraser, and her husband, Scott Fraser. These studies have clearly demonstrated that individual cells of the neural crest are indeed multipotent, not only before they depart from the neural crest itself but also while they migrate through an embryo. Together with Nicole, Taube Rothman and I went on to show that crest-derived cells are still multipotent even after they have finished migrating and have made their grand entrance into the developing bowel.

The technique that revealed that the gut is colonized by multipotential precursors is called *back-transplantation.* Nicole had used it for the first time in a different context, and we adapted it to the developing bowel. For back-transplantation, an organ, in our case the gut, that has already been colonized by cells from the neural crest is removed and grafted back into a younger host embryo. This procedure determines the degree to which postmigratory neural crest cells retain the developmental potential of their ancestors. Once they have migrated, do their fates become specified, and do they lose their wanderlust? Alternatively, are their options still open; can they still "go abroad" and wander?

In her earlier experiments, Nicole had shown that after back-transplantation some of the neural crest–derived cells in a target organ of the older donor recover an urge to travel and migrate a second time through the tissues of the host embryo. The primitive environment of the younger host embryo brings out this potential. In our experiments, we back-transplanted segments of gut from a quail donor into a chick host. In some experiments, we even back-transplanted a chimeric bowel in which only the neural crest–derived cells were quail while the rest were chick. The experimental design enabled us to identify the crest-derived cells from the donor's gut, no matter where they migrated or what they became in the host embryo.

Taube placed the back-transplants of gut next to the neural tube, so that the crest-derived cells in the older bowel would find themselves in a neural crest migration pathway in the younger embryo. Nicole was willing to go along with us and see what would happen, but she did not expect crest-derived cells, which were already giving rise to ganglia in the quail gut, to do much in the host embryo, even after back-transplantation. In a manner suitable for science fiction, however, crest-derived cells that had previously colonized the quail bowel *came out of it* after back-transplantation and migrated again in their embryonic chick host. Where they went, moreover, depended on where we placed the donor's gut. If the quail bowel was grafted to the vagal or sacral regions of the host embryo, the bowel-derived crest cells would follow the appropriate trails and colonize the host's gut. In contrast, if the donor's bowel was grafted to any other region of the embryo, none of the bowel-derived crest cells migrated to the host's gut. Instead, they followed pathways that were appropriate for the level of the chick embryo at which the graft was placed. Depending on where we placed the back-transplanted bowel, therefore, we could induce crest-derived cells to migrate from it to the host's sensory ganglia, sympathetic ganglia, peripheral nerves, and the adrenal. In other words, the crest cells that colonize the gut obviously learn nothing from their trip to the bowel. Instead, they remain multipotent and able to give rise to many types of nerve cell, even to some that they do not produce when they stay in the gut.

The collaborative experiments that I carried out with Taube Rothman and Nicole Le Douarin to investigate the developmental potential of the cells that form the enteric nervous system were lots of fun to plan, do, and analyze. The paper describing our results, however, was probably the single most painful one that I have ever helped to write. As was usual in transatlantic collaborations before the advent of E-mail, we sent many drafts of manuscripts back and forth across the ocean. Eventually, when we were close to something that seemed right to all of us, I happened to be invited to a meeting in Europe. That provided me with the opportunity to visit Nicole's laboratory and get together with her to finish the paper. Initially, all went well. Nicole and I met, reached agreement in a short conference, and I went off to write the final version. Writing, however, turned out to be a problem.

Both Nicole and I used Macintosh computers. Her computers, however, had French keyboards with letters in what were, for me, the wrong places. The solution was to use an American Macintosh that one of her postdoctoral fellows had brought with him to France. That computer was located in the animal room, where it was keeping track of the sounds made by a very strange population of chimeric birds that Nicole and her postdoctoral fellow had produced. They were studying the origin of the nerve cells responsible for birdsong. Nicole and her postdoctoral fellow had grafted parts of the presumptive brain from quail donors into chick embryos so that they could trace the migrations of nerve cell precursors. In this case, the birds were allowed to hatch. The animals looked like chickens but they chirped like quails. As time went on, the quail cells began to be rejected by the chicken's immune system. In contrast to most mature nerve cells, those which are responsible for birdsong can be replaced. As the quail nerve cells were rejected, therefore, new cells were substituted for them, but the new nerve cells were all of chicken origin. While they were in the process of rejecting the quail nerve cells they were born with, the birds alternated quail chirps with cock-a-doodle-dos that were distinctly chicken. Eventually, once all the quail cells were disposed of, the birds clucked away in a manner that would have been appropriate in any barnyard.

I went up to the animal room to work on the birds' Macintosh and discovered a problem. I am terribly allergic to chickens. People have often asked me how I managed to survive to adulthood with my ethnic background and that particular allergy. Still, chickens or no chickens, I had a job to do and I was going to get it done, even if it killed me, which it nearly did. I thought I was in an avian variety of hell, and when I finished, I was nearly unrecognizable. The paper, however, was an important one and was published without difficulty.

The demonstration that the crest-derived cell population that gives rise to the enteric nervous system is multipotent implies that the bowel itself must play an important role in determining what becomes of the crest cells in its wall. In essence, the neural crest–derived cells are analogous to the first-year medical students I teach at Columbia. Like the neural crest émigrés in the gut, this student population is multipotent. It gives rise to many different types of medical specialists as well as to family doctors, although almost the entire set of students becomes physicians. Despite the fact that they are multipotent, the crest-derived cells that colonize the gut are similarly restricted, in the sense that their developmental horizons are limited to the nervous system. They appear, for example, to have lost the ability found in other cells of the neural crest to give rise to pigment cells or bone; nevertheless, as we have seen, many different types of nerve cell, even some that would be inappropriate for the bowel, can develop from the enteric population of neural crest–derived émigrés.

The choices of medical specialties open to medical students become progressively more restricted as their training proceeds. Options decline as some fields attract their interest and further study, while other fields, which strike them as repellent, are neglected. The choices students make are heavily influenced by the faculty of the college and by other more advanced students who serve as role models. Similarly, the developmental choices made by neural crest–derived cells in the gut are also influenced by a "faculty" and "other students." For the neural crest–derived cells, the faculty is represented by non-neural-crest-derived cells of the bowel, which provide them with instruction in the form of chemical signals called *growth factors*. These growth factors are essential for the development and even the survival of enteric nerve cells and glia (the gluelike cells that support the nerve cells of the brain and gut). The supporting gel that cells live in within the fetal gut (called the *extracellular matrix*) also participates in the instruction of crest-derived cells. Eventually, if the instruction provided by the "faculty" of the bowel is complete and not distorted by a congenital defect, enteric nerve cells begin to differentiate in the wall of the gut. These precocious nerve cells, like the more advanced medical students, make important contributions to the development of laggards. Through their activity and the actions of their neurotransmitters, these early-developing nerve cells exert powerful influences on the developmental decisions made by the cells that follow in their wake.

Revisionism

One of the great certainties of science is that any major advance will be challenged by revisionists. This process is really a very positive one, because it ensures that concepts will be rigorously tested before they are accepted into the canon of received wisdom. Faith has no place in scientific progress. Nicole Le Douarin's discoveries about the origins of the enteric nervous system were no exception. Very soon after she published her observations, they were challenged by Allen and Newgreen, two investigators in Australia. These authors pointed out that Nicole's experiments with quail-chick chimeras lacked a control. Nicole could see where quail cells migrate in chick embryos, and she could see where chick cells migrate in quail embryos. She could not, however, do the control studies to see where chick cells migrate in chick embryos and quail cells migrate in quail embryos. Nicole's work was thus based on the untested assumption that a foreign crest cell migrates in a host embryo just as the host's own crest cells would have done had they not been replaced. Allen and Newgreen were particularly upset about Nicole's identification of the sacral crest as a source of precursors for the enteric nervous system. Their own work had revealed that enteric nerve cells can be recognized in a smooth oral-to-anal progression that they attributed to the progress of vagal crest cells down the gut. In recent years, Don Newgreen has been absolutely relentless in his attacks on the idea that the sacral neural crest contributes to the formation of the enteric nervous system.

The role of the sacral neural crest is an important issue because the hindgut, a region of the bowel that these cells help to colonize, is the part of the bowel that is most frequently affected by birth defects of the enteric nervous system. Hirschsprung's disease, for example, is a common congenital problem in which the terminal bowel contains no enteric ganglia at all (it is thus *aganglionic*). Hirschsprung's disease occurs in about one of every five thousand live births. Since there are four million births per year in America, that means that every year about eight hundred Americans are born with a life-threatening defect. Because these children have no intrinsic nerve cells in their terminal gut, the propulsive reflexes that depend on these cells are missing from this region of their bowel. As a result, their colon becomes obstructed and dilates massively above the aganglionic zone. Another name for Hirschsprung's disease is thus *congenital megacolon.* If left untreated (and sometimes even if it is), the condition is fatal. The dilated bowel may perforate, allowing infection to spread through the patient. Treatment consists of surgically removing the aganglionic portion of the gut, which can be done if the aganglionic zone is not too large, and pulling the normal gut through to the anus. Clearly, in order to determine why the terminal gut becomes aganglionic in Hirschsprung's disease, it is important to know whether sacral crest cells normally participate in the colonization of this region of the bowel.

In order to test the validity of Don Newgreen's objections to the work of Nicole Le Douarin, it is necessary to be able to trace the migrations of crest-derived cells by a technique that does not involve the construction of quail-chick chimeras. I thus set out to do this together with Howard Pomeranz, a student in Columbia's combined M.D.-Ph.D. program, who did his graduate research with me. Howard was a sensitive human being with a literary bent and a fearless willingness to undertake almost any experiment, no matter how difficult it might appear to be. Howard and I were joined originally by Bob Payette, a board-certified neurologist and an experienced investigator who had previously worked at the University of Pennsylvania. Bob had found life in Philadelphia to be insufferably boring. He always said that he became a member of my laboratory because he liked the research that I did, but I suspected that the fact that I worked in New York was at least as important. Bob was a great help in providing Howard with research training. Unfortunately, Bob's untimely death prevented him from seeing the completion of the project. Taube Rothman took Bob's place and did the critical manipulations of chick embryos.

We used three different methods to follow the migrations of cells from the sacral neural crest. One technique was simply to trace the cells that we identified immunocytochemically with antibodies that specifically recognize cells of neural crest origin. In subsequent experiments, we injected a fluorescent dye, called *DiI,* into the vagal and sacral regions of the neural crest of chick embryos. DiI inserts itself into the plasma membrane of cells, stays there, and provides a convenient cell marker. Finally, we injected a genetically engineered retrovirus into the neural crest of chick embryos. The virus was altered so that it was able to invade cells that were exposed to it but not replicate within these cells or come out of them. The virus was also engineered to contain a gene for a bacterial enzyme, *beta galactosidase,* that we could detect in tissue sections. By demonstrating the presence of the beta galactosidase, we could positively identify infected cells. Since the virus, which cannot spread, was injected into the sacral crest, the presence of infected cells in the bowel would prove that sacral crest cells migrate there.

Every one of these three techniques confirmed that Nicole had been right. Sacral crest–derived cells do indeed migrate to the gut. The immunocytochemical marker revealed a trail of cells between the sacral crest and the hindgut, and after the injections of either DiI or the retrovirus into the sacral crest, there were always labeled cells in the hindgut. Our studies in chicks were confirmed and extended to mice by Marianne Bronner-Fraser, who also injected DiI into the sacral crest. More recently, Nicole Le Douarin and Alan Burns, a young Northern Irish expatriate working with her as a postdoctoral fellow, revisited the question of the sacral neural crest migration and found that the number of sacral cells that participate in forming the enteric nervous system of the terminal bowel is actually very large. The sacral cells, however, do not arrive until relatively late in the course of development. In fact, they allow their vagal compatriots to traverse the entire bowel and get into the colon before them. This tardiness accounts for the difficulty encountered by Don Newgreen and others in finding the sacral component of the enteric nervous system. The entry of the sacral cells was masked by the presence of vagal cells. It is impossible to distinguish a vagal from a sacral neural crest cell unless one marks them before they mix.

The recent studies of the contribution of the sacral neural crest to the formation of the enteric nervous system can, in a sense, be thought of as the rediscovery of the wheel. Nicole Le Douarin had really made the seminal observation before any of these investigations were undertaken. Still, even in retrospect, I think it was important to confirm Nicole's conclusions. If this particular wheel had not been rediscovered, we could never have come up with a molecular explanation that accounts for the pathogenesis of Hirschsprung's disease. That is another story that I will come to, but since the bottom of the bowel is totally aganglionic in patients with the condition, it is obvious that the lesion will not occur unless *two* streams of neural crest–derived émigrés fail, not just one. To explain the origin of Hirschsprung's disease, both vagal and sacral cells have to be taken into account.

Mimicry

Once I had become convinced that the microenvironment of the bowel plays a crucial role in the formation of the enteric nervous system, I stopped thinking about other factors that might also be important. That turned out to be a mistake. Nothing in biology is simple, and anything as complicated as the second brain should have been expected to depend on more than a single set of factors. The first clue that the development of the enteric nervous system was going to depend on something other than what the gut alone can provide was an observation made independently by Gladys Teitelman, an investigator who had previously spent a year in my laboratory, and by Philippe Cochard, a French colleague of Nicole Le Douarin, who was then on leave doing research at Cornell. The discovery was that for a brief window of time the primitive gut of fetal rats transiently contains cells that look for all the world as if they are sympathetic nerve cells. These cells contain the sympathetic postganglionic neurotransmitter norepinephrine. Since norepinephrine is a member of a class of chemicals called *catecholamines,* the *t*ransiently *c*atecholamine-containing cells of the developing bowel were named TC cells for short. There are no catecholamine-containing cells in the adult gut. Where TC cells came from and why they disappeared were, at the time of their discovery, major mysteries.

Gladys Teitelman and I began to collaborate again to try to discover what TC cells were up to. Gladys is the kind of person who makes me think that we in America owe a great deal to the odious despots of the world. Every time some tyrannical dictatorship becomes intolerable somewhere on earth, another infusion of talent comes to America. Gladys was driven out of Argentina by the military junta that appeared to be running what was, in essence, a reverse Westinghouse (now Intel) Science Talent Search. The best scientists were persecuted so that the survivors were forced to emigrate to America and Europe. When Gladys first joined my laboratory, the Argentinean military was making people "disappear" in what has come to be known as the "dirty war" (as if wars are sometimes clean). Gladys left Argentina when the military began to move against her family. As far as I am concerned, this kind of transfer of people is just another illustration of the old saying, the rich (us) get richer and the poor (Argentina) get poorer.

Gladys and I soon found that the TC cells in the fetal rat gut, as well as their counterparts in the fetal mouse bowel, express a set of molecules that are only found in nerve cells. That observation told us that the TC cells are members of the nerve cell fraternity (or sorority). Since all of the nerve cells in the gut are derivatives of the neural crest, we concluded that TC cells or their progenitors must get to the bowel by migrating there from the neural crest. My first thought was that the TC cells might actually have been sympathetic nerve cells that made a wrong turn during their migration from the neural crest, became lost, and wound up in the gut. If that were so, then TC cells might disappear because the enteric microenvironment is hard on strangers and kills them off. The body contains many tough neighborhoods, where cells that do not belong or that fail to follow local rules are simply murdered. Cell death is a fact of life and is especially common during development. As it turned out, however, my first thought was quite wrong. I was thus forced to think again.

Gladys and I began to think that an untimely demise is not the reason that TC cells disappear when we found that they proliferate. Sympathetic nerve cells do not proliferate. In fact, with rare exceptions (including the nerve cells responsible for birdsong and the sensory nerve cells that detect odors), nerve cells do not divide or proliferate in any way. The last division of a precursor of a nerve cell is thus considered that nerve cell's birthday. Once a typical nerve cell is born, it is irreplaceable and has nothing to look forward to but death. Gladys went on to study other things, but I was joined in my investigation of TC cells by a particularly adept graduate student, Greg Baetge. Greg and I not only confirmed Gladys's demonstration that TC cells proliferate, but we actually caught several of them *in flagrante,* shamelessly undergoing *mitosis* (cell division). The fact that TC cells divide indicated that they are not nerve cells. Since they are dividing cells from the neural crest that express nerve cell markers, we proposed that TC cells are nerve cell precursors. It also seemed very unlikely that cells about to die would be reproducing. We began to suspect that TC cells are progenitors of the enteric nervous system that disappear because they, or their successors, lose the sympathetic nerve cell properties that enabled them to be identified in the first place. Our suspicion was thus that the TC cells go on to give rise to enteric nerve cells that are not remotely like any that are found in the sympathetic nervous system. To test this idea, we would need a new marker that persists after the sympathetic nerve cell characteristics of TC cells are lost.

Molecular Persistence

At this point in our studies, we caught a break. We were using methods of molecular biology to look not only at the enzymes that TC cells use to make their neurotransmitter but also at the RNA transcripts that direct the biosynthesis of these enzymes. We found that the transcripts were present in the cells for only a very short period of time. TC cells thus are "flashers," like the people I meet every now and again when I get off the "A" train at the 168th Street subway stop. They expose themselves briefly and then cover up again. The protein encoded by the transcripts that are expressed so evanescently by TC cells continues to be present long after the RNA that encoded it disappears. The persistence of this protein allows the successors of TC cells to be identified, even after they have, so to speak, changed their occupation. It was as if the TC cells had indelibly stained themselves by what they were doing, so that they could be spotted forever after. In the same way, you can recognize painters by the paint on their hands after they are no longer painting, and first-year medical students by the cadaverous odor they carry around after they have finished dissecting. We could thus follow the progress of cells that were once TC cells, even after they no longer made the enzymes that enabled them to mimic sympathetic nerve cells.

One enzyme in particular, known as *DBH* (the acronym for dopamine beta-hydroxylase),[2] is present in TC cells and is never totally lost from their progeny. DBH is the final enzyme in the chain needed to make norepinephrine. DBH continues to be present in certain nerve cells of the adult enteric nervous system, even in fully mature animals. We found DBH, for example, in mature serotonin-containing nerve cells. Since none of the adult enteric nerve cells that contain DBH make norepinephrine, there is nothing in them for DBH to do. The cells cannot even produce the molecule that DBH works on. We concluded that the DBH in adult enteric nerve cells is a functionless remnant of their historical past, a memento of their prior existence. Our observations thus showed that TC cells do not die but disappear because they are precursors that differentiate into enteric nerve cells that do not contain norepinephrine (or any other catecholamine). It was not obvious, however, why the development of enteric nerve cells should be so tricky as to include the presentation of one face and then another.

Greg Baetge went on to show that the TC cells that one finds in the bowel are actually part of a larger population of vagal crest–derived émigrés. TC cells thus are found studded all along the route of vagal crest cell migration to the gut. In fact, they outline the vagal pathway. While the TC pioneers enter the bowel early and make good headway down the gut before they begin to lose their sympathetic markers, other TC cells migrate at a more leisurely pace and are overtaken by the down-growing fibers of the vagus nerves. Eventually, however, continued migration cleans out the vagal pathway, and all of the TC cells enter the bowel.

After Greg finished his graduate training with me, he went on to become a pathologist in southern California, which is where he came from and where he always dreamed of returning. After Greg left, an Israeli postdoctoral fellow, Eran Blaugrund, arrived to continue the project. Eran had been steered in my direction by Chaya Kalcheim, another Argentinean émigré and an Israeli developmental biologist whom I met through Nicole Le Douarin. Chaya is a great scientist, and she also treats her friends the way politicians treat big donors. I must have once unknowingly done something nice for Chaya. She is too good to me for me to believe that she does not think she is paying back an imaginary debt. In any case, Eran was a gift from Chaya.

Eran ostensibly came to work with me to receive training, which was a bit of a joke because he arrived fully trained. He was not in my labora-

[2] Norepinephrine is made from the dietary amino acid tyrosine. You get that mainly by breaking down the protein that you eat. *Tyrosine* is converted by an enzyme called *tyrosine hydroxylase* to a molecule by the name of L-DOPA (DOPA is an abbreviation for *dihydroxyphenylalanine*). L-DOPA is the molecule that has revolutionized the treatment of Parkinson's disease. It enabled rigid people to move again. L-DOPA is converted to a product called *dopamine* by another enzyme, *L-amino acid decarboxylase*. Dopamine is used by nerve cells in the brain as a neurotransmitter, but in sympathetic nerve cells it is the immediate precursor of norepinephrine. DBH converts dopamine to norepinephrine.

tory for more than a week before he was teaching people how to do things better. The gut was new for him, as were TC cells, but as far as I was concerned, I had acquired an experienced collaborator, not a trainee. Eran's goal was to try to learn why at least some of the cells that participate in the formation of the enteric nervous system divert so much energy to the imitation of sympathetic nerve cells, only to tire of the charade after a few days and give it all up. Imagine a transvestite going to similar lengths to perfect the costume, makeup, and demeanor of the opposite sex, and then, having briefly done so, forgetting about it.

At the time Eran and I set out to analyze the peculiar behavior of TC cells, we knew that they gave rise to enteric nerve cells, but we did not know how many. We guessed that only some enteric nerve cells develop from TC progenitors, but for all we knew at the time, the entire enteric nervous system might have done so. The observation that only some mature enteric nerve cells are branded with DBH, the marker left over from the TC phase of their progenitors, suggested that only a subset of enteric nerve cells have TC ancestors. Still, since we had no idea how indelible a tattoo DBH expression was, it was possible that the mature nerve cells that did not express DBH might have succeeded in erasing (repressing) it. We decided to try to find out whether sympathetic mimicry was a quirk displayed by a few eccentric precursors or a peculiarity of the entire set of crest-derived émigrés in the bowel.

A Precursor of Everything?

While we were studying the universality of the TC origin of enteric nerve cells, a paper appeared, by Carnahan, Anderson, and Patterson, that suggested that the entire enteric nervous system, the sympathetic nervous system, and the central core of the adrenal gland (the *adrenal medulla*) are all derived from a common embryonic precursor. They called this postulated Abraham of a cell, whose progeny would be as numerous as the stars in the sky and the sand at the sea, the *sympathoadrenal-enteric* progenitor. David Anderson and Paul Patterson are to developmental neurobiology what E. F. Hutton was to the financial world in the old television commercials. When they speak, people listen. The hypothetical sympathoadrenal-enteric progenitor immediately dominated our thinking, and we set out to determine whether it really exists.

The proposal that enteric and sympathetic nerve cells are ancestrally linked was based on the observation that TC cells mimic more of the properties of sympathetic nerve cells than just the machinery involved in making and using norepinephrine. TC cells and sympathetic nerve cells also wear the same uniform. As they develop, cells express a distinctive set of proteins in their plasma membrane, called *differentiation antigens,* because antibodies to these proteins are made when they are introduced into an animal of a different species. The particular set of differentiation antigens that cells present to the outside literally constitutes their uniform and allows members of the same tribe to be distinguished. Carnahan, Anderson, and Patterson had followed the appearance and disappearance of a series of differentiation antigens on the surfaces of sympathetic nerve cells. These proteins had seemed at first to be specific for sympathetic and adrenal precursors, but the new paper by Carnahan, Anderson, and Patterson now revealed that they had found that the same differentiation antigens are also expressed by TC cells in the fetal bowel. These differentiation antigens, moreover, are not only shared by TC and sympathetic nerve cells but appear and disappear on both cell types at the same times during development. The enteric and sympathetic cells, therefore, do more than just wear the same uniform. They also put it on and take it off in synchrony. It was the shared uniform and the synchronized robing and disrobing that suggested to Carnahan, Anderson, and Patterson that TC cells and sympathetic nerve cells belong to the same cellular army.

Eran succeeded in getting TC cells to go through their routine in tissue culture. He removed the gut from fetal rats, used appropriate enzymes to dissolve the "glue" (extracellular matrix) that holds cells together, and cultured the resulting suspension of "dissociated" cells. Sure enough, some of the cultured cells acquired the characteristics of sympathetic nerve cells and then lost them, even though the cells were growing all by themselves in a defined solution. This phenomenon was interesting by itself because it demonstrated that the transience of TC cells is the result of properties that are intrinsic to the cells themselves. TC cells do not need to receive signals from elsewhere in the fetus, or even the bowel, to lose their sympathetic appearance. They are programmed to imitate sympathetic nerve cells for a finite period of time and will simply stop doing so when this time expires.

Eran went on to use antibodies that recognize the sympathetic set of differentiation antigens on the surfaces of TC cells as selective agents of death. A group of blood proteins called *complement* punches lethal holes in the plasma membrane of cells that have antibodies bound to their surfaces. Complement ignores cells to which antibodies are not attached. Eran thus added both antibodies and complement to his cultures and waited for the TC cells to die. Eran's approach succeeded. All of the cells in his cultures that even began to express the differentiation antigens that marked them as TC cells were instantly killed. If all of the cells of the enteric nervous system were to have been derived from TC cell precursors, then no nerve cells should have been able to develop in the cultures that were exposed to antibodies and complement. In fact, however, despite the treatment, nerve cells continued to develop, albeit in reduced numbers. These observations suggested that there are at least two lineages of enteric nerve cell precursors. One might be common to the sympathetic nervous system, but the other is clearly not.

Our studies on the development of TC cells prompted me to call David Anderson to discuss the issue with him. He had, at that time, been studying a genetically altered mouse in which a gene that is essential for the development of the sympathetic nervous system was knocked out. This gene is called *mash–1*.[3] Since sympathetic nerve cells depend on *mash–1*, the *mash–1* knockout mice are born with almost no sympathetic nervous system. The *mash–1* knockout mice all died at birth. David had noted that there are nerve cells in the gut of the *mash–1* knockout animals, and that they develop during fetal life, although their initial appearance is two days late. Obviously, the fact that enteric nerve cells are present in mice in which the development of the sympathetic nervous system is lost does not support the idea that enteric and sympathetic nerve cells arise from a common progenitor. Conceivably, the mutation might not be lethal to the postulated sympathoadrenal-enteric precursor but only to the sympathetic nerve cells that arise from it at a later stage in development. It was thus possible to wiggle around the inconvenient facts with an explanation (or rationalization). After some thought, however, David agreed with me that he had an interest in further study of the particular set of enteric nerve cells that develop in the bowel of *mash–1* knockout mice.

After discussing Eran's data, David agreed to collaborate with us to test the hypothesis that there are multiple lineages of enteric nerve cell precursors. I already knew from studies carried out with Taube Rothman and an associate, Tuan Pham, that enteric nerve cells of different types arise in a reproducible sequence. Serotonin-containing cells were among the earliest to develop, while those containing a peptide called CGRP[4] were the last. The concept that we set out to evaluate was that the delay in the development of enteric nerve cells that David had seen in the *mash–1* knockout mice was not the result of a general delay occurring in mice that were abnormal but rather the result of the specific loss of enteric nerve cells that develop early. The selective wipeout of the early-

[3] Mash–1 gene is an acronym signifying that it is the mammalian homologue of a fruit fly gene called *achaete-scute*. It is customary to write the names of genes in italics.

[4] CGRP is an abbreviation for *calcitonin gene related peptide*. Calcitonin is a hormone produced in the thyroid gland. It is encoded by the same gene that also encodes CGRP. The RNA transcript yields two different products by a process called *alternative splicing*.

born subset of enteric nerve cells would create a delay in the appearance of nerve cells in the fetal gut. Although the "early bird" nerve cells would fail to appear, the late-developing nerve cells would show up on schedule, creating the appearance of a developmental delay.

We wanted to identify the source of the early-born nerve cells that we proposed would be missing from the guts of *mash–1* knockout mice. TC cells mimic the properties of sympathetic nerve cells; therefore, they were our prime candidates. Since serotonin-containing nerve cells develop early and contain DBH when they grow up, we postulated that serotonin-containing nerve cells develop from TC cell precursors. In contrast, since CGRP-containing nerve cells are all born late, we proposed that none of them arise from TC cells. Our hypotheses could thus easily be tested. They predicted that *mash–1* knockout mice would lack enteric serotonin-containing nerve cells but possess a normal complement of nerve cells that contain CGRP. Very few of my hypotheses have been confirmed by the actual experiments quite as precisely as these.

When we examined the *mash–1* knockout mouse gut, there were, as predicted, no TC cells and no serotonin-containing nerve cells; however, CGRP-containing nerve cells were present. We also found that the *mash–1* gene was selectively expressed by TC cells in the normal fetal bowel. Eran then went back to his cultures and found that the antibodies and complement he had used to destroy TC cells had prevented the development of serotonin-containing but not CGRP-containing nerve cells. These studies thus established that there are multiple (at least two) lineages of precursors contributing to the development of the enteric nervous system. Lineage and the enteric microenvironment thus both sculpt the enteric nervous system.

Environmental factors work only on cells that are prepared to respond to them. Before a cell can be affected by any type of signal, for example, it has to express receptors on its surface, and these receptors have to be coupled to an internal transduction mechanism. The lineages to which cells belong determine the set of receptors they express and the transduction pathways to which these receptors are linked; therefore, cell lineages decide which signals can influence nerve cell precursors.

Fishing

To study the microenvironmental signals of the fetal bowel in action, it is necessary to extract the neural crest–derived cells from the gut after they colonize it and to separate them from their non-neural-crest-derived neighbors. The isolated crest-derived cells can then be cultured in defined medium[5] (which does not itself contain any potentially confounding substances) and exposed to signaling molecules. The goal of all these manipulations is to isolate the precursors of the enteric nervous system from uncontrolled outside influences. If cells of various types were to be allowed to interact, there is no way to know what they might say to one another. One cell could easily slip a surreptitious developmental message to another without being detected. Surprising results are fine in science, but mistakes are made when experimental subjects are allowed to interact in unknown ways. Research subjects, cells included, have to be treated like the citizens of a totalitarian state. To keep control of the population, it is critical to isolate individuals from any input except that provided by the people in charge. Unknown signals are the enemies of scientists in the same way that fax machines and the Internet are the enemies of autocrats.

Isolating neural crest cells is easy if one is satisfied with obtaining them before they migrate to the organs they colonize. One has only to remove the closing neural tube and grow it in tissue culture. The neural crest cells then separate themselves by migrating away from the neural tube along the surface of the culture dish. After the exodus of the neural crest cells, the neural tube can be removed from the culture with a fine tungsten needle, and one is left with almost nothing in the culture but neural crest cells. These neural crest cells, however, are not the same as those that colonize the gut. The cells that have completed their migration through an embryo to enter the bowel are both older and more experienced than their premigratory predecessors. The postmigratory cells should be called neural crest–derived rather than neural crest cells because they have undergone changes that have either been programmed by genetic instructions to occur spontaneously as the cells mature or to occur in response to signals that the cells encounter along their migratory route. To study the effects of microenvironmental signals on the development of enteric nerve cells, therefore, it is necessary to fish the neural crest–derived cells out of the bowel itself.

Finding a relatively simple method to isolate neural crest–derived cells from a recently colonized segment of fetal gut was a daunting problem that impeded progress for several years. The neural crest–derived cells migrate through the fetal bowel mix with other cell types, and, if not for molecular markers, it would be impossible to distinguish them from their non-neural-crest-derived neighbors. Molecular markers, however, do exist.

[5] *Media* are the fluids within which cells are grown in tissue culture. Cells often require complex additions to tissue culture media, such as blood serum and chick embryo extract (prepared by putting chick embryos in a blender). Blood serum and chick embryo extract contain many unknown substances.

The differentiation antigens, which comprise the distinguishing uniform that cells wear, allow neural crest–derived cells to be identified in the wall of the fetal gut. Eran Blaugrund had used antibodies to the differentiation antigens expressed by TC cells to select them for death. I now decided to use similar antibodies to select the whole population of enteric neural crest–derived cells for a life in tissue culture.

I began the experiments to isolate neural crest–derived cells from the fetal bowel together with Howard Pomeranz and Taube Rothman, and finished them with Alcmène Chalazonitis. Alcmène had been an independent research scientist at the Albert Einstein Medical School before joining my laboratory. In contrast to the graduate students and postdoctoral fellows with whom I usually work, Alcmène was already an established investigator. She thus came to collaborate, not to be trained. The method we used was to pick out a differentiation antigen that was expressed in the gut only by cells of neural crest origin. We then dissociated the fetal bowel with enzymes to produce a suspension of individual cells and exposed them to antibodies that specifically recognized the antigen we had selected. This treatment decorated the surfaces of neural crest–derived cells with antibodies. The next step was to add a second set of antibodies that reacted not with cells but with the antibodies we had just attached to differentiation antigens on cell surfaces. These second antibodies were coupled to magnetic beads. When we subsequently applied a magnet to the cell suspension, the neural crest–derived cells, which now contained a coating of magnetic beads, were attracted to it. The amazing thing about this process of magnetic *immunoselection* was that it actually worked. We wound up with two pots of cells. One, the "immunoselected" pot, contained an almost pure population of neural crest–derived cells. The other, the "residual" pot, contained a mixture of all the cells that were left after the neural crest–derived cells were removed by the magnetic field. The "residual" population of cells was thus neural crest–depleted.

Sure enough, when we cultured the cells in the two pots that we obtained by immunoselection from developing mouse, rat, chicken, or quail gut, nerve cells and glia developed only in the cultures of immunoselected cells. The cultures of residual cells contained smooth muscle. A few years later, Jun Wu, a graduate student who is still working with me, found that interstitial cells of Cajal also developed exclusively in cultures of residual cells. The experiments with Jun confirmed that the interstitial cells of Cajal, which had only recently been shown to be real, are not neural crest derivatives. Interstitial cells are present in the residual cell cultures because they come from the same family of progenitors that also produces smooth muscle. We were not the first to draw this conclusion. Kent Sanders and Nicole Le Douarin had each suggested it earlier. Kent's suggestion was based on his observation that a variety of molecules that are thought to be specifically expressed by smooth muscle are also found in interstitial cells of Cajal. Nicole's conclusion was based on experiments with quail-chick chimeras. When she replaced the vagal neural crest of a chick embryo with that of a quail, all of the nerve cells that developed in the gut above the umbilicus were quail cells, but all of the interstitial cells of Cajal were chick.

Immunoselection provided Alcmène with the tool she needed to investigate the molecular meaning of the term "enteric microenvironment." The neural crest–derived cells that she immunoselected from the gut could be exposed to molecules that she suspected might be growth factors for enteric nerve cells, and the response of enteric neural crest–derived cells to these factors could be unambiguously determined. Alcmène could also directly examine the growth factor receptors that enteric neural crest–derived cells express because she had a purified batch of these cells to analyze; furthermore, she could distinguish factors that enteric neural crest–derived cells produce in order to communicate with one another from factors produced by non-neural-crest-derived cells of the bowel wall. For the first time, it had become possible to "tap into" the molecular conversations between the cells that influence the development of the second brain.

Signals and Knockouts

Alcmène began her studies by investigating members of a family of growth factors called *neurotrophins*. The very first growth factor of any type ever identified, *nerve growth factor* (aka *NGF*), is a neurotrophin. Rita Levi-Montalcini and Stanley Cohen received the Nobel Prize for their contributions to the discovery of NGF. The neurotrophin family is known to contain at least five members: NGF, *b*rain *d*erived *n*eurotrophic *f*actor (*BDNF*), and *n*euro*t*rophins–3, –4/5, and –6 (*NT–3, NT–4/5,* and *NT–6*). NT–6 has only recently been discovered and was not known to exist at the time of Alcmène's study. All neurotrophins bind to two kinds of receptor. One of these receptors, called in the trade, $p75^{NTR}$, is nondiscriminating in its taste for neurotrophins and is happy to allow any of them to bind to it. The second kind of receptor is more selective, binds neurotrophins with higher affinity, and displays preferences for particular neurotrophins. The selective molecules are called Trk receptors.[6] There

[6] The Trk receptors are all enzymes that have the ability to add phosphate groups to the amino acid tyrosine in proteins. An enzyme that adds phosphates is called a *kinase*. The Trk receptors are thus *re*ceptor *ty*rosine *k*inases; hence the name Trk.

are three Trk receptors. TrkA likes NGF, TrkB prefers either BDNF or NT–4/5, while TrkC favors NT–3.

What turned Alcmène on to the neurotrophins was that Greg Baetge had previously discovered that all of the neural crest–derived cells that colonize the fetal bowel express $p75^{NTR}$. In fact, Alcmène found that antibodies to $p75^{NTR}$ are superb probes for the immunoselection of crest-derived cells from the fetal gut. The expression of $p75^{NTR}$ by the enteric population of crest-derived cells suggested that at least some of these cells were likely to respond to one or more of the neurotrophins; however, because $p75^{NTR}$ is nonselective, its expression says nothing about which neurotrophin(s) might affect the development of enteric nerve cells or glia.

Although Alcmène tried all of the neurotrophins except NT–6 (which had not yet been identified), only NT–3 was actually found to exert an effect on the development of the neural crest–derived cells that she immunoselected from the fetal gut. NT–3 promoted both the development of nerve cells and glia in the cultures, while NGF, BDNF, and NT–4/5 did nothing that could be detected. More enteric nerve cells were produced in the presence of NT–3, but NT–3 did not cause precursor cells to proliferate; therefore, NT–3 must have increased the proportion of precursors that chose to differentiate as nerve cells or glia. Alcmène also found that nerve cells that were induced to develop by NT–3 became addicted to it. Once these cells responded to NT–3, they became so dependent on getting an NT–3 "fix" that they dropped dead if NT–3 was no longer provided to them. NT–3, therefore, is a survival factor as well as a factor that promotes the development of enteric nerve cells and glia.

Alcmène went on to carry out similar experiments that established the efficacy of several additional growth factors. The most powerful of these was one called *g*lial cell line *d*erived *n*eurotrophic *f*actor, or *GDNF.* This factor was originally discovered as a product secreted by a cultured line of glial cells that had been derived from a malignant tumor. The action of GDNF that first attracted attention to it was an ability to promote the development and survival in tissue culture of midbrain cells that contain the neurotransmitter *dopamine.* Since the degeneration of these cells is the cause of Parkinson's disease, anything that enhances the ability of midbrain dopamine-containing nerve cells to survive is viewed as a cause célèbre. As I noted earlier, Parkinson's disease may affect the enteric nervous system as well as the brain. In fact, the telltale lesions, called *Lewy bodies,* that enable neuropathologists to make a definitive tissue diagnosis of Parkinson's disease are found not only in the brain but in the bowel as well. Since midbrain dopamine-containing and enteric nerve cells each degenerate in Parkinson's disease, we reasoned that the enteric nervous system might contain at least some cells that are related to the dopamine-containing nerve cells of the midbrain. If that were to be true, then the good things that GDNF does for nerve cells of the brain it might also do for nerve cells of the gut.

It took Alcmène and me some time to get around to investigating the enteric actions of GDNF. By the time we did so, we already knew that GDNF was going to turn out to be a major player in the development of the enteric nervous system. Vassilis Pachnis, working across the street from us at Columbia in the laboratory of Frank Costantini, had produced mice that carried a targeted mutation in a gene called *c-ret.* At the time Vassilis and Frank produced these knockout animals, they knew that *c-ret* encoded a cell-surface receptor, Ret (Ret, like the Trk receptors, is a receptor tyrosine kinase); however, the molecule (ligand) that bound to Ret and activated it was unknown. Despite the fact that they did not know what its ligand might be, Vassilis and Frank were interested in Ret, because mutations in the human *RET*[7] gene cause terrible clinical problems. These mutations, which allow the receptor to become spontaneously active (that is, Ret turns on even in the absence of a ligand to stimulate it), are found in people who suffer from, and often die of, multiple cancers of the endocrine glands. Vassilis and Frank hoped to gain insight into the normal function of Ret in development by determining what goes wrong when Ret is absent. The knockout mice that lacked Ret died at birth and had no nerve cells (at all) below the esophagus and the immediately adjoining region of the stomach. (Mice that lack Ret also have no kidneys.) The *c-ret* knockout mice thus established that the stimulation of Ret is absolutely essential for the development of all enteric nerve cells, except for the tiny population that comes from the truncal neural crest and colonizes the esophagus and a bit of the adjacent region of the stomach.

The functional ligand that stimulates Ret was subsequently discovered by several groups of investigators to be GDNF.[3] A targeted mutation that prevents the expression of GDNF thus produces exactly the same lethal effects as a mutation that knocks out Ret. Either way, therefore, whether you take out the ligand or its receptor, the enteric nervous system is doomed. While these observations clearly show that the development of the enteric nervous system is GDNF-dependent, the studies of knockout animals do not demonstrate why that is so. The cells in the developing gut that express Ret are all of neural crest origin, suggesting that these cells

[7]The animal gene is called *c-ret*, while the human gene is capitalized as *RET*.

[8]GDNF does not actually bind directly to Ret. To stimulate Ret, GDNF requires a coreceptor, a friend that helps out. Such a coreceptor, which is not an integral component of a cell membrane, is called an *alpha component*. More than one type of coreceptor exists that helps ligands to stimulate Ret. The one that seems to be most important for GDNF is a molecule called *GFR alpha-1*.

themselves are likely to be the targets of GDNF. In fact, David Anderson had used antibodies to Ret to immunoselect neural crest–derived cells from the fetal mouse bowel. Many of the cells that antibodies to Ret immunoselected were multipotent, indicating that they were not very advanced in their state of differentiation.

The hypothesis that Alcmène and I set out to test was that GDNF is required to stimulate the development of the earliest and most primitive of the vagal and sacral neural crest–derived precursors of the enteric nervous system. If a common progenitor of everything has to be stimulated by GDNF in order to differentiate or survive, then nothing in the way of an enteric nervous system would be expected to develop the absence of GDNF or Ret. Our hypothesis would thus account for what happens in the bowel of GDNF or Ret knockout mice.

Alcmène immunoselected neural crest–derived cells from fetal mice at a variety of ages, cultured the cells, and exposed them to GDNF. The results were spectacular. When Alcmène first examined a dish of GDNF-treated cells, she gasped. None of us had ever seen anything like it. NT–3 had promoted the development of nerve cells, but its effects paled in comparison to those of GDNF. The culture dish contained a virtual lawn of nerve cells.

The reason that the effect of GDNF was so striking turned out to be that it does something that NT–3 and other growth factors that we had previously investigated do not. Like NT–3, GDNF promotes the development and survival of enteric nerve cells, but unlike NT–3, GDNF also causes the neural crest–derived precursors of these cells to proliferate. As a result, the GDNF-treated population as a whole becomes larger and acquires more nerve cell progenitors. These effects dwarf the actions of the other growth factors, which enhance differentiation of enteric nerve and glial cells at the expense of proliferating precursors.

Alcmène found that the effects of GDNF are highly age-dependent. GDNF is very effective early in development, but its efficacy declines when it is applied to cells from older fetuses. At the earliest times, when the effect of GDNF is strongest, enteric crest-derived cells are unable to respond at all to NT–3. The early precursors of the enteric nervous system thus respond to GDNF before they become responsive to NT–3. In fact, if the progenitors do not first see GDNF, they will never be able to respond to NT–3. Eventually, once the cells acquire the ability to respond to NT–3, the effects of GDNF and NT–3 add to one another, so that the effect of exposing cells to a combination of these two factors is greater than that of either one alone. These observations were all nicely compatible with our starting hypothesis.

It is striking how different the effects on the gut of knocking out GDNF or Ret are from the effects of knocking out *mash–1*. The loss of

GDNF or Ret is a catastrophe that prevents the development of even a single nerve or glial cell anywhere in the gut below the esophagus and adjoining stomach. The loss of *mash-1*, however, selectively deletes only one enteric nerve cell lineage and allows the glial cells to develop at will. Even more restricted effects are produced by the knockout of some of the other genes that are involved in the construction of the enteric nervous system.

An example of a more restricted effect is seen when targeted mutations are made that knock out genes encoding receptors for a class of growth factors called *neuropoietic cytokines*. The knockout of the neuropoietic cytokine receptor causes the specific absence of just the motor nerve cells that excite or inhibit smooth muscle.[9] (I hate the name "neuropoietic cytokine," which is difficult to pronounce and hard to remember, or even to spell. There is a logic to the nomenclature, but since it is arcane, technical, and not very interesting, I prefer to forget about it.) As far as a newborn mouse is concerned, the difference in effects between these various gene knockouts is highly academic. What happens to the mouse is always the same. Its gut blows up soon after the animal is born and the mouse dies. Either the enteric nervous system is up and running on the day of birth or an animal will not survive for more than a few hours. On the other hand, as far as scientists are concerned, the differences between the effects of the different knockouts and the actions of the growth factors are highly informative.

The Ballet

What the research on the actions of the various growth factors suggests is that the development of the enteric nervous system is like a ballet combination. (This ballet and the effects of making mistakes are shown in the

[9] The identity of the actual ligand that stimulates the neuropoietic cytokine receptor upon which the development of enteric motor nerve cells depends is unknown. The neuropoietic cytokine receptor has three components, one alpha, which binds to a ligand, and two betas, which are integral membrane proteins and do the signaling. The alpha component of this receptor is known to be critical, because the development of enteric motor nerve cells is prevented when only the alpha component is deleted. A molecule called ciliary neurotrophic factor (or CNTF for short) binds to the alpha component, activates the neuropoietic cytokine receptor, and promotes the development of enteric nerve cells and glia. CNTF, however, can be knocked out in mice without affecting the development of the enteric nervous system, and about 2 percent of the otherwise normal human population of Japan happens to lack CNTF. A molecule other than CNTF must therefore do the stimulating of the neuropoietic cytokine receptor in the developing enteric nervous system.

COLONIZATION OF THE GUT AFTER THE KNOCKOUT of GDNF or Ret

Precursor cells still migrate to
the gut from the neural crest.

Vagal and sacral crest–derived cells die
because they cannot survive unless they are stimulated by GDNF.
The gut becomes aganglionic below the esophagus and adjacent stomach.

COLONIZATION OF THE *MASH-1* KNOCKOUT MOUSE GUT

Precursor cells migrate to the gut from the neural crest;
the population is multipotent.

Vagal and sacral crest–derived cells proliferate
and populate the bowel below the esophagus.

Within the gut, cells express Ret.

They get their "GDNF fix" and thus
they survive.

All of the *mash-1*-dependent cells die;
therefore, the gut lacks all of the
(TC cell–derived) nerve cells in this lineage.

The *mash-1*-independent
lineage develops on schedule.

Identity?

CGRP-containing nerve cells

COLONIZATION OF THE MOUSE GUT AFTER THE KNOCKOUT OF THE NEUROPOIETIC CYTOKINE RECEPTOR

Precursor cells migrate to the gut from the neural crest;
the population is multipotent.

Vagal and sacral crest–derived cells proliferate
and populate the bowel below the esophagus.

Within the gut, cells express Ret.

They get their "GDNF fix" and thus
they survive.

The *mash-1*-dependent
lineage survives.

The *mash-1*-independent
lineage survives.

Seritonin-containing nerve cells

There are
no motor
nerve cells.

Identity?

CGRP-containing nerve cells

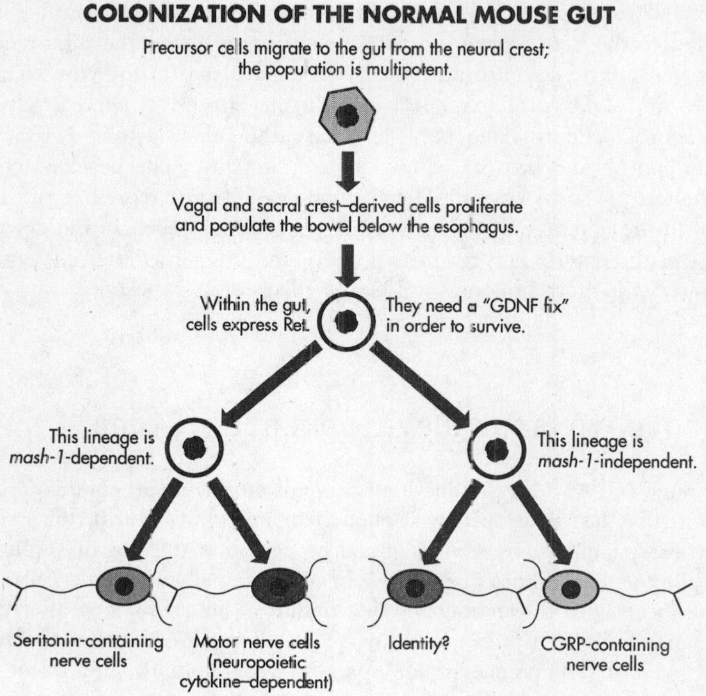

COLONIZATION OF THE NORMAL MOUSE GUT

Precursor cells migrate to the gut from the neural crest;
the population is multipotent.

Vagal and sacral crest–derived cells proliferate
and populate the bowel below the esophagus.

Within the gut,
cells express Ret.

They need a "GDNF fix"
in order to survive.

This lineage is
mash-1-dependent.

This lineage is
mash-1-independent.

Seritonin-containing
nerve cells

Motor nerve cells
(neuropoietic
cytokine–dependent)

Identity?

CGRP-containing
nerve cells

illustration.) The sequential order of the steps is as important as the form of the steps themselves. The early founders of the enteric nervous system express Ret when they enter the gut and need to be caressed by GDNF, which is produced by non-neural-crest-derived cells in the wall of the gut. The founders perish if GDNF is taken away from them, and if the stimulation provided by GDNF is absent, there will be no enteric nervous system at all. If the first step (GDNF stimulation) is botched, therefore, it no longer matters what comes next.

When the founding neural crest–derived precursors are suitably stimulated by GDNF, they go on to sort themselves into two distinct cell lineages, one of which must express the *mash–1* gene, while the other does not have to. This is the next step in the combination. If the *mash–1* gene is knocked out, then TC cells do not appear, and all of the early-developing nerve cells that arise from TC cells are lost. Late-developing nerve cells, however, go on to differentiate on schedule. When *mash–1* is missing, there are thus no serotonin-containing nerve cells (and probably no motor nerves to the smooth muscle), but there are nerve cells that contain CGRP and there are glia. The *mash-1* lineage then appears to sort itself into still more restricted sublineages. One of the sublineages becomes dependent on stimulation of the neuropoietic cytokine receptor and forms the motor nerve cells that talk to smooth muscle. This is a third step in the combination. NT-3 and TrkC probably effect even smaller groups of nerve cells than the neuropoietic cytokine because, unlike the other factors, NT-3 and TrkC can be knocked out without lethally impairing the development of the enteric nervous system. Their actions would thus represent still later steps in the enteric nervous system's developmental ballet. In the development of the enteric nervous system, therefore, the earlier a given signal is required, the more profound will be the effects of its knockout.

Serotonin as a Developmental Signal

Although it is obvious that life itself depends on having an enteric nervous system that has developed well enough to function at birth, the enteric nervous system of a newborn animal or person is still not fully mature. Development continues postnatally. In mice, new enteric nerve cells continue to be born throughout the first month of life. How long after birth humans continue to add enteric nerve cells has never been determined, but given the difference in life span between humans and mice, one mouse month would be the equivalent of a little more than three human years. The postnatal acquisition of new enteric nerve cells means that an infant's nervous system is still plastic and developing. It is therefore possible that the early experiences of a young bowel can affect the "personality" of the second brain that matures within it.

Although I have never seen a study that proves the point, it is often said that children with intestinal colic grow up to be adults with the irritable colon syndrome. It is difficult to get reliable information about the relationship between the experience of the early gut and the behavior of the mature bowel. Still, recent observations of the effect of serotonin on the development of enteric nerve cells have given me a molecular reason to take seriously the idea that what happens to a gut during childhood may affect the kind of bowel it grows up to be.

The early development of serotonin-containing enteric nerve cells raises the possibility that serotonin could be a growth factor as well as a conventional neurotransmitter. That possibility was strengthened by electron micrographs taken years ago by Diane Sherman of the developing guinea pig enteric nervous system. These pictures revealed that early-developing enteric nerve cells actually form synapses on the dividing precursors of enteric nerve cells. The serotonin-containing EC cells of the gut's lining also develop before the bulk of enteric nerve cells are born. Enteric nerve cells that develop late, therefore, are probably exposed to serotonin as they differentiate. To be affected by serotonin, however, these developing nerve cells would have to express appropriate receptors.

As I have already said, enteric serotonin receptors are things that I have been studying for a long time. Most recently, I have been collaborating with a young colleague, Elena Fiorica-Howells, in an effort to clone the gene that encodes the still-elusive 5-HT_{1P} receptor. Elena is nothing if not resourceful. Although she has not yet been able to clone the 5-HT_{1P} receptor, she has, while trying to do so, cloned several other serotonin receptors from isolated myenteric ganglia. One of these, the 5-HT_{2B} receptor, provided an interesting surprise. This receptor had previously been thought to be expressed in the gut only by the smooth muscle in a particular region (the fundus) of the stomachs of mice and rats. It was thus hard to take the 5-HT_{2B} receptor seriously, because there is no equivalent region in the human stomach. The fundus of the rat and mouse stomach is a specialized compartment where symbiotic bacteria live and digest the cellulose that these animals eat. Despite the prevailing belief that the 5-HT_{2B} receptor is a restricted muscle receptor, Elena cloned it from nerve cells of the small and large intestines of guinea pigs, as well from those of mice and rats. Since the 5-HT_{1P} receptor was, for Elena, a lemon, she decided to use the 5-HT_{2B} receptor to make lemonade.

Elena obtained antibodies from a French collaborator, Luc Maroteaux (who had also cloned the mouse 5-HT_{2B} receptor), to locate the 5-HT_{2B} receptor in tissues by immunocytochemistry, and she produced molecular probes so that she could use in situ hybridization to locate the cells in the gut that synthesized the 5-HT_{2B} receptor. The results of her experiments could not have been predicted. In the mature rodent bowel, the 5-HT_{2B} receptor was indeed found in ganglia of the myenteric plexus, but only in a very small number of cells. The receptor could be detected in no more than one cell in every four or five adult myenteric ganglia. In fetal mice, however, Elena found that the 5-HT_{2B} receptor is amazingly abundant and is produced in many cells in virtually every ganglion.

Detailed studies showed that the 5-HT_{2B} receptor is "developmentally regulated." This means that its expression rises and falls during the course of development, like the Roman Empire during the course of history. The 5-HT_{2B} receptor makes its debut in the gut of fetal mice on the fourteenth day of gestation, blazes away on days fifteen and sixteen, and then fades away to adult levels of expression by the time the mice are born, about eighteen days after conception. These times are interesting because of what is happening simultaneously with enteric sources of serotonin. The 5-HT_{2B} receptor is at its height when the last of the enteric serotonin-containing nerve cells are born (on gestational day fifteen) and when the serotonin-containing EC cells first appear (on gestational day sixteen). The timing of its expression thus suggests that the 5-HT_{2B} receptor in the fetal bowel might well occupy itself with developmental actions of serotonin.

Elena and I decided that the preliminary data were good enough to see whether serotonin and the 5-HT_{2B} receptor really do influence the development of enteric nerve cells. We were also encouraged to do so by data that Luc Maroteux had acquired from studies of cranial neural crest–derived cells (the region in front of the vagal crest) and the cardiovascular system, which suggested that 5-HT_{2B} receptors might exert developmental effects. Elena isolated neural crest–derived cells from the fetal mouse bowel at ages when she had found that the 5-HT_{2B} receptor is present, and cultured them in defined media in the presence or absence of serotonin. Sure enough, serotonin acted like a growth factor and promoted the development of enteric nerve cells in her cultures at least as well as the well-known growth factor NT–3 had done in the prior experiments of Alcmène Chalazonitis; furthermore, Elena could mimic the effects of serotonin with a drug called DOI (a derivative of LSD), which is known to be an agonist that stimulates the 5-HT_{2B} receptor. Elena could also block the effects of serotonin and DOI with 5-HT_{2B} antagonists (methysergide or ritanserin). These observations are compatible with the idea that serotonin is a growth factor that influences the development of the enteric nervous system.

Many investigators had previously suggested that serotonin might affect the development of the brain. The evidence in favor of these suggestions is strong but rather indirect. Elena's work is therefore important because it is the first direct demonstration that serotonin really does promote the development of at least some vertebrate nerve cells. The idea that serotonin might be a growth factor as well as a neurotransmitter thus passes out of the realm of suspicion and into the world of established actions. We now know what serotonin *can do* to developing nerve cell precursors. Given the presence of serotonin in the fetal gut and the developmental regulation of 5-HT_{2B} receptors, its seems likely that promotion of the development of enteric nerve cells is something that serotonin actually *does* in real life. That conclusion, however, remains to be proven.

Neurotransmitters are released by nerve cells that are stimulated. Since a neurotransmitter, serotonin, may well affect the development of enteric nerve cells, it follows that whatever activates the serotonin-containing nerve cells (or other developmentally active neurotransmitters) in the immature bowel is likely to exert effects, for better or for worse, on the development of the enteric nervous system. The experience of an infant's bowel determines how active various enteric nerve cells become, and when these cells are turned on or off. By thus regulating the activity of enteric nerve cells and the release of developmentally active neurotransmitters, like serotonin, experience may alter the course of enteric nervous system development. It follows that it would probably be wise to treat the immature gut well. Unfortunately, no one now knows what the immature enteric nervous system considers to be good treatment. Happily, that question provides yet another reason for more research to be done.

Lineage

The story of the development of the enteric nervous system is strikingly like a chronicle of immigration to America. Irving Howe could have written a sequel to *The World of Our Fathers* about it, although he might have had a bit of trouble with the title. *The World of Our Fathers' Bellies* has an unfortunate ring to it, and *The Guts of Our Fathers* carries a different meaning. The precursors of enteric nerve cells are all émigrés from a faraway place. The neural crest is their old country. What urge drives these intrepid cells to migrate to the promised bowel? It is unclear. Could they be trying, in an inchoate way, to escape the tyranny of the brain and its vassal, the spinal cord? Are the neural crest–derived émigrés searching for the liberty and freedom of the gut where the enteric nervous system can operate independently of central nervous system control? The routes that the émigrés follow from the neural crest to the bowel are defined, but they are long and tortuous. New cells are born and old cells falter during the journey to their destination in the gut. The colonization of the bowel is thus filled with peril, and cell death stalks the enteric ghettos established by the newly arrived immigrants.

As the neural crest–derived émigrés mature, they interact with the environment that they find in the lower east colon (and the rest of the bowel). The progeny that develop from the original immigrants are irrevocably altered by this interaction, and they are quite different from their parents. Still, nature as well as nurture comes to the fore. Lineage, no less than the environment of the gut, plays a role in the development of the younger generation.

Italian Americans, Polish Americans, African Americans, Asian Americans, and Irish Americans are all Americans, and they are vastly different from the Italians, Poles, Africans, Asians, and Irish of their former lands. The American experience affects even the first generation born in the New World; nevertheless, each of these groups of hyphenated Americans is also different from the others. Lineage matters. Immigrants reflect their roots because parents transmit a heritage from one generation to the next. So it is with neural crest–derived émigrés in the bowel. They too reflect both the effects of their environment and the influence of their heritage. It is this ferment of interacting forces that produces the incredible diversity of the nerve cells of the gut, which work together in the rich, multicultural, and polyglot melting pot of the enteric nervous system.

11

LOCATION, LOCATION, LOCATION

THE MOST SATISFYING thing about a long and arduous journey is not the first step but the last. So too for the cells that come to the bowel from the neural crest. The last step is often the most troublesome, and in some people[1] it is never taken. For the Israelites who escaped from Egyptian bondage, for example, it surely must have been nice to put the pyramids behind them. Still, once the hoopla of the breakout from Egypt was over, what the Israelites had to look forward to was forty years of wandering through the Sinai desert. If you have never seen the Sinai, let me tell you that you would find forty years of wandering in it a trying experience. The joy of the parents in starting the Exodus from Egypt, therefore, must have been like nothing in comparison to the jubilation of their children in taking possession of Canaan. For neural crest cells, the terminal colon is Canaan.

Every year at Passover, Jews are obligated to retell the story of the Exodus. There is one seder table where a biological parallel to the Exodus from Egypt comes to mind. At my seder table, the exodus from the neural crest of the progenitors of the second brain is remembered. My concern is not for the escape of these precursors from the bondage of the neural folds, or even about their wandering though the wilderness of the embryo. Instead, I will concentrate on the designated sites between the layers of intestinal muscle that lie at the end of their journey and that have been given to the descendants of the original émigrés to found their own nervous system.

[1]Those with Hirschsprung's disease, also called congenital megacolon, which is the congenital absence of ganglion cells in the terminal colon.

The Long March

Most of the time the colonization of the bowel by cells from the neural crest goes as smoothly as the colonization of Canaan went for the Israelites. In fact, the colonization of the gut usually goes even more smoothly, because there are no enteric Canaanites that have to be displaced, and there is no need for a neural crest–derived Joshua to lead a cellular army of conquest. Cell death is important in the formation of some organs such as the kidney, where each final kidney is the successor to two predecessors (as if the highest civilization displaces two earlier ones).[2] In some people and in some animals, however, the exodus from the neural crest to the bowel fails, and when it does, the failure occurs at the end of the trip, in the terminal colon. Like Moses at Nebo, the ill-fated neural crest–derived cells approach, but cannot reach, their destination. The failing vagal neural crest–derived cells enter the colon, but they stop migrating long before they reach its lower end. At the same time, the sacral neural crest–derived cells arrest their migration at the colonic border and do not even enter the bottom of the bowel. When the exodus from the neural crest is incomplete, the terminal colon becomes aganglionic. As we have already seen, the absence of ganglia in a segment of gut is a catastrophe that, unless a surgeon intervenes, is incompatible with life.

An intact and functioning enteric nervous system is an absolute requirement for intestinal propulsion; therefore, when a region of the bowel has no ganglia, it obstructs the flow of intestinal contents. As surely as if a dam had been constructed inside the gut, fecal material backs up above the aganglionic zone, causing the overlying bowel to dilate. Aganglionosis equals obstruction, no matter whether the bowel is congenitally aganglionic, as it is in patients with Hirschsprung's disease, or whether ganglia die as they do in patients with Chagas' disease. Hirschsprung's disease, or congenital megacolon if you hate eponyms, is not yet epidemic, but as I noted earlier in discussing the sacral region of the neural crest, Hirschsprung's disease is a relatively common birth defect. A variety of animals, including mice, rats, rabbits, and horses, are born with the same problem as patients with Hirschsprung's disease. These "animal models" are named for their appearance (spotted), their fate (death), and their species.

Mutated genes are the source of the trouble for all of the animals, who inherit their aganglionosis from their parents. Mutated genes are also the source of the trouble in Hirschsprung's disease, but in contrast to the animals, patterns of inheritance are not always clear, and the faulty gene is only sometimes

[2]During development, a primitive *pronephrous* forms and is replaced by a *mesonephrous* and finally by the *metanephrous,* which is our kidney. Bits of the earlier kidneys are retained, but most of them die off like the ancient Canaanites.

known. Hirschsprung's disease may run in families, but it also has an annoying tendency to appear spontaneously, like a random audit by the Internal Revenue Service. Moses unquestioningly obeyed the command of the Eternal to stay out of Canaan. I, on the other hand, have devoted a great deal of time to asking why cells from the neural crest are denied entrance into the terminal bowel in humans with Hirschsprung's disease and in the mice that model that condition.

Colonization of the Colon

My first studies of the problem of congenital megacolon were carried out with Taube Rothman and utilized a strain of mouse called "lethal spotted." These animals had been discovered at the Jackson Laboratory in Bar Harbor, where the massive breeding of mice and the perceptive observation of their offspring provides many animal models of human disease. Lethal spotted mice are not monsters from outer space who kill people. Quite the contrary, it is they who die. The "lethal" in the name "lethal spotted" reflects the fact that, in contrast to patients with Hirschsprung's disease, it is not customary to rescue the animal by cutting out its aganglionic terminal colon. The cells that color a mouse's coat, like enteric nerve cells, are derived from the neural crest; therefore, the association of spots and aganglionosis is not that surprising.

Lethal spotted mice inherit both their spots and their megacolon as a recessive trait. What this means is that each of two possible genes[3] at the

[3]The lethal spotted gene was called *ls* in the days before anyone knew what protein the gene encoded. Genes are strung out along a chromosome, like a string of pearls in a necklace. Since chromosomes come in matched pairs, there can be two genes (or *alleles*) at every chromosomal site (or locus), one for each of the chromosomes of the pair. The two alleles at each locus may or may not be the same. If they are the same, each allele directs the synthesis of the same product. If the two alleles are not the same, then each allele directs the synthesis of its own product in one cell or another. In this case, however, either product is only half as abundant as that produced when both alleles are identical. One gene of a locus may become mutated and encode a protein that either cannot be produced or which is unable do its job if it is produced. If half the usual amount of protein is adequate to maintain whatever function the protein is needed for, the mutation will be inconsequential. The normal gene covers up for its abnormal partner and is said to be dominant. This is the case for lethal spotted mice. If the genetic makeup of the animals is *ls*/+, meaning there is one mutated and one wild-type allele, the mice are fine. To acquire spots and aganglionosis the mice have to be *ls*/*ls*. An *ls*/+ parent can contribute, as a matter of chance, either an *ls* or a + gene to its offspring; therefore, you would expect only ¼ of the pups (the unlucky winners of the sweepstakes) who get the *ls* gene from each parent to become abnormal. On the other hand, if either parent is +/+, then none of the offspring will have the disease, no matter what the genetic makeup of the other parent might be. In contrast, when *ls*/*ls* mice breed with one another, all of the offspring are abnormal.

chromosomal sites that deal with this issue must be mutated in order to make a mouse inherit the condition. If either gene is *wild-type* (not mutated) in nature, the mice are normal. If you breed two affected lethal spotted mice with one another, neither the mother nor the father has a normal gene to contribute to their pups. Both genes in both parents are mutated. One hundred percent of the offspring will thus be affected by the disease. It is difficult, but possible to keep lethal spotted mice alive long enough to get them to breed. When sex is concerned, a mouse's fortitude is truly admirable. Animals with a colon so dilated that they can hardly walk will nevertheless be amorous and breed.

The beauty of the recessive inheritance, from an experimental perspective, is that every single fetus that results from a coupling of lethal spotted parental mice can safely be presumed to be a developing lethal spotted mouse. Early stages in the development of aganglionosis can therefore be studied in fetal animals. It is not necessary to wait for the bowel to become aganglionic to know that it is going to become so; moreover, the location of the region of the gut that will become abnormal is also known, even before the abnormality occurs.

The guarantee of transmission of the lethal spotted trait to all of the offspring of lethal spotted parents absolutely was what allowed us to consider investigating what goes wrong in the formation of the enteric nervous system in these mice. Back in the days when we started these experiments, we needed the guarantee because we did not know the product of the gene that was mutated in the mice, and we had no molecular probes to use to identify affected animals. Of course, once a mouse acquires spots, the identification of affected animals is easy; however, coat color is acquired postnatally, and there are no spots on the skin of a fetal mouse. We also had no molecular markers to use to identify enteric nerve cells or their precursors. As a result, for us to know that a cell was a nerve cell, it was actually necessary to wait for it to mature and become obviously recognizable by its form or neurotransmitter. We thus devised an indirect assay to detect the presence of neural crest–derived precursors of enteric nerve cells in the fetal gut.

Our method of precursor detection was to grow the bowel in tissue culture, wait a week, and then determine whether or not nerve cells developed in the culture dish. The concept was simple. As admittedly migratory as neural crest–derived cells are, they cannot cross a room and leap into a culture dish; therefore, if nerve cells developed in our cultured segments of gut, their precursors must have been in the bowel at the time the original segment was removed from the animal. By dividing the fetal mouse gut into many small segments and culturing each of them in a separate dish, we could locate the level of nerve cell precursors in the bowel at any age.

Our indirect assay revealed that the colonization of the mouse bowel by nerve cell precursors was a much faster process than Nicole Le Douarin's studies of chick-quail chimeric embryos had led us to expect. It did not seem to take the mouse's neural crest much more than a day to colonize the entire gut. That explosive day, moreover, occurred very early in gestation, beginning on about the middle of the ninth day. Once we had satisfied ourselves that the assay worked, Taube and I went on to compare the colonization of the fetal bowel by nerve cell precursors in normal and lethal spotted mice.

The oral part of the gut of the two types of mouse appeared to be colonized at more or less the same time. The major difference between lethal spotted mice and their normal colleagues appeared to be that the last two millimeters of the lethal spotted gut did not appear ever to become colonized. Nerve cells developed in the cultures obtained from every other segment of the gut, but the cultures of the last two millimeters never yield any. We began to suspect that the problem faced by lethal spotted mice, and by analogy, patients with Hirschsprung's disease, might have at least as much to do with their colon as with their neural crest. We also thought that it was likely that the precursor cells simply stayed out of what we called the "presumptive aganglionic zone"; nevertheless, our experiments did not rule out the alternative possibility that neural crest–derived cells do enter this zone, but that after doing so they die.

Long after our indirect experiments were published, Raj Kapur, a pediatric neuropathologist at the University of Washington in Seattle, introduced a bacterial gene encoding an enzyme, beta-galactosidase (this is the same enzyme that Howard Pomeranz, Bob Payette, and I had used to identify virally infected cells) into a strain of mice. Mice that express foreign genes, like this one, are called *transgenic* animals, and the gene that is experimentally introduced into a transgenic animal is called a *transgene.* The bacterial beta-galactosidase that Raj introduced to his transgenic mice was under the control of a promoter that induced nerve cell precursors to express the bacterial protein while migrating down the gut. Raj used the promoter for DBH, which you may recall is the enzyme that enteric nerve cells derived from TC cells never succeed in repressing. The DBH-beta galactosidase transgenic mice turned out to be really marvelous animals for the study of the development of the enteric nervous system. The beta-galactosidase that vagal crest–derived cells express makes them as easy to recognize in the bowel as a McDonald's restaurant on a highway in the desert.

By crossing his DBH-beta galactosidase transgenic mice with lethal spotted animals, Raj was able to produce lethal spotted pups with labeled vagal neural crest–derived cells. He could thus follow the migration of vagal neural crest–derived cells down the lethal spotted gut. Raj found that the migration of beta galactosidase–labeled neural crest–derived cells in the lethal spotted bowel remains normal until the cells reach the colon. After the cells enter the large intestine, however, their progression slows down, becomes erratic, and finally stops, well short of the anus. Both our data and Raj's thus seemed to suggest that the neural crest–derived cells of lethal spotted mice were not themselves abnormal. If they had themselves been defective, something should have gone wrong with their migration through the small intestine. It also seems unlikely that the enteric nervous system of the small intestine would develop normally, as it does, both in patients with Hirschsprung's disease and in lethal spotted mice, if the neural crest–derived precursors of nerve cells were substantially defective. Raj's observations, like ours, thus pointed toward something being very wrong with the colon of lethal spotted mice.

The Matrix

We began to look at the lethal spotted colon. Bob Payette, the neurologist from the University of Pennsylvania who had joined my laboratory, decided that the smart money was on an examination of the stuff through which cells migrate, the extracellular matrix. His reasoning seemed sound to me. If cells had trouble getting into a piece of the body's turf, then it was certainly possible that the terrain might be rough. In particular, Bob thought that it would be interesting to investigate some of the molecules of the extracellular matrix to which cells were known to adhere. Logically enough, these substances are known collectively as *adhesion molecules.* Bob's first experiment turned up something interesting. A very large adhesion molecule, by the name of *laminin,* was extremely abundant in the terminal region of the lethal spotted colon. A segment-by-segment comparison with the normal bowel revealed not only that laminin accumulated in the region of the lethal spotted gut that became aganglionic, but the distribution of the laminin within this region was also abnormal. The accumulated laminin was concentrated in the outer portion of the wall of the gut, which put it right in the path of neural crest–derived cells attempting to colonize the bowel. The excess of laminin, furthermore, could be detected even before nerve cells were due to appear in the terminal colon. We began to wonder whether an overabundance of laminin might help to keep the neural crest–derived cells out of the terminal colon.

Bob's results were intriguing, but they were not free of problems. We had not expected to find an excess of an adhesion molecule in the presumptive aganglionic gut. Since the lethal spotted trait was inherited as a recessive, we expected that the genetic defect would cause whatever the product of the mutated gene might be to be deficient. While we realized that the loss of a regulatory factor might indirectly cause another protein to become overly abundant, we were very uncomfortable with the idea that an excess of an adhesion molecule could be the cause of the failure of neural crest cell migration.

Migrating neural crest–derived cells stick to the structural elements of the extracellular matrix, mostly collagen fibers (the biological rope that holds tissues together), because the cells are glued to them by adhesion molecules. The failure of neural crest cell migration would thus be the kind of effect you would anticipate from a deficiency, not an excess, of an adhesion molecule. If cells cannot get a purchase of the ground over which they have to crawl, they are likely to slip and slide and get nowhere. Such nonadherent cells would be expected to behave like cars without chains or snow tires on the road after an ice storm. The effects of an excess of an adhesion molecule were more difficult to envision. We wondered whether the cells that were migrating through the lethal spotted gut became stuck too tightly to their supporting fibers. The accumulated laminin might thus have turned the extracellular matrix of the lethal spotted colon into something like flypaper, which trapped the cells after they hit it and would not let them move on. The concept of the extracellular matrix as flypaper, however, was not very compelling. In my experience, flypaper never works. Some flies stick to it, but most get by and continue to do their annoying flylike things.

Bob immediately discovered more than just conceptual problems with the laminin-as-the-cause-of-aganglionosis hypothesis. His first problematic discovery was that laminin was not the only molecule that accumulated excessively in the presumptive aganglionic region of the lethal spotted colon. He found that the lethal spotted colon also contains an excess of every other molecule that, like laminin, is a constituent of a structure called a *basal lamina.* Basal laminae are relatively amorphous bands of material that separate connective tissue from the cells that line the body's cavities and that encircle muscle cells. The lining of the intestine thus sits on a basal lamina, and smooth muscle cells of the gut are each wrapped in one. Basal laminae also surround blood vessels and can act as filters, which hold back cells and large molecules while allowing liquids and small molecules to pass through them. The defect in the lethal spotted animals, therefore, appeared to be one of basal lamina regulation, not just laminin production.

Bob's second problematic discovery was that laminin and the other constituents of basal laminae did not just accumulate in the lethal spotted colon. They were equally abundant throughout the pelvic region of the mice. The colon seemed almost to be involved as an innocent bystander. It had too much basal lamina material because it happened to run through a zone where the massive accumulation of that stuff was the fashion. Since laminin accumulated outside of the lethal spotted colon as well as in it, and the overabundance of laminin in lethal spotted mice was not specific to that molecule, we began to think that the laminin abnormality might just be one of a package of defects in a strange mouse. Bob's finding thus might not be causally related to the development of aganglionosis. I was, however, not quite ready to forget about laminin. Certainly, bad things sometimes occur more as a result of an accident than by design. Conceivably, therefore, the defect in the gut of lethal spotted mice could be the adventitious result of a regional fault in the regulation of the production of a group of molecules, one of which, laminin, happens to interfere with the migration of neural crest–derived cells. Although this speculation, however, kept the theory alive, it was clear that unless new evidence was uncovered in a hurry, Bob's hypothesis that laminin might contribute to the development of aganglionosis was going to be abandoned.

While we were thinking about what seemed to be nonspecific aspects of the laminin excess in lethal spotted mice, Betty Hay, then the chair of the anatomy department at Harvard, threw more cold water on the idea that an overabundance of laminin could be causally linked to a failure of neural crest migration. Betty had just published a paper that demonstrated that laminin actually promoted the migration of primary neural crest cells away from the closing neural tube. Betty had been one of my teachers in medical school, and I was therefore very embarrassed when she told me that my suggestion that laminin might contribute to the failure of neural crest cells to colonize the bowel was ludicrous. If anything, Betty thought an abundance of laminin in the lethal spotted gut would be more likely to promote the migration of neural crest–derived cells than to stop it. I hate to be told that my ideas are ludicrous (even when they are); therefore, I pointed out to Betty the truism that the neural crest–derived cells that enter the gut, which were the cells we were investigating, are not the same cells as those in the process of leaving the neural tube, which were the cells that she was studying. The premigratory neural crest cells have not had the opportunity to learn from the information imparted to neural crest–derived cells as they migrate to and within the bowel. Our cells were thus older and wiser than her cells. Laminin might thus exert effects on neural crest–derived cells in the gut that are completely different from those Betty had observed. Betty was not persuaded, and neither, despite my brave talk, was I. Before I was going to go to the mat for laminin, I would need some additional data.

The new data that I wanted was some solid information about the responses to laminin of the neural crest–derived cells that had actually colonized the gut. I also wanted to know if the crest-derived cells in the bowel actually come into contact with laminin as they migrate through the wall of the gut. Immunoselection provided an opportunity to find out what laminin can do to neural crest–derived cells that have entered the bowel, and the offer was one that I could not afford to turn down. I thus suggested that Alcmène Chalazonitis get together to do the experiments with Virginia Tennyson, who was then collaborating with Taube Rothman and me to study lethal spotted mice. Virginia had been at Columbia for many years and had welcomed me to the school when I was recruited. Since some of the older faculty members of the Department of Anatomy were unenthusiastic about my devotion to research and frankly hostile, Virginia's welcome won her a very special place in my heart. When the funding for her own research on the development of the dopamine-containing nerve cells of the brain dried up, I was very happy to have Virginia join forces with me to study the gut.

Alcmène and Virginia used a novel primary antibody to immunoselect cells from the fetal mouse gut. We had been given this antibody by Hynda Kleinman, a research scientist at the National Institutes of Health. The antibody binds to a receptor for laminin that Hynda had recently discovered on the surfaces of nerve cells. Hynda called this receptor *LBP110.* The initials LBP stand for *l*aminin *b*inding *p*rotein, while the number 110 refers to its molecular weight (which is measured in units called kilodaltons).

Howard Pomeranz and Bob Payette had previously used Hynda's antibodies to the LBP110 receptor in an immunocytochemical study of tissue sections and found that the antibodies bound to the surfaces of neural crest–derived cells in the gut. Interestingly, the antibodies did not recognize cells in the premigratory neural crest, and they did not even react with the neural crest–derived cells that were in the process of migrating to the bowel. It was only after the neural crest–derived cells entered the gut that they acquired the LBP110 receptor and could be seen by Hynda's antibodies.

Alcmène and Virginia now found that the antibodies to the LBP110 receptor could be used to immunoselect cells from the gut, and the cells that the antibodies picked out were precursors of nerve cells. When the immunoselected cells were grown in tissue culture following immunoselection with antibodies to the LBP110 receptor, nerve cells selectively developed in the cultures of immunoselected cells. In contrast, cultures of residual cells, the cells that were not selected by the antibodies to the LBP110 receptor, contained few or no nerve cells. Taken together, these observations confirmed that the neural crest–derived cells that migrate to the bowel really are different from cells of the premigratory neural crest, just as I had maintained to Betty Hay that they might be. The set of neural crest–derived cells that were our concern, those in the gut, expressed the LBP110 receptor, while the neural crest cells that were Betty's concern, those in the premigratory neural crest, did not. More importantly, the acquisition of the LBP110 receptor expression only after neural crest–derived cells enter the bowel suggested that laminin might affect the enteric population differently from the way it affects their predecessors.

The Flypaper Is Really a Growth Factor

When Alcmène and Virginia exposed the neural crest–derived cells they had immunoselected from the fetal gut to laminin, the effects were dramatic. Laminin strongly promoted the development of nerve cells. At the time Alcmène and Virginia began their experiments, laminin was not thought of as a growth factor. Nevertheless, if laminin was not a growth factor, it was giving a pretty good imitation of one. When we reported the observation, critics sniffed at it and said that laminin, which is an adhesion molecule, probably just allowed nerve cell precursors to stick to the floor of the culture dish. In the absence of laminin, the critics postulated, the nerve cell precursors might just wash away. In their view, the reason we saw more nerve cells when laminin was present in the cultures of immunoselected cells was that laminin induced nerve cell precursors to hang on, not to differentiate. To cope with this criticism, we changed the design of the experiments to allow the cells to become adherent to the culture floor before exposing them to laminin. Cells adhere nicely to another molecule, a synthetic polymer, *polylysine.* We thus cultured immunoselected cells on polylysine for twenty-four hours. Following this "adhesion interval," we washed away all of the loose cells and then added laminin to the medium in a soluble form. Laminin again promoted the development of nerve cells, and its effects were just as spectacular as when laminin was present at the start of the experiment and a component of the floor of the cultures.

Laminin exerted another effect that we thought was unlikely to reflect its role as an adhesion molecule. Laminin, even when it was added to immunoselected cells that had already adhered to the culture dish, stimulated the cells to express the immediate early gene, *c-fos.* As discussed previously, Fos, the protein encoded by *c-fos,* is involved in turning on a cell's genetic apparatus when the cell is appropriately stimulated. Nerve cells, as we have seen, express Fos when they become physiologically active, and almost all cells express Fos when they respond to a growth factor. The expression of Fos induced by laminin, moreover, showed all of the characteristics expected of a response to a growth factor. Fos expression could be detected in the nuclei of cells exposed to laminin shortly after exposure, and their expression of Fos was quite transient. As I said, laminin was either a growth factor or a superb impersonator of one.

By this time we had developed a healthy suspicion that the LBP110 receptor is responsible for mediating laminin's effects on the development of nerve cells. To determine whether the LBP receptor on the surfaces of the neural crest–derived cells actually mediates promotion of the development of nerve cells by laminin, we took advantage of the molecular information that Hynda Kleinman had obtained about the LBP receptor-laminin interaction. Laminin comes in many varieties. In fact, there is not one laminin molecule but several. The various laminin molecules (laminin–1, laminin–2, etc.) are modular assemblies of three different classes of subunit, called alpha, beta, and gamma. The subunits of each class are assigned numbers to keep track of them, so that there is an alpha1, an alpha2, a beta1, a beta2, a gamma1, etc. The type of laminin in the fetal bowel was not known when Alcmène and Virginia were doing their experiments, but they did know that the type they were adding to the immunoselected cells was laminin–1, which is composed of an alpha1, a beta1, and a gamma1 subunit. The presence of the alpha1 subunit was significant because Hynda Kleinman had demonstrated that this is the subunit that contains the binding site for the LBP receptor. The LBP110-binding site is a sequence of five amino acids, isoleucine, lysine, valine, alanine, valine. Since the chains of amino acids in protein and peptide tend to be very long, a single-letter amino acid code has been worked out to avoid the tedium of writing out the names. According to the code, the isoleucine, lysine, valine, alanine, valine sequence of the LBP110-binding sequence is IKVAV.

At first, Alcmène and Virginia tried to mimic the effects of the whole laminin molecule by adding a peptide with the IKVAV sequence. They hoped that the binding site alone would suffice to activate the LBP receptor and promote nerve cell differentiation. In fact, nothing happened after cells were exposed to the IKVAV-peptide. The cells simply sat there and looked stupid. Very few nerve cells developed, and Fos did not turn on. The IKVAV-peptide was thus not an agonist. Alcmène and Virginia then added the IKVAV-peptide together with laminin–1, and the cells still sat there looking stupid. Once again, very few nerve cells developed and Fos did not turn on. This time, however, laminin–1 was present and should have promoted the development of nerve cells and activated Fos. That laminin did *not* do so in the presence of the IKVAV-peptide showed that the peptide competitively prevented the larger molecule from binding to and stimulating the receptor. None of the cells, furthermore, fell off the culture dish when they were exposed to the IKVAV-peptide. The peptide thus did not interfere with the adherence of neural crest–derived cells but only with their developmental response to laminin–1. In contrast to the IKVAV-peptide, control peptides (comprised of a nonsense sequence of amino acids or the sequence of laminin alpha1 adjacent to IKVAV) were unable to interfere with the effects of laminin–1. An antibody to laminin alpha1, which covered the IKVAV region, also antagonized the effects of laminin–1. These observations established that laminin–1 is a growth factor, that it powerfully promotes the development of enteric nerve cells, and that it acts through an interaction of its alpha1 subunit with the LBP receptor at the cell-surface.

Having established that laminin can promote the development of enteric nerve cells, it became important to know whether nerve cell precursors actually get to see laminin at the time they differentiate in the bowel. To determine whether the migrating neural crest–derived cells within the gut make contact with laminin while they migrate, Virginia Tennyson and Diane Sherman carried out an electron microscopic study. They needed the high resolution to be able to visualize laminin and to distinguish real contact from close appositions. They used immunocytochemistry with two sets of antibodies to identify simultaneously extracellular matrix molecules and neural crest–derived cells. Each antibody was labeled with gold particles of different sizes so that they could be distinguished from one another when viewed in the electron microscope. The differentiation antigens selectively expressed by cells of neural crest origin provided the marker that enabled these cells to be recognized. Virginia and Diane demonstrated that neural crest–derived cells really do make contact with tufts of laminin, which they found was not restricted to formed basal laminae in the developing gut. Using Hynda Kleinman's antibodies to the LBP110 receptor, they also found that there are clusters of the LBP110 receptor at many of the sites where the neural crest–derived cells made contact with laminin.

The observations that laminin can promote the development of enteric nerve cells, and the fact that enteric nerve cell precursors migrating in the gut encounter laminin, convinced me to take seriously the accumulation of laminin that occurs in lethal spotted mice. I must admit, however, that the ability of laminin to promote the differentiation of nerve cells rather than to inhibit the migration of precursors confused me for a long time. The confusion ebbed when I realized, in one of those sudden moments of revelation that cause me to be religious, that the differentiation of neural crest–derived cells into nerve cells is a great way to stop these cells from migrating. Neural crest–derived cells are great travelers, while nerve cells are as sedentary as cells get to be. Nerve cells plant their bodies in the turf and send axons and dendrites out exploring. The cells themselves have to stay where they are because they are targets that other nerve cells are looking to innervate. It is amazing enough that the trillions of nervous connections that exist in the brain get made when nerve cells sit still. I cannot imagine how axons could find their targets if they were allowed to move.

In any case, when a wandering neural crest–derived cell becomes a nerve cell and sends out an axon, it is as if the cell were dropping an anchor. I also realized that a migrating cell had better reach its destination before it does this. If a cell differentiates too soon, the region that it was supposed to move into is going to miss it. An overabundance of laminin in the lethal spotted gut, therefore, might contribute to aganglionosis by pushing neural crest–derived cells to commit themselves to becoming nerve cells before their time. In other words, there may be too much of a good thing in the lethal spotted gut. Differentiation may thus occur at the expense of continued migration. Migration would thus not be completed, and the last part of the bowel to be colonized would become aganglionic.

While I was trying to imagine a possible mechanism that might relate the laminin excess that Bob Payette had discovered to the development of aganglionosis in lethal spotted mice, several investigators reported that the very same defects of the extracellular matrix that we had found in lethal spotted mice also occurred in the bowel of human patients with Hirschsprung's disease. Laminin and other molecules found in basal laminae were thus overly abundant in the aganglionic zones of men (and women too) as well as mice. Of the basal lamina molecules, only laminin has been shown to be a powerful promoter of nerve cell differentiation. The fact that it is not the only molecule that accumulates in the abnormal gut thus does not mean that its accumulation is trivial. The fact that laminin also accumulates outside of the bowel is also important. Sacral crest–derived cells see this external laminin and have to negotiate their way through it just to reach the gut. The pattern of laminin accumulation in lethal spotted mice and patients with Hirschsprung's disease therefore positions the molecule to interfere simultaneously with both of the streams of neural crest–derived cells (vagal and sacral) that head for the colon. Laminin had begun to look like a pretty good villain in the Hirschsprung's plot.

Taube Rothman and Jingxian Chen made laminin look even more villainous. They showed that each of the three subunits of laminin–1 (alpha1, beta1, and gamma1) are synthesized with excessive zeal in the gut of lethal spotted mice. These molecules do not just accumulate, for example, because the gut's garbage-disposal system is out of order and the stuff hangs around too long. The colon of lethal spotted mice actually makes too much laminin-1. The problem is one of excessive biosynthesis. In addition, laminin-1 expression is developmentally regulated. It is high when enteric ganglia are forming, but then it drops off as fetuses age. This pattern is consistent with the idea that laminin-1 is provided to nerve cell precursors when they need it. At every age, however, the critical alpha1 subunit is far more abundant in the colon of lethal spotted than in normal mice. Laminin-1, moreover, does not accumulate in the gut when the bowel lacks neural crest-derived cells for reasons that are unrelated to the lethal spotted gene. It is not seen, for example, in the gut of Ret knockout mice, which have no neural crest-derived cells at all. The excessive production of laminin-1 that characterizes the lethal spotted and Hirschsprung's bowel, therefore, is a primary result of the genetic defects and is not a secondary consequence of the absence of neural crest-derived cells from the bottom of the gut.

The Genes

My enthusiasm for the idea that laminin plays a role in causing the terminal bowel of lethal spotted mice and patients with Hirschsprung's disease to become aganglionic had been following the trajectory of a roller coaster. I had enjoyed some nice highs, but our work on the effects of laminin alpha1 was followed by a real low. This low came from a discovery so unexpected that the information even shocked its discoverer. The gene that is mutated in lethal spotted mice was identified, and its product turned out to be a signaling molecule that had never previously been suspected of playing a role in development.

The ability to knock out virtually any gene an investigator chooses to delete has had a revolutionary effect on developmental biology. For a long time birds and worms had seemed to be the animals of choice for studies of developing systems. Bird embryos are accessible and amenable to surgical manipulation, while worms are simple, have small nervous systems, and every one of their nerve cells is known. The ability to pick the characteristics one wants an animal to inherit, however, is at least as potent an advantage as any of these properties. Mice have thus acquired an admiring constituency of scientists. The gene knockout technique, however, is full of surprises. Many genes that were confidently thought to be absolutely essential for life, or at least for some critical function, have been knocked out with no discernible consequences. The ability of genes to compensate for the loss of their fellows is prodigious. Take out one gene and others may find a different way to accomplish what the knocked-out gene used to do. In other instances, compensation does not occur, but a function is lost that was not previously suspected to require the expression of the knocked-out gene. Experiments that involve playing games with mouse genes are thus not good for people who like to live predictable lives.

The studies that revolutionized research on the cause of Hirschsprung's disease were experiments in which a gene that was knocked out to study blood pressure regulation unexpectedly produced mice with congenital megacolon. Masashi Yanagisawa, working at the University of Texas Southwestern Medical Center in Dallas, had been studying a group of signaling peptides called *endothelins*. These peptides, which are chains of twenty-one amino acids, come in three flavors: endothelin–1 (*ET–1*), endothelin–2 (*ET–2*), and endothelin–3 (*ET–3*). The endothelins act on two cellular receptors, noted in scientific shorthand as ET_A and ET_B. The ET_A receptor prefers to be stimulated by endothelin–1, accepts endothelin–2, but ignores endothelin–3. The ET_B receptor, on the other hand, is more egalitarian and shows no favoritism for one endothelin over another; ET_B will let itself be stimulated by any of the endothelins. As a result of these properties of the receptors, endothelin–3 is useless unless it has ET_B receptors to work on.

At the time that Masashi Yanagisawa was contemplating knocking out the genes that encode the endothelins and their receptors, the peptides were known to be produced by the cells that line blood vessels,[4] and they were known to cause the smooth muscle that surrounds blood vessels to contract. In fact, it was the endothelins' outrageously potent ability to constrict blood vessels that made them interesting to the biomedical community. This interest was enhanced by the discovery that the endothelins also stimulate the heart to beat faster and harder. Masashi was thus interested in the cardiovascular effects of the endothelins, and he hoped that the knockout of the peptides and/or their receptors would provide insight into the role they play in the physiology of the cardiovascular system, particularly in the regulation of blood pressure.

The knockout mice that Masashi produced never had a chance to manifest cardiovascular problems. The knockout animals all had such severe developmental problems that they did not survive long enough. Among the knockout mice that Masashi produced were a set of animals that lacked endothelin–3 (the endothelin that acts only on ET_B receptors) and another that lacked ET_B (the endothelin receptor that allows itself to

[4]The cells that line blood vessels are called endothelial cells. The endothelins, therefore, received their name from their origin in endothelial cells.

be stimulated by endothelin-3). To his great surprise, animals lacking either endothelin-3 or ET_B were born with spots and a lethal aganglionosis of the terminal colon. Masashi is not a person who lets unexpected results get him down. What began as an investigation of the cardiovascular system, therefore, turned very quickly into an investigation of the neural crest and its ability to colonize the gut.

The Hirschsprung's disease-like syndrome brought about by knocking out endothelin-3 and the ET_B receptor would have been striking if Masashi had published these observations and did nothing else. He did not, however, stop there. Masashi went on to investigate the strains of mice in which similar abnormalities occur naturally as a result of inherited birth defects. One of these was the mouse I had been studying for years, the lethal spotted.

Masashi's additional studies revealed that lethal spotted mice carry a mutation that prevents them from making active endothelin-3. The synthesis of endothelins begins with the production of a large protein, called a *preproendothelin,* which contains 203 amino acids. This preproendothelin is cut down in cells to yield an intermediate peptide, called a big endothelin, which has thirty-nine amino acids. The big endothelin is not itself active. To get to the active twenty-one amino acid peptide, the big endothelin has to be pared back even further, this time by a novel membrane-bound *endothelin-converting enzyme.* Why evolution has made the manufacturing of endothelins such a complicated process is unclear, but as you might imagine, a chain of events as complex as this one can be disturbed at multiple sites.

The mutation in lethal spotted mice causes a tiny mistake to be made in the sequence of preproendothelin. When the big endothelin-3 intermediate is produced from the abnormal preproendothelin in cells of lethal spotted mice, one of its thirty-nine amino acids is the wrong one. The amino acid, arginine, is replaced by another amino acid, tryptophan, near the terminal end of the molecule. Remember that proteins and peptides are written in a language that is spelled out in amino acids. If there is a mistake in the amino acid sequence, then a word (the protein or peptide) is misspelled. The substitution of a single amino acid may or may not have detrimental consequences. It depends on whether or not the mistake changes the meaning of the word or prevents its recognition. The same thing is true of spelling words in English. The substitution of a single letter may be benign, but it can also be disastrous. Consider, for example, the effect of substituting the letter "e" for the letter "i" in the word "six."

Because of the placement of the tryptophan where the arginine should be, the misspelled big endothelin-3 in lethal spotted mice cannot be recognized by the endothelin-converting enzyme, and thus the unfortunate big endothelin–3 is not turned into active endothelin–3. Lethal spotted mice are thus deficient in endothelin–3. Their big endothelin–3 is useless to them because it cannot stimulate the ET_B receptor (big endothelin–3 could not have stimulated the ET_B receptor even if it had the correct sequence; the physiological rule is convert it or forget it!). It makes no difference to a mouse's terminal colon whether endothelin–3 is lacking because the animal cannot produce it from big endothelin–3 or because the gene encoding preproendothelin–3 is knocked out. The net result of both genetic abnormalities is the same: The coats of the mutant mice are spotted and the terminal bowel is aganglionic.

Still another natural mutant was investigated by Masashi and his group. This mouse, called *piebald* lethal, was found to carry a mutation that prevented it from producing ET_B receptors. The final appearance of these mutant mice is identical to that of ET_B knockout animals. For that matter, these mice also look very much like the endothelin–3 knockout and lethal spotted mice. To get the colon to be colonized by cells from the neural crest, therefore, ET_B receptors on some cell, somewhere, at some time during development have to be stimulated by endothelin–3. Which cells require the stimulation of the ET_B receptors and why were not made clear by Masashi's experiments.[5]

Since the lethal spotted and piebald mice are animal models of Hirschsprung's disease, it was logical to determine whether the human condition, like that of the mice, is due to the loss of either endothelin–3 or the ET_B receptor. Masashi and his colleagues found that the genetic defect in some patients with Hirschsprung's disease is essentially the same as that of piebald lethal mice. These patients lack ET_B receptors. More recently additional patients have been found who lack endothelin–3. Unfortunately, neither the mutations in genes encoding ET_B nor those which encode endothelin–3 account for even a majority of cases of Hirschsprung's disease. Hirschsprung's disease evidently can occur when any one of a number of different genes is mutated. The gene encoding ET_B is only one of the culprits that have turned up in patients. Others are *RET, GDNF, SOX10*, and *DLX–2* (the names of many genes are arcane initials). It is not clear how or why mutations of all of these unrelated genes produce the same syndrome. In fact, it was not even clear at the

[5]Since ET_B receptors are able to respond equally well to all of the endothelins, moreover, the question arises as to why endothelin–1 or endothelin–2 do not compensate for the loss of endothelin–3. ET_B receptors are intact in endothelin–3 knockout and lethal spotted mice; therefore, if endothelin–1 or endothelin–2 could get to the ET_B receptors of the mutant animals, there would be no need for endothelin–3. Circulating endothelin–1 or endothelin–2 would have rescued the colon. That they do not do so implies that endothelins do not circulate in fetal mice.

time Masashi published his revolutionary discoveries in three back-to-back papers in *Cell* (with a cover picture in color showing very cute spotted mice) what role(s) endothelin–3 and ET_B receptors might play in the formation of the enteric nervous system.

Neural Crest or Colon, Where Is the Lesion?

Masashi was struck by the coincidence that the coats of all of the animal models of Hirschsprung's disease were spotted. The pigment that colors an animal's coat is called *melanin.* It is produced by cells called *melanocytes.* Like enteric nerve cells, melanocytes are derivatives of the neural crest. One of the defined pathways that leave the neural crest leads to the skin, and the cells that follow this pathway give rise to melanocytes. Masashi thus concluded that endothelin–3 is required for the development of both the enteric nervous system and melanocytes. His hypothesis, which was viable when he put it forward in 1994, was that endothelin–3 and the ET_B receptor are each expressed by migrating neural crest–derived cells. Mashashi proposed that these cells secrete endothelin–3, which then turns around and stimulates their own ET_B receptors. He postulated that the neural crest–derived cells require their ET_B receptors to be activated in order for them to survive and develop as enteric nerve cells or melanocytes. This kind of self-stimulation is called an *autocrine* effect.

I have never been comfortable with the concept of autocrine actions. They sound too much like something my parents told me would put hair on my palms. Still, in the absence of any knowledge to the contrary, the "autocrine hypothesis" was very attractive. It accounted for both the aganglionosis and spots that occurred in animals when either endothelin–3 or ET_B receptors were missing. On the other hand, there was one real problem that was not accounted for by the "autocrine hypothesis." If neural crest–derived cells require the stimulation of ET_B receptors in order to survive and differentiate into enteric nerve cells (or glia, which are also missing from the aganglionic segments of terminal colon in lethal spotted mice), then why was the defect so geographically localized when endothelin–3 or ET_B receptors were missing? Even in piebald lethal mice and patients with Hirschsprung's disease, who have no ET_B receptors at all, enteric nerve cells develop very nicely in the small intestine and first part of the colon.

I vividly remember my first reading of Masashi's papers. My initial feeling was one of unrestrained admiration for the elegance of the work and the clarity of the information. If there was a loose end, or a string Masashi left untied, I could not see it. My second emotion was depression, because I had worked on the lethal spotted mouse for years without coming close to identifying the gene that is mutated in the animals. Everybody knows that they have limitations, and those of us who are older scientists know that our efforts are going to be surpassed by our younger colleagues. In fact, we devote a great deal of time to training people who we actually hope will go on to surpass us. Still, Masashi's three papers in *Cell* hit me like Mozart's *Don Giovanni* hit Salieri in Peter Shaffer's play *Amadeus*. My third attitude was to forget the personal nonsense, go on, and try to relate the new findings of Masashi Yanagisawa to what I had previously learned about the development of congenital megacolon in lethal spotted mice and patients with Hirschsprung's disease.

The failure of Masashi's published hypothesis to account for the normal development of enteric ganglia everywhere in lethal spotted mice and patients with Hirschsprung's disease except the terminal colon bothered me. I decided to call him and ask him how he explained it. I assumed that he would have devoted considerable thought to this problem and would have an explanation, which he might not have wanted to publish, perhaps because the ideas were too speculative. I had never met Masashi before, and I did not know what to expect. He turned out to be very communicative and happy to discuss common problems. He did not, however, know why enteric nerve cells should be able to develop in the small intestine and upper colon in animals that did not have ET_B receptors. He pointed out that his working hypothesis was put forward simply as an idea to be tested. It was not intended to be the last word but rather a useful framework to use in designing future experiments. We agreed, however, that whatever explanation of the effect of ET_B receptor stimulation was ultimately accepted would have to account for the sharp localization of the lesion in endothelin–3/ET_B-deficient animals.

After Masashi's papers were published, Nicole Le Douarin began to explore the effects of endothelin–3 on neural crest cells. She worked, as was her usual custom, with avian neural crest cells. When she added endothelin–3 to neural crest cells growing in tissue culture, the results were stunning. The cells proliferated wildly in response to endothelin–3 and covered the floors of their culture dishes with jet black carpets of melanocytes. Nicole thus proposed that the major effect of endothelin–3 on neural crest cells is to get them to proliferate. Since isolated neural crest cells responded to endothelin–3, she concluded that they must express ET_B receptors. Although most of the vastly enlarged population of neural crest–derived cells generated by endothelin–3 were not nerve cells, Nicole thought that the effect she had observed provided a good explanation for the generation of an aganglionic terminal colon in mice or humans who lack either endothelin-3 or ET_B receptors.

Nicole's idea was that a critical mass of vagal neural crest-derived cells is needed to get the population as far as the terminal colon, which lies at the end of their road. If the population that enters the gut is too small, as she postulated that it would be in animals that were unable to use the endothelin-3/ET_B system to stimulate the proliferation of neural crest émigrés, then the cell mass would peter out before reaching the terminal colon. This idea made good use of the observation that endothelin-3 does indeed cause neural crest cells to proliferate, but like Masashi Yanagisawa's earlier hypothesis, it did not explain all of the data. Some of the data that seemed to me to be least compatible with Nicole's concept were her own.

The idea that neural crest cells might not make it all the way to the terminal colon if there were not enough cells in the émigré population was fine as long as it was applied only to the vagal set of these cells. When the sacral set of neural crest-derived cells was included, however, Nicole's hypothesis became less attractive. Sacral neural crest-derived cells enter the bowel at its bottom end. If they were to reach the gut and not get to the finish line of their migratory route, they would fail to ascend to the umbilicus. There might thus be a gap in the innervation of the colon between the points where the two populations of neural crest-derived cells become exhausted and stop migrating. In the absence of the endothelin-3/ET_B system, however, a gap in the enteric nervous system is not what is found. The aganglionic region is always at the very bottom of the bowel. Nicole's hypothesis, therefore, did not take into account the contribution made by the sacral neural crest to the formation of the enteric nervous system. Since she had, in fact, been the person who discovered this contribution, I was surprised to see her espousing a theory that did not account for it.

Nicole went on, in subsequent studies, to discover that birds have a special subtype of ET_B receptor, which has not yet been found in mammals, that works especially well on cells in the melanocyte lineage. This existence of this receptor probably explains why she produced so many melanocytes and not nerve cells when she added endothelin-3 to cultures of avian neural crest cells. Nicole has also recently documented, once again (in collaboration with Alan Burns), the importance of the sacral region of the neural crest in the formation of the enteric nervous system. The major problem, however, with Nicole's hypothesis attributing aganglionosis of the terminal colon to an inadequate number of neural crest-derived émigrés is that her hypothesis leaves no room for the involvement of any tissue except the neural crest. It has now, however, become quite clear that the aganglionosis of the gut in both lethal spotted (endothelin–3-deficient) and piebald lethal (ET_B-deficient) mice is not just the fault of the neural crest. The colon is also involved.

Taube Rothman and I had actually begun to investigate the question of whether the aganglionosis of the terminal colon of lethal spotted mice is the result of defective neural crest–derived cells or a bad colon years before Masashi Yanagisawa identified the genetic lesion. In collaboration with a postdoctoral fellow, Janet Jacobs-Cohen, we grew segments of gut together with neural crest cells in coculture. We employed a variety of sources of both normal neural crest cells and those from lethal spotted mice. The idea behind our experiments was simple: to determine what neural crest cells would do when tempted with a nice tidbit of gut that was itself empty of cells of neural crest origin. Would normal or lethal spotted neural crest cells colonize such a segment of bowel? Could either of these neural crest cells colonize the presumptive aganglionic bowel of lethal spotted mice?

The results of our studies showed that virtually any source of the neural crest cells that we picked would colonize a segment of normal gut whenever both were present in the same culture dish. There were no species boundaries operating in culture. Quail neural crest–derived cells were happy to colonize a segment of mouse bowel, and mouse neural crest–derived cells were equally delighted to enter a chick or quail gut. We could even use a lethal spotted mouse's stomach or small intestine as a source of neural crest–derived cells to colonize a chick hindgut. In contrast to the ability of almost every kind of neural crest to contribute cells that enter a normal bowel, no source of neural crest émigrés, from anywhere, provided cells that would enter the terminal region of the lethal spotted gut. These observations strongly suggested that the neural crest–derived cells of lethal spotted mice could behave normally when they are confronted with a normal bowel, but that something is radically wrong with the colon of lethal spotted mice.

Taube Rothman went on to test the lethal spotted mouse colon in a second and completely different way. She had, as we have already discussed, previously collaborated with Nicole Le Douarin to use the technique of back-transplantation to evaluate the developmental potential of cells from the neural crest. Those studies had revealed that when back-transplanted neural crest–derived cells that had already colonized the bowel found themselves in a neural crest migration pathway in a younger embryo, they would leave the gut and migrate again through the tissues of their new host. Taube now carried this technique one step further. She back-transplanted the colon from both normal and lethal spotted mice into younger quail embryos. DNA is distributed so differently in the nuclei of mouse and quail cells that it is easy to distinguish a mouse cell from a quail cell in the resulting chimeric embryos. Quail cells, in particular, have no reason to be affected by a mouse's genetic abnormality.

One of the possibilities that the back-transplantation of mouse gut into a younger host was designed to evaluate was whether or not neural crest–derived cells enter the terminal colon of lethal spotted mice and die when they get there. We reasoned that it might be harder for neural crest–derived cells to survive in the terminal colon than in the more cushy regions of the rest of the bowel. If so, then we might be able to rescue these cells by providing them with a way to escape from the colon by leaping into the embryonic environment of a younger quail embryo. In fact neural crest–derived cells departed from the back-transplanted segments of normal mouse colon and migrated to nerves and ganglia in the quail hosts; however, no cells ever left the lethal spotted colon. More interestingly, neural crest cells from the quail host migrated into and through the back-transplanted colon of a normal mouse, but the colon from a lethal spotted mouse stopped the quail cells in their tracks. The quail's neural crest cells formed giant ganglia immediately outside the lethal spotted colon, but they never entered or migrated through it.

The results of our coculture experiments and those involving back-transplantation are incompatible with Masashi Yanagisawa's initial working hypothesis that endothelin–3 is an autocrine growth factor that is both produced and required by neural crest–derived cells. If neural crest–derived cells are the critical source of endothelin–3, normal neural crest cells would have been able to colonize the lethal spotted colon (in coculture or after back-transplantation). Our research showed that even normal neural crest–derived cells could not enter the lethal spotted colon. Our experiments thus suggest that the colon becomes intrinsically abnormal when endothelin–3 is deficient.

Two additional experiments, one that I carried out with Taube Rothman and Dan Goldowitz, a collaborator who is now at the University of Tennessee, and another by Raj Kapur, left no doubt that the defect in mice that lack either endothelin–3 or the ET_B receptor is not just a problem of the neural crest. These studies used what are called *aggregation chimeric* mice. The last chimeras that we spoke about were beasts that contained cells from two different species. Nicole Le Douarin made these chimeras by combining cells from quail and chick embryos. Aggregation chimeric mice are made by combining the embryos of two different strains of mouse. To make these chimeras, mice are mated, and just enough time is allowed for the eggs of the mother to be fertilized and to divide once or twice. The early mouse embryos are then flushed out of the oviducts of the maternal mice. The embryos that one obtains thus consist of two or four cells. Their protective capsule is dissolved off (with an appropriate enzyme), and two embryos are pushed together with a micromanipulator. The embryos fuse together, forming a single chimeric embryo that contains four or eight cells. This is the "aggregation" referred to in the term "aggregation chimera." When the fused embryos are implanted into a surrogate mother, they go through development and are born as the offspring of four parents (a tetraparental mouse). Each one of the fused embryos had a mother and a father; therefore, since the resulting pup is the product of two embryos, it had two mothers and two fathers. This is not science fiction. The cells of each embryo cooperate in making the aggregation chimeric mouse pup, which is thus a mosaic of the cells of each of the original embryos.

In our case, we fused lethal spotted mice with another strain that does not have congenital megacolon. The two strains were selected so that we could recognize the cells of each strain (with an endogenous marker) in the resulting chimeric animals. Almost none of the chimeric mice acquired megacolon. The colons of the chimeric animals did not become aganglionic. The most striking observation, however, was that lethal spotted nerve cells (each of which carried two out of two mutated genes) were found in every enteric ganglion of the chimeric animals, even in the most terminal region of the bowel. Mutant neural crest–derived cells can thus colonize the terminal colon without difficulty, as long as at least some of the cells of the bowel into which they migrate are normal.

Raj Kapur obtained the same results with aggregation chimeras constructed by fusing lethal spotted mouse embryos with those of his DBH-beta galactosidase transgenic animals. He used the beta galactosidase to identify the cells of the normal embryos. Even more interestingly, Raj also fused piebald lethal mouse embryos with those of the DBH-beta galactosidase transgenic animals. The difference in Raj's two experiments is that the lethal spotted mice lack endothelin–3, while the piebald lethal mice have no ET_B receptors. Masashi Yanagisawa had argued that the aggregation chimeras that we and Raj Kapur had studied initially, which involved fusions of normal and lethal spotted mice, did not rule out a defect limited to the neural crest. He suggested that neural crest cells march through embryos side by side. Since aggregation chimeric mice are mosaics and neural crest–derived cells stay close together, the secretion of endothelin–3 by the normal neural crest–derived cells in the chimeric animals might rescue their lethal spotted endothelin–3-deficient neighbors.

At the time these studies were carried out, which cells actually make endothelin–3 was unknown; therefore, the possibility that endothelin–3 secretion by normal neural crest–derived cells could rescue lethal spotted cells in chimeric animals could not be refuted. The importance of Raj Kapur's second study with the ET_B-deficient piebald lethal mice was that it is impervious to this criticism. What Raj's study of chimeras constructed with piebald lethal mice showed was that neural crest–derived cells could colonize the entire bowel even if they lack ET_B receptors. Cells that lack ET_B receptors cannot respond in any way to endothelin–3; therefore, there is no possibility of their being rescued by their neighbors' endothelin–3. It is impossible to explain these data without postulating that something goes wrong in the colon when endothelin–3 is not available or when ET_B receptors are not there to be stimulated.

Our observations on the overproduction of laminin in the lethal spotted gut, and the observations of others that laminin also is overproduced in the aganglionic bowel of patients with Hirschsprung's disease, are, of course, direct demonstrations of something that has indeed gone very wrong in the colon when endothelin–3 is not available or when ET_B receptors are not there to be stimulated. The colon is thus not a passive receptacle for cells from the neural crest but an active participant that determines whether or not it will allow itself to be colonized.

Up to this point in the story, Raj Kapur and I had been in reasonable agreement about what is going on in endothelin–3/ET_B-deficient animals. We still concur on the basic observations, but there are some important nuances of difference in how we interpret the data. Raj and I agree that the bowel does not become aganglionic in endothelin–3/ET_B-deficient animals simply because neural crest–derived cells cannot give rise to enteric nerve cells unless their ET_B receptors are stimulated by endothelin–3. We also agree that the aganglionosis cannot be explained by an abnormality that is limited to the neural crest. In the language of science the aganglionosis is not *neural crest-autonomous*. We have, however, advanced different hypotheses to explain what else, besides neural crest–derived cells, is involved.

Raj's hypothesis is that neural crest–derived cells are normally stimulated by endothelin–3 (via ET_B receptors) to put out a signal that prepares the colon for their entry. I like to think of Raj's suggestion as the heraldic hypothesis. It turns the large intestine into a fine place, staffed by cells whose function is, upon receipt of the news of the imminent arrival of the delegation from the neural crest, to prepare the bowel for a royal entrance. The colonization of the colon by neural crest–derived cells is thus seen as a grand event, like the storied meeting in 1520 of Henry VIII and Francis I on the Field of the Cloth of Gold. It would have been unthinkable for either the king of England or the king of France to arrive at the meeting site in the absence of a suitable welcome. This welcome would have been triggered by a discreet advance message of imminent arrival sent out by the advancing royal parties.

In Raj's idea, the neural crest–derived cells blow their own horns and send the message that allows the colon to prepare for their arrival. Raj also requires that the ET_B receptors on the neural crest–derived cells be stimulated in order for them to do so. The nature of the signal that the neural crest–derived cells transmit to the colon is unknown, as is the nature of the welcome that the bowel makes ready. Still, it is unthinkable, in Raj's hypothesis, for neural crest–derived cells to migrate through a colon that has not been suitably prepared to receive them. I have to hand it to Raj. Who else could conceive of a drama in which the colon plays the role of the Field of the Cloth of Gold?

My ideas are much more plebeian than Raj's. No heralds are needed by the colon to announce the coming of the neural crest. In my view, the signal that the bowel requires to allow itself to be colonized by cells from the neural crest is endothelin–3, and it makes the endothelin–3 all by itself. I believe that endothelin–3 acts on ET_B receptors, which are expressed in the colon by enteric smooth muscle precursors (as well as by cells from the neural crest). In response to endothelin–3, these cells reduce their biosynthesis of components of basal laminae, including laminin–1. In the absence of this effect of endothelin–3, too much laminin–1 is made, and because of the alpha1 subunit of laminin–1, the accumulated laminin–1 provides neural crest–derived cells with too strong a stimulus to differentiate. The neural crest–derived cells thus turn into nerve cells before they have finished colonizing the bowel. Vagal neural crest–derived cells stop migrating in the upper part of the colon, and sacral neural crest–derived cells stop migrating before they even enter the gut. The critical function of endothelin–3, therefore, is to dampen the reception that the colon provides for the neural crest–derived cells.

I see endothelin–3 as a factor that is needed for restraint and moderation, not a royal welcome. In the absence of endothelin–3, the exuberance of the neural crest–derived cells, presumably thrilled by the splendid microenvironment that they encounter in the colon, drives them into a diverting frenzy of differentiation. As a result of this binge, the precursor cells find that they all have become nerve cells before they have finished their mundane march to the end of the bowel. By then, of course, it is too late. The terminal bowel is irreversibly aganglionic.

Actions of Endothelin-3

The power of molecular biology has generated a wave of enthusiasm that has sometimes transcended the bounds of reality. Many people, including a few scientists who should have known better, actually believed that the discovery of the gene or genes responsible for causing a birth defect would carry with it the solution to the problem. Solutions, however, have proven to be elusive. To understand why a birth defect arises, it is necessary, but not sufficient, to know the identity of the abnormal gene. The mechanism of gene action, which is often difficult to work out, is equally important, and is not revealed by the knowledge alone of the identity of the gene or even its product. The finding that mutations in genes encoding endothelin–3 and ET_B receptors cause congenital megacolon illustrate these limitations very well. Although Masashi Yanagisawa made a major contribution to biology and medicine when he made these discoveries, the breakthrough did not by itself reveal what endothelin–3 is required to do in order for the colon to become colonized by cells from the neural crest. Certainly, it would be impossible to understand why congenital megacolon occurs without knowing that endothelin–3 and ET_B receptors play a critical role in the colonization of the bowel by cells from the neural crest; nevertheless, just learning that genes encoding endothelin–3 and ET_B receptors are involved in the process did not reveal exactly what it is that they do.

Masashi Yanagisawa and Raj Kapur have collaborated to try to gain some insight into what endothelin–3 actually does for the colon. In their new work, they have used Raj's transgenic mice, in which the DBH promoter is employed to induce vagal neural crest–derived cells to express specified genes in the gut of transgenic mice. The first experiment of Masashi and Raj was to employ the DBH promoter to drive the expression of endothelin–3 in vagal neural crest–derived cells within the bowel of lethal spotted mice. The experiment worked. The vagal neural crest–derived cells of lethal spotted mice carrying a DBH-ET–3 transgene did indeed express endothelin–3; moreover, the expression of the transgenic endothelin–3 prevented the colon of the lethal spotted mice from becoming aganglionic. Despite the expression of endothelin–3 in their gut, however, the transgenic lethal spotted mice did not lose their spots. The importance of the results of these experiments is that they established that the bowel is the site of action of endothelin–3. The DBH promoter does not turn on until the vagal neural crest–derived cells enter the bowel; moreover, the fact that the melanocytes remain abnormal, despite the introduction of a normal source of endothelin–3, indicates that the effects of the endothelin–3 that have been targeted to the gut are quite local. This experiment also conclusively rules out the possibility, suggested by Nicole Le Douarin, that the role of endothelin–3 is to stimulate the premigratory neural crest cell population to expand enough to colonize the entire bowel.

A second experiment by Masashi Yanagisawa and Raj Kapur was to use the DBH promoter to direct the expression of ET_B receptors to vagal neural crest–derived cells in the gut of mice that were genetically unable to produce the receptor. Once again, the introduction of a transgene prevented the occurrence of aganglionosis of the colon. The earlier transgenic experiments with endothelin–3 had shown that the locus of endothelin–3's action is in the bowel; however, since endothelin–3 can move around within the gut, the correction of an endothelin–3 deficiency by targeting the ligand to the bowel does not establish which enteric cell is at fault. In contrast, the correction of an ET_B deficit by targeting ET_B receptors to the neural crest–derived cells that colonize the gut certainly suggests very strongly that ET_B receptors on neural crest–derived cells have to be stimulated in order for them to colonize the colon. Curiously, the technique of transgenically directing ET_B receptor expression to neural crest–derived cells is much better at curing the aganglionosis of ET_B-deficient mice than of ET_B-deficient rats. In any case, as a result of these most recent experiments with transgenic animals, we now know that endothelin–3 acts in the bowel, and that it must act on neural crest–derived cells. To explain why the aganglionosis associated with an endothelin–3/ET_B deficit is not neural crest autonomous, therefore, it is necessary to postulate either that the neural crest–derived cells send a signal to the colon in advance of their arrival (Raj's heraldic hypothesis) or that another cell in the colon in addition to neural crest–derived cells must also be stimulated by endothelin–3 (my plebeian hypothesis).

To determine what endothelin–3 does for the neural crest and for the colon, I took an approach that was quite different from that of Masashi Yanagisawa and Raj Kapur. I decided to determine what endothelin–3 actually does, both to a purified population of neural crest–derived cells immunoselected from the developing mouse gut, and to the residual population of cells that is left after the neural crest–derived cells are removed (by immunoselection). These experiments were carried out with Jun Wu, the graduate student who had previously used immunoselection to confirm that the interstitial cells of Cajal are not of neural crest origin.

Jun is from the People's Republic of China and is a great example of the benefits that accrue to America when we ignore national boundaries in science and education. We have been experiencing a decline in the number of Americans selecting careers in biomedical research. This has created a gap that has been filled by China, among other countries. Jun may or may not return to China, but I have benefited greatly from having had the opportunity to work with him, and our American students have received an equal benefit from having had the opportunity to interact with him. Of course, Jun has been trained as a scientist by me, but if he continues to make fundamental discoveries of biomedical importance, what difference does his nationality make? Our government continues to restrict its support of training to American citizens or people with green cards. This policy flies in the face of the international character of science and is increasingly counterproductive. The noncitizens whose training we attempt to prevent are very likely to be the people upon whom American science will depend in the future.

The effects of endothelin–3 on enteric cells from the neural crest turned out to be not at all what anyone predicted, and, in fact, they were counterintuitive. Since the development of nerve cells does not occur in a region of the bowel when endothelin–3 is deficient, the most obvious explanation of the developmental role of endothelin–3 would be for it to be an essential signal that promotes the differentiation and/or survival of enteric nerve cells. Endothelin–3 might, for example, be anticipated to act like GDNF and stimulate the proliferation of nerve cell precursors. Alternatively, endothelin–3 might be predicted to act like NT–3 and enhance differentiation and/or survival of mature nerve cells. In actuality, however, endothelin–3 was found to do none of the things a reasonable person might have expected of it.

The first surprise in Jun's experiments was his demonstration that isolated enteric nerve cells develop happily in the total absence of endothelin–3, and equally happily in the presence of an antagonist of the effects of endothelin–3 on ET_B receptors. Masashi Yanagisawa's original hypothesis that endothelin–3 is required for the development of enteric nerve cells was thus shown to be wrong. Endothelin–3 is not required for these nerve cells to develop, and neither is the stimulation of the ET_B receptor. On the contrary, Jun found that endothelin–3 itself actually inhibits the development of nerve cells in a very nice concentration-dependent manner. What is more, Jun also showed that compounds that mimic the actions of endothelin–3 on the ET_B receptor (act as agonists) are also highly effective inhibitors of the differentiation of enteric nerve cells in tissue culture. In contrast to their effects when applied by themselves to cultures of enteric nerve cell precursors, which is to do nothing, ET_B antagonists completely block the effects of endothelin–3 on enteric nerve cell development.

When I first saw Jun Wu's data, I assumed that endothelin–3 must be preventing the development of nerve cells because it was stimulating the proliferation of their precursors. If precursor cells kept on dividing, we would not see them developing as nerve cells. Since Nicole Le Douarin had found that endothelin–3 exerts just this effect on cells of the premigratory neural crest, it was only natural to assume that endothelin–3 would do the same thing to the postmigratory neural crest–derived cells that Jun had fished out of the bowel. Fortunately, we did not satisfy ourselves with making that assumption and actually investigated what endothelin–3 was doing. In fact, endothelin–3 did not increase the rate of proliferation of neural crest–derived precursors, nor did it stimulate any other cells in cultures obtained from the fetal gut to proliferate. To get a sensitive look at proliferation, we measured the uptake of two molecules that become incorporated into the DNA of cells that proliferate, radioactive *thymidine* and *bromodeoxyuridine.* The uptake of neither molecule was altered in any way by endothelin–3. Endothelin–3 thus appears to act as a brake on differentiation. The effects of endothelin–3 on postmigratory neural crest–derived cells, like those of laminin–1, are thus different from the effects of endothelin–3 on the premigratory cells of the neural crest itself. The neural crest–derived cells that finally arrive in the gut are not the same as their predecessors. In any case, Jun's data reveal that endothelin–3 is probably a restraining molecule that holds neural crest–derived cells back and prevents them from differentiating before their time.

Jun used a specific smooth muscle marker, a molecule called *desmin,* to investigate the effects of endothelin–3 on cultures of residual cells. In contrast to its effects on the development of enteric nerve cells, Jun found that endothelin–3 promotes the development of smooth muscle. Whereas endothelin–3 decreases the number of nerve cells that develop in cultures of immunoselected cells, it increases the number of smooth muscle cells developing in cultures of residual cells. Perhaps because endothelin–3 enhances the maturation of muscle cell precursors, which are secretory, into smooth muscle cells, which concentrate their energy on contraction, endothelin–3 also decreases the biosynthesis of subunits of laminin–1, including that of the nerve cells' development-promoting alpha1 subunit.

In keeping with his observation that smooth muscle cell development is promoted by endothelin–3, Jun found that the ET_B receptor is expressed by cells in the residual population as well as by neural crest–derived cells. This observation was confirmed by Jingxian Chen, who further demonstrated that the ET_B receptor is expressed in the bowel of fetal and newborn Ret knockout mice. Since there are no cells of neural crest origin in the gut of Ret knockout mice at all, the presence of the ET_B receptor in the Ret knockout bowel proves that ET_B receptors are expressed in the gut by non-neural-crest-derived cells. Cells of the bowel that are from the neural crest and cells that are not from the neural crest thus both express the ET_B receptor.

Cheryl Gariepy, a postdoctoral fellow working in the laboratory of Masashi Yanagisawa, has recently confirmed (by using in situ hybridization) that endothelin–3 and ET_B receptors are expressed in the fetal bowel by non-neural-crest-derived cells. Cheryl found that RNA transcripts encoding preproendothelin–3 are actually found preferentially in nonneuronal cells in the wall of the fetal bowel. In contrast, RNA transcripts encoding the ET_B receptor are expressed in the fetal gut most strongly by neural crest–derived cells; nevertheless, shortly afterward they appear in the émigrés from the neural crest, transcripts encoding the ET_B receptor are also expressed by cells that are not of neural crest origin.

Jun Wu's experiments, and the location of RNA encoding endothelin–3 and ET_B receptors, provided a dramatic confirmation of what had previously been my suspicion about the cause of aganglionosis. When endothelin–3 or ET_B receptors are absent, the neural crest–derived cells that "attempt" to colonize the colon are presented with a double whammy. On the one hand they are deprived of a restraint (the ability of endothelin–3 to inhibit their premature differentiation), while on the other they encounter a major promoter of nerve cell development (the alpha1 chain of the excessively expressed laminin–1) as they enter the endothelin–3/ET_B-deficient colon. Confronted by these conditions, I believe that the nerve cell precursors are unable to keep themselves from differentiating into nerve cells.

The consequence of the premature development of nerve cells is that the bowel is left incompletely colonized. Since the neural crest–derived cells also cease to divide as they differentiate, the total number of cells from the neural crest within the endothelin–3/ET_B-deficient colon is also diminished. The overproduction of laminin–1 occurs outside the bowel as well as in it. As a result, there are out-of-place (*ectopic*) ganglia in the pelvis of lethal spotted mice that do not exist in the pelvises of their normal littermates. Some of these ectopic ganglia actually fuse with ganglia of the myenteric plexus in the colon just above the aganglionic zone. I interpret the ectopic ganglia as evidence for sacral neural crest–derived cells stopping and differentiating into nerve cells before reaching their targets in the bowel.

My hypothesis for the origin of aganglionosis in endothelin–3-deficient animals remains to be verified, and it should not be taken as gospel, but it has several advantages. One is that the hypothesis accounts for all of the data. Another is that all of its tenets are supported by actual observations. Endothelin–3 really does inhibit the development of nerve cells while promoting that of muscle cells. Endothelin–3 actually does decrease the secretion of laminin–1, probably by enhancing the conversion of secretory muscle cell precursors to mature contractile muscle cells. Laminin–1 does, in fact, promote the development of enteric nerve cells.

The aganglionosis that occurs in the bowel of patients with Hirschsprung's disease is a more complex issue than that which occurs in the animal models of the condition. When mutations in genes encoding endothelin–3 or ET_B receptors are at fault, it seems likely that the pathogenesis of the human congenital megacolon is similar, if not identical, to the animal defects associated with the same abnormalities. The other genetic deficits that have been linked to Hirschsprung's disease, however, defy any kind of facile explanation.

Defects in Ret or GDNF are superficially explainable, because this receptor and its ligand are required for the survival of enteric nerve cell precursors in the bowel. On the other hand, when Ret or GDNF are knocked out in mice, the result is not merely a colon that is aganglionic, but the entire bowel below the esophagus and adjacent stomach. In order for any abnormality to be seen in the animals, moreover, the genetic abnormality has to involve both the gene inherited from the mother and the gene inherited from the father. In humans, only one of the *RET* genes is ever abnormal (if both are, the defect is probably so lethal that afflicted individuals do not even survive long enough to be born) and the abnormality is limited to the colon. Obviously, more needs to be learned about the human disease, which probably arises through a variety of mechanisms.

The more we learn about Hirschsprung's disease, as the complex facts discussed immediately above illustrate, the more it is apparent the condition is misnamed. Hirschsprung, when he described the problem, saw only an end result, a bowel that lacked ganglia at its bottom end. He had no idea what caused that to happen. He did the only thing he could: He described what he saw. It now turns out, however, that there probably is not just one Hirschsprung's disease but an array of Hirschsprung's disease*s*. All of these conditions lead to a similar end result, because the repertory of migrating neural crest–derived cells is limited. The terminal colon is to them what California was to the forty-niners, a destination that might not be reached for a variety of causes. The reasons why neural crest–derived cells may not reach the terminal bowel may vary, but if the cells do not get there, the end result is always the same; ganglia will not be where they should be. Cells can die en route, become distracted, or get killed on arrival. Whatever the cause, a colon without a nervous system is a megacolon.

Hirschsprung's disease illustrates a truism about any kind of trip. When you travel it is wise to know where you are going, how to tell if you have arrived, and when to stop when you get there. The neural crest–derived cells that colonize the bowel are no exception. They need to find the right sites to establish the ganglia they will form, and they need to know when they reach these locations. If the neural crest–derived cells stop migrating in the wrong place, the colon is just as lost as if these cells were to drop dead. In the affairs of the colon and their nervous control, everything is retail. That means that nothing is more important to a colon than having its ganglia form in the right places; in other words, location, location, location.

12

THE STATE OF THE BOWEL

PROGRESS MADE IN understanding both the second brain and its development has been, in recent years, extremely rapid. Although I say this, you might, if you were a contrarian, have reason to take issue with me. I have recently reviewed the world's literature on Hirschsprung's disease for a technical article that I wrote on that subject. Having looked at all of the information that is now available from human patients with congenital megacolon and the animal models of the condition, I still find myself unable to provide a satisfying answer to the question: What makes the terminal colon become aganglionic? Superficially, the answer is simple. Genes have gone bad and are not delivering the information they are supposed to. That simple answer, however, is no answer. I can name a variety of genes that, when mutated, lead to Hirschsprung's disease. In fact, I have just done so above. Getting from the genes to the aganglionic colon, however, is the problem. We still cannot do it.

We come closest to knowing what is going on when the mutated genes that are associated with Hirschsprung's disease are those that normally encode endothelin–3 or ET_B receptors. But even with these genes, we have only conflicting theories to explain what happens, not an established set of facts. As we have seen, moreover, the mechanism that makes the colon aganglionic when endothelin–3 or ET_B receptors are lacking is unlikely to be operative when the mutated gene is *RET* or *GDNF.* Since, in contrast to endothelin–3 and ET_B receptors, *RET* and *GDNF* expression really are required by neural crest–derived émigrés to survive in the bowel, it seems likely that a loss of their function causes the precursors of enteric nerve cells to be abnormal in some way; nevertheless, much less is known about why the colon becomes aganglionic when *RET* or *GDNF* are mutated than when the mutated genes are *ET–3* or ET_B. In the case of *RET* or *GDNF,* we can only speculate as to what exactly goes wrong in the gut, why the lesion in the human bowel is restricted to the terminal colon, and why humans and animals are affected differently by mutations of the same genes. Even less is known about the actions of the other genes that have been found to be associated with cases of Hirschsprung's disease.

It is easy to look at this plaintive summary of the limitations of current knowledge and wonder what we scientists do with our time. Billions of public dollars have been poured every year into biomedical research, and a simple question like why some people are born with no nerve cells at the end of their gut still cannot be properly answered. A more perceptive look at the same set of facts, however, reveals that things are really pretty wonderful and rapidly becoming more so.

A written word was probably first carved in a stone near the Euphrates River by a Sumerian ancestor of ours over five thousand years ago. That act can be said to have initiated the history of the Western World. A long time transpired, however, between the initial carving of words in stone and the systematic provision by our predecessors of a written account of the events they witnessed. The Bible is the written record on which Judeo-Christian civilization is based. Written Hebrew, the language of the Bible, can be traced back to about the tenth century B.C.E.[1] The Pentateuch probably did not exist as we know it until after the Babylonian captivity (538 B.C.E.). To get from the first human word to the Bible, therefore, it took our ancestors much more than two thousand years. About the same amount of time expired between the completion of the Bible and the paper, published in 1954 by Yntema and Hammond, showing that the enteric nervous system is formed by émigrés that migrate to the bowel from the neural crest. That manuscript was to the second brain what the Sumerian word carved in stone was to Judeo-Christian civilization, the beginning of its history.

Comparatively speaking, we scientists are not doing so badly. It did not take us two millennia to get from the start of modern research on the second brain to the identification of genes that control its formation and that, when mutated, cause disease. It took us only forty years. In the scant four-year period since the discovery of the role of endothelin–3 in the formation of the enteric nervous system, we have learned how it affects cells from the neural crest and how its actions change after neural crest–derived cells colonize the bowel. Furthermore, while we still do not know precisely how abnormalities of endothelin–3 or ET_B receptors cause defects in the formation of enteric ganglia, we at least have testable hypotheses to account for them. Of even greater importance, the seemingly unfathomable complicated process of development of the second brain has been decoded sufficiently to understand it as an orderly mechanism. We have learned that

[1]I use the notation B.C.E. (before the common era) instead of the more frequently employed B.C. (before Christ) to avoid the religious connotation of the latter term. For the same reason, I will use the designation C.E. (common era) instead of A.D. (*anno domini*; in the year of our Lord).

the bowel is colonized both by specified and uncommitted nerve cell precursors, that these colonists are tutored by signals provided by the enteric microenvironment, and that their responses are determined, in part, by the heritage the various lineages of neural crest–derived cells bring to the gut.

In short, even though unknown worlds still exist, progress made in understanding the development of the second brain cannot be described as slow. Scientists have been making good use of their time. But will there be a return on the public investment? Surely no one can, or should, question whether the public is better off for having supported biomedical research in general. It certainly is. Compared to the cost of fending off the former Soviet Union during the cold war, the cost of fending off disease has been utterly trivial. I admit that the Soviets were a scary bunch, but more Americans are killed every year by colon cancer, germs, heart attacks, and strokes than by the combined efforts of the Soviets throughout the cold war period. A society that could afford to spend hundreds of billions of dollars every year to protect itself from the potential threat of the Soviet Union can easily afford to spend a tiny fraction of that amount to protect itself against the much more immediate threat of disease.

In fact, the dollars spent by the United States government to gain protection from illnesses of various types have been extremely good investments. Aside from the lives they have saved and the suffering they have prevented, the dollars spent, particularly on basic biomedical research, have probably made a substantial profit. Health care is not cheap. Every illness prevented by a scientific advance is thus money in the bank for the nation.

When I was a child, my parents lived in fear of poliomyelitis. Every summer it seemed that there would be a polio scare. A nearby children's camp would be closed by a polio outbreak, or a rumored epidemic would force a trip to be canceled. The president, whom they admired more than any man on earth, had been crippled by it. The hideous polio epidemics that once were real are no more. Compare the cost of distributing polio vaccine to what the cost of polio would be like today, if its associated iron lungs, hospitalization, quarantine, braces, rehabilitation, aid to the handicapped, and untimely funerals were still needed. Compare the cost of ulcer surgery to that of Prilosec and Zantac, and compare the cost of even the medical treatment of ulcers, heartburn, and stomach cancer to a one-time course of antibiotics to eradicate *Helicobacter pylori.* Fundamental discoveries, as opposed to technical advances in treatment, save money.

In addition to the monetary savings (and better life) brought by progress in biomedical science, the investment our government has made in supporting the work of its scientists has spun off the world's leading biotechnical industry. This industry is beginning to put out useful products and more will certainly come, but its existence has already produced large numbers of jobs and great wealth. Justification of biomedical research in general is thus not hard. It is not a controversial subject. Since no one is immortal or invulnerable, everyone benefits from biomedical research, and there is no downside. Progress in health care is not a zero-sum game, in which one person's gain is another's loss. Make an advance and it becomes available for everyone. Even politicians like biomedical research. After all, what kind of a constituency do germs have? No one is for disease.

In contrast to biomedical research in general, it is often difficult to see the good that comes of basic research in particular. Fundamental advances are the result of tiny increments in knowledge. Little steps add up, but they are not by themselves spectacular, and even the people who are taking the little steps may not know where they lead. In fact, those of us who explore things that are unknown at the beginning of our studies never know where our research is going. That is the nature of being unknown. It is much different in the field of applied research, where specific needs provide the motivation for doing the work and where nothing new has to be discovered to get it done. Applied research consists of making use of available knowledge to accomplish desired ends. It is thus unlikely, for example, that a person trying to design a better artificial heart valve would be asked about the utility of the endeavor. It is obvious that heart valves are often damaged by various disease entities and have to be replaced. There is a risk in heart valve replacement that probably would be reduced by providing surgeons with better valves. Basic scientists thus often find themselves being asked to give a rationale for their work, while applied scientists rarely face this problem.

My own life's work on the second brain has been quite basic in nature. People in the field have never had much trouble understanding why I do what I do. As I have acknowledged above, they have not always been happy with my observations, and colleagues have sometimes hated my conclusions, but they have never questioned the utility of my effort. In contrast, many of the people I have really loved, including my own father, have expressed grave doubts about its value.

My father never quite forgave me for doing research instead of curing the sick. He frequently sat down with me, usually after a Sunday brunch, and asked me the question: "So, Michael, what have you discovered lately?" Telling him was a major problem. Explaining that he had a brain in his belly that I was trying to understand got me nowhere. My experience of trying to convince a skeptical world that there are at least three neurotransmitters (including serotonin) in the peripheral nervous system, not just the accepted two, left him cold. Sequential gene expression in the development of the enteric nervous system was a subject I would not even think of discussing with him. Whenever I would relate the results of a particular study, my father would listen attentively and then ask: "So,

Michael, what disease is that going to cure?" It got to the point where I would look for someplace to hide whenever a conversation with my father began with the words "So, Michael, . . ."

My father died a little over a year ago. He passed away quietly, at home, in his bed, surrounded by his family, just as he would have wished. He was ready to go at the time. I think he waited for his first great-grandson to be born, to hold the child himself, and to see a Democratic president get reelected one more time (the first one since the beloved Roosevelt). At this point in writing this book, my father comes to mind. I know that if he were here to do so, he would carefully read what I wrote, and he would ask me a question. It is a hard question that would begin: "So, Michael. . . ." I do not have fill in the rest, you just know what would follow. I feel that I owe my father something of an answer.

I did not intend this book to be a "how to" document, explaining to readers how to cope with a variety of gastrointestinal complaints. In that sense, like my research, no disease of a kind that I could explain to my father is going to be cured by what can be found in these pages. On the other hand, there is a disease that I very much hope my literary effort will attack. It is the disease of despair that is faced by so many of the victims of an unmanageable gut. I really feel for these individuals, whose tortured existence has for so long brought out the worst in the very people to whom they have turned for help.

There is something about gastrointestinal problems that has, until recently, turned off both the sympathy and the empathy of many physicians. The history of psychiatric illness has been marked by a similar attitude. Many of the problems faced by the brain in the head have been dismissed because no anatomical or biochemical abnormalities have been identified as their cause. Unlike the heart, which visibly beats, we cannot see the brain think; therefore, when thought is disturbed for reasons that are not apparent, there is a tendency to blame the thinker for not doing it right. It has often been just like that with the bowel. When a cause for gastrointestinal malfunction is not obvious, there is a tendency to deny that the problem is real, or if that denial becomes untenable (as it may in the face of visible difficulties), to attribute the problems of the gut to bad thoughts emanating from the brain. This attribution again provides a rationale for blaming the thinker.

I would like the readers of this book to realize that the rediscovery of the brain in their bowel is a breakthrough for hope. The focus of scientific attention on the second brain holds within it a great potential, some of it already realized, for understanding how to treat and prevent gastrointestinal disease. The realization that an independent center of integrative nervous activity lurks inside the abdomen has become a magnet for attracting good research. That magnetic attraction has turned a forgotten field into a "hot" arena of discovery. In the ferment of work thus inspired, there are scores of cures for diseases that have not yet been discovered and which exist today only as miscellaneous syndromes or symptom complexes.

Perhaps the greatest potential of all is the excitement generated by the new information about the development of the second brain. I truly believe that Hirschsprung's disease is the tip of a very large iceberg. Congenital megacolon is obvious, because it produces an aganglionic zone that can be seen. There is no doubt, however, that there are also many more conditions that are not so easy to recognize. For example, we are just beginning to appreciate that too many ganglia, or ganglia of the wrong kind, can be just as damaging to the ability of the gut to function as no ganglia at all. We call these emerging problems *neurointestinal dysplasia*. Little is currently known about the cause(s) of neurointestinal dysplasia, but that is coming soon.

Functional bowel diseases, including the irritable colon, which may afflict 20 percent of Americans, may well have developmental origins. The ability of neurotransmitters such as serotonin to affect developing enteric nerve cells provides a mechanism by which the experience of an impressionable primitive gut may affect the personality of the mature bowel that arises from it. The concept of the inflammatory bowel diseases (ulcerative colitis and Crohn's disease) passed during the course of my lifetime from mental illnesses to autoimmune conditions. The inflammatory bowel diseases may also have a developmental basis.

The nervous system works so closely with the immune system in defending the gut that a developmental interaction between the nervous and immune systems, each of which secrete chemicals that affect the other, should be anticipated. There are, for example, more nerve cells in segments of gut affected by inflammatory bowel disease. Since nerve cells do not proliferate, how can this happen? Are these additional nerve cells the products of precursors that are retained in the enteric nervous system of adults, or were they preexisting? Could it be that too many nerve cells in particular regions of the gut is a cause of inflammatory bowel disease? Obviously, I cannot answer these questions, but the fact that I can now pose them are reasons why the explosion of discovery about the second brain carries so much hope for making the lives of people with troubled bowels so much better.

If I could ever again sit down after a Sunday brunch with my father I would finally answer his question. The minute he said to me: "So, Michael, . . ." I would interrupt and reply, using the name by which he was known in the family. "Bomber," I would tell him, "I have cured no one in my lifetime. That is something I leave to Anne, your daughter-in-law, the pediatrician. What I have done is something just as good. I have done my bit to made it possible to find many cures for many people." Deep down where it counts, I know he would finally have been satisfied.

ENDNOTE

Animals in Biomedical Research

I WOULD LIKE TO explain why biomedical scientists are willing to kill animals. None of us relishes suffering, and we have no wish to be the agents of the unnecessary death of any sentient being. We thus apply rigorous standards of care and compassion to the way we treat research animals. They live in air-conditioned spacious quarters and eat food that they love and that is good for them. Animal care and use committees have been established at all federally funded research facilities to make sure the standards are enforced.

In contrast to biomedical scientists, disease places no limits on the degree of suffering it causes, either in animals or in humans. Cancer that has spread to bones, for example, may cause them to break without respite or healing, and each break is agonizing. Ulcers give rise to pain, and when they perforate the bowel, the resulting infection, shock, and torment is heartrending. What drives scientists to experiment with animals, therefore, is simply the very real compassion they feel for their fellow human beings. Studies done with animals, moreover, have been, and continue to be, both necessary and effective. The major advances in medicine that we now take for granted are all due to animal experimentation. Almost no disease can be treated without making use of agents or procedures developed as a result of research done on animals.

People who object to animal research blind themselves to its benefits. They feel for the animals, but they do not feel for people. In crusading for animal rights, activists often accuse scientists of "specieism," a term they equate with racism or sexism. Simply to use the term "specieism," however, is to announce that its user is unable to understand that human life is sacred. Since research on animals alleviates disease and human suffering, demands that this research stop are really proposals to expand disease and promote human suffering. That is unacceptable. Adolf Hitler opposed animal research, which for him was morally consistent. His alternative was Dachau. Since Hitler's minions were tried at Nuremberg, a code of ethical decency has been established. Article Three specifically forbids that experiments be carried out on people that are not based on results previously obtained with animals. Products or procedures, therefore, are not applied to human beings until they have acquired at least some track record for safety.

I have never observed moral consistency in animal rights crusaders. Those who oppose animal research should forgo the use of the products and techniques that come from it. Activists owe it to their constituency to announce and carry out a personal boycott of, for example, antibiotics, cancer therapy, painkillers, vaccines, and heart surgery, all of which have been developed through the use of animal experimentation. If anyone has done this, I have yet to hear of it. If an animal rights person were to renounce the benefits of animal experimentation, I would regard such a person as incredibly stupid, and I would anticipate that he/she would die prematurely and probably painfully; nevertheless, I would mourn for him/her as one who has lived a truly righteous and consistent life.

The use of animals to make human lives better has been considered moral throughout history. The human use of animals is found in ancient sources and is explicitly acknowledged in the Bible. "God said: Let us make humankind, in our image, according to our likeness! Let them have dominion over the fish of the sea, the fowl of the heavens, animals, all the earth, and all crawling things that crawl about the earth!" (Genesis I: 26). This is a mandate that should not be abused by cruelty, but neither should it be forgotten. Animals work with and for us. Horses pull carriages and carry riders. Oxen plow fields. Cowboys round up beef cattle, and most of us like lamb chops. Vegetarians who do not wish to eat meat make a personal statement by not doing so, and this decision harms no one. This is admirable. In contrast, attempts by activists to prevent research on animals are harmful to others, and thus are not admirable. Fundamentalism, which extends one's personal beliefs to others, is no better when it is practiced in the name of animals than when it is practiced by Iranian ayatollahs in the name of God.

It is sometimes suggested that computers or tissue culture cells be used instead of animals. These ideas are put forward by people who know nothing about research or biology. Computers manipulate information that is put into them; they do not, like scientists, explore the unknown. Computers are useful for making models that can be tested by actual experiments, they are wonderful for finding facts buried within huge compendia of data, and they are spectacular for manipulating numbers, but they do not invent or discover.

Tissue culture cells are also helpful, but even they are ultimately derived from people or animals. Worse, they do not think like a brain, pump like a heart, run like a jogger, or digest like a gut. Cells aggregate to form tissues, and tissues aggregate to form organs; therefore, studies limited to single cells in tissue-culture dishes cannot reveal the workings of tissues or organs in living bodies. For me, in particular, there is no enteric nervous system in a tissue culture cell. The gut and the enteric nervous system, therefore, can only be studied where they are, in animals and people. We learn from the problems of people, and we do experiments with animals to solve them. The goal, which has been, and is continuing to be, achieved to an amazing degree is to improve the human condition.

To determine what a substance does to the gut, it is necessary to isolate the bowel to be sure that the molecule under investigation actually acts directly on the gut itself. If, instead, the compound were to be administered to an intact animal, the molecule can spread all over the body. This dissemination makes possible all kinds of indirect effects. Hormones, which can influence the gut, can be released in distant organs and reach the bowel via the circulation. Nerves can be stimulated anywhere and directly or indirectly affect the gut. Compounds can also affect the behavior of the bowel by changing the availability of its blood supply, for example, by increasing or decreasing the heart rate or by contracting or dilating blood vessels. Scientists cannot deal with this kind of complexity because it defies clear interpretation of the phenomena we observe. We like experiments to be simple so that we can establish causality with the odds of being correct in our favor. We introduce a perturbation and note a response. The fewer things there are in a system to be affected by our perturbation, the more clear the effect of that perturbation is likely to be.

Animal experimentation, therefore, is a necessary and moral part of life. Just as we depend on animals for food and labor, so too we depend on them for knowledge of how to improve the health of humankind. We do not yet know enough about how to prevent or cure many of the diseases that afflict us. Until we do, we must continue to rely on animals to learn what we need to know. The rediscovery of the second brain was a relatively recent event. Its benefits are just now beginning to unfold. To keep the discoveries coming and to exploit their promise, it will be critical to test new drugs, procedures, and experimental hypotheses on animals. Fortunately, the pets and farm animals that have entered into partnership with us humans, and who depend on us to take care of their problems, will derive as much benefit as we will from the advances made by biomedical scientists. There is no reason to believe that humans are unique in having an enteric nervous system that malfunctions. Veterinarians are in the loop and see to it that our new drugs and procedures are available to pets and farm animals. Animal experimentation must thus continue to be carried on. Both humans and animals depend on it.

INDEX